AF361695

10th Anniversary of Cells—Advances in Cell Cycle

10th Anniversary of Cells—Advances in Cell Cycle

Editor

Zhixiang Wang

MDPI • Basel • Beijing • Wuhan • Barcelona • Belgrade • Manchester • Tokyo • Cluj • Tianjin

Editor
Zhixiang Wang
Department of Medical
Genetics
University of Alberta
Edmonton
Canada

Editorial Office
MDPI
St. Alban-Anlage 66
4052 Basel, Switzerland

This is a reprint of articles from the Special Issue published online in the open access journal *Cells* (ISSN 2073-4409) (available at: www.mdpi.com/journal/cells/special_issues/advances_cell_cycle).

For citation purposes, cite each article independently as indicated on the article page online and as indicated below:

LastName, A.A.; LastName, B.B.; LastName, C.C. Article Title. *Journal Name* **Year**, *Volume Number*, Page Range.

ISBN 978-3-0365-5216-3 (Hbk)
ISBN 978-3-0365-5215-6 (PDF)

Contents

Editorial

Editorial to Summarize the Papers Published in the Special Issue "10th Anniversary of Cells—Advances in Cell Cycle"

Zhixiang Wang

Signal Transduction Research Group, Department of Medical Genetics, Faculty of Medicine and Dentistry, University of Alberta, Edmonton, AB T6G 2H7, Canada; zhixiang.wang@ualberta.ca

Citation: Wang, Z. Editorial to Summarize the Papers Published in the Special Issue "10th Anniversary of Cells—Advances in Cell Cycle". *Cells* **2022**, *11*, 2437. https://doi.org/10.3390/cells11152437

Received: 22 July 2022
Accepted: 2 August 2022
Published: 5 August 2022

Publisher's Note: MDPI stays neutral with regard to jurisdictional claims in published maps and institutional affiliations.

To celebrate its 10th anniversary, the prestigious journal *Cells* launched a series of Special Issues in 2021. This Special Issue entitled "10th Anniversary of Cells—Advances in Cell Cycle" was launched together with other sister Special Issues under the umbrella "10th Anniversary of Cells". I am honored to be invited to serve as the academic editor for this Special Issue. This Special Issue attracted the attention of many scientists in the cell cycle field and consists of 10 high quality papers, including 4 research articles and 6 scientific reviews—a great success. The four research articles focus on various important topics of the cell cycle using a broad range of model organisms, including yeast, sea urchins, green algae, and human cancer cell lines.

The first research article published in this issue is focused on the cell cycle progression of green algae *Desmodesmus quadricauda* [1]. The cell cycle progression of D. quadricauda consists of cell growth, duplication, and division, which is usually examined by a biochemical analysis of macromolecules (RNA, protein, DNA, and starch). However, such analyses are rather time consuming, complicated, and require a large quantity of initial samples. Czech scientist Bišová and her team aimed to find a more efficient way to analyze these large biomolecules. In their research, they used two independent methodologies to examine the dynamics of starch, lipid, polyphosphate, and guanine pools during the cell cycle of the synchronized green algae: a conventional biochemical analysis of cell suspensions and confocal Raman microscopy of single cells. They concluded that confocal Raman microscopy can detect even low levels of macromolecules naturally present in the cells during their vegetative development. Confocal Raman microscopy is especially suited for the detection of polyP, lipids, and guanine crystals within cells [1].

Mitosis is the most dynamic period of the cell cycle, involving a major reorganization of virtually all the cell components. Mitosis is divided into prophase, prometaphase, metaphase, anaphase, and telophase. The progression of the cell cycle through these mitotic subphases is tightly regulated by complicated molecular mechanisms. Mitosis is also the most fragile period of the cell cycle. Therefore, most cancer drugs are designed to specifically target mitotic cells. The next two research articles published in this Special Issue focused on the mitotic phase of the cell cycle [2,3].

The research article by Bari et al., aimed to develop a novel therapeutic agent against Glioblastoma multiforme (GBM) by arresting the cancer cells in mitosis [2]. GBM is the most malignant and frequent human brain tumor, representing more than 60% of all brain tumors in adults [2]. Despite the great advances made in understanding the molecular alterations that occur in Glioblastoma (GB), there is no definitive cure, and mortality is still very high. Therefore, the development of new therapeutic agents is clinically urgent. The Italian research team previously led by Dr. Tata showed that the arecaidine propargyl ester (APE), an orthosteric agonist of M2 muscarinic acetylcholine receptors (mAChRs), arrests the cell cycle of glioblastoma (GB) cells and reduces their survival. They showed that M2 agonist treatment causes an arrest in cell cycle progression with an accumulation of cells during the pro-metaphase/metaphase transition, which causes a significant increase in abnormal mitosis and multipolar mitotic spindle formation [2]. These findings highlight the M2 muscarinic receptor as a new strategic therapeutic target in GBM therapy.

Vannini et al., studied the regulation of the mitotic exit in the budding yeast *Saccharomyces cerevisiae* [3]. The mitotic exit is critical for the successful completion of a cell division cycle. The mitotic exit network (MEN) is a Ras-like signal transduction pathway that promotes mitotic exit in anaphase. A crucial step in MEN activation is the activation of Cdc15, which requires the association of Cdc15 with spindle pole bodies (SPBs) and the presence of the Tem1 GTPase and the Polo kinase Cdc5. However, it was previously unclear how Cdc15 associated with SPBs. In this research, a collaborative team led by Dr. Seshan identified a hyperactive allele of NUD1, nud1-A308T, that recruits Cdc15 to SPBs in all stages of the cell cycle in a CDC5-independent manner. They further showed that nud1-A308T leads to the early recruitment of Dbf2-Mob1 during metaphase [3]. Their findings highlight the importance of scaffold regulation in signaling pathways to prevent improper activation.

In the fourth research article, Limatola et al., studied the role of the vitelline layer (VL) of sea urchin (*Lytechinus pictus*) eggs in fertilization [4]. Sea urchins have been a model organism in cell cycle research and contributed to the discovery of cyclins [5]. In this study, the research team led by Italian scientist Santella attempted to partially disrupt the VL with a reducing agent, dithiothreitol (DTT), and then observed its effects on fertilization. They showed that DTT treatment did not elevate the fertilization envelop, but instead caused a few anomalies at fertilization, including compromised Ca2+ signaling, the blocked translocation of cortical actin filaments, and impaired cleavage. The authors concluded that the fertilization envelop is not the decisive factor preventing polyspermy, and that the integrity of the VL is crucial to the egg's fertilization response [4].

This Special Issue also published six comprehensive high quality review articles. These reviews covered a broad range of topics: the impact of 5-Bromo-2′-deoxyuridine (BrdU) on the proliferative behavior of cerebellar neuroblasts [6], the regulation of the cell cycle by telomerase [7], growth factors [8], heat shock transcription factors [9], Cyclin-Dependent Kinases and CTD Phosphatases [10], and the effects of low-dose radiation on the cell cycle [11].These reviews focused on various model organisms, including mammalian cells, plant cells, yeast, and parasites of the *Leishmania genus*.

As a marker for DNA synthesis, BrdU has generated important insights into the cellular mechanisms underlying the proliferative behavior of neuroblasts. BrdU labeling is the most widely used procedure for studying cell-cycle phases and their durations, as well as to identify dividing neuroblasts and follow their fates. Even though BrdU has demonstrated toxicity, its detrimental effects on the proliferation and viability of different cell types have been frequently neglected. In a focused review article that included numerous data from his own lab, Dr. Martí-Clúa evaluated the potential negative impact of using BrdU to study the proliferation of neuroblasts [6]. He found that incorporation of BrdU into newly synthesized DNA may lead to inaccurate results. Thus, caution should be exercised when interpreting the results obtained using BrdU. This is particularly important when high or repeated doses of this agent are used.

Leishmaniasis are a group of common poverty-related endemic diseases. They cause a wide spectrum of clinical manifestations, and approximately one million new diagnoses are expected yearly in East Africa, the Indian subcontinent, and Latin America. Thus, novel treatments against leishmaniasis are urgently needed. For this purpose, molecular biology studies on Leishmania spp. have been conducted to elucidate the different aspects of parasite biology. Among these studies, trypanosomatid telomere biology has generated great interest in the scientific community, as telomeres are essential for genome stability and cell cycle progression. In this comprehensive review, Dr. Cano and her colleagues aim to shed light on what we know about the phenomena behind telomere maintenance and how it impacts the parasites' cell cycle and survival [7]. Their review covers the knowledge available so far on the Leishmania spp. cell cycle and telomere homeostasis, with an emphasis on the remaining gaps and the advances reached in the last few years. Among the impressive progress made on understanding the biology of these parasites are the facts that they remain in the G0 state during their infective stages and the remarkable divergence of their telomeric shelterin-like complex relative to mammals. However, these aspects are only a few examples of how this subject can be linked in the future to leishmaniases treatment and how far scientists are from a deeper knowledge of these peculiar eukaryote parasites.

A review I authored myself focuses on the regulation of cell cycle progression by growth factor (GF)-induced cell signaling [8]. The driving force of cell cycle progression are GF-initiated signaling pathways that control the activity of various Cdk-cyclin complexes. While the mechanism underlying the role of GF signaling in the G1 phase of the cell cycle progression is well understood, little is known regarding the function and mechanism of GF signaling in regulating other phases of the cell cycle, including the S, G2, and M phases. In this review, we briefly discuss the history of cell cycle research and the process of cell cycle progression through various phases. The emphasis in this research is on the role of signaling pathways activated by GFs and their receptors (mostly receptor tyrosine kinases) in regulating cell cycle progression through various phases. As summarized in the review, the accumulated results so far suggest that GF signaling may regulate cell cycle progression throughout all phases of the cell cycle, but further research is needed to sustain these findings and to uncover the underlying molecular mechanisms [8].

Heat shock transcription factors (HSFs) have been noted as critical proteins for cells' survival against various stresses; however, recent studies suggest that HSFs also have important roles in cell cycle regulation-independent cell-protective functions. During the cell cycle progression, HSF1 and HSF2 bind to condensed chromatin to provide immediate precise gene expression after cell division. In a comprehensive review, Dr. Hayashida and his colleagues discussed the function of these HSFs in cell cycle progression, cell cycle arrest, gene bookmarking, mitosis, and meiosis [9]. They briefly described the early essential findings in cell cycle studies and the discovery of the heat shock response and the essential functions of HSFs. They further describe all of the important discoveries to date regarding the function of HSFs in cell cycle regulation, including the interaction between HSF1 and p53 as well as HSFs and WD40 repeat proteins [9].

The precise control of transcription is crucial for the synthesis of many phase-specific proteins and for the orderly progression of the cell cycle. Dr. Zheng wrote a timely review for this Special Issue to discuss this important topic in cell cycle research [10]. This comprehensive review highlights highly conserved transcriptional regulators that are shared in budding yeast (Saccharomyces cerevisiae), Arabidopsis thaliana model plants, and humans, which are separated by more than a billion years of evolution. In addition, the functions of plant-unique regulators in relation to cell division are also discussed. The regulators discussed in the review include structurally and/or functionally conserved regulators such as cyclin-dependent kinases (CDKs) and RNA polymerase II C-terminal domain (CTD) phosphatases. The nature of the classical versus the shortcut models of Pol II transcriptional control is also discussed. These CDKs and CTD phosphatases have conserved domains across the three eukaryotic kingdoms, but some of them have also evolved with unique structures and functions in each of the kingdoms. This review is an insightful future perspective regarding the precise control of transcription in cell cycle regulation [10].

This Special Issue is concluded by a well-written comprehensive review regarding the effects of low-dose radiation on the cell cycle [11]. Cells exposed to ionizing radiation undergo a series of complex responses, including DNA damage, reproductive cell death, and altered proliferation states, which are all linked to cell cycle dynamics. Ionizing radiation has been used in cancer treatment for more than one hundred years and is currently a standard option in treating 20–60% of all new cancer cases. Through the years, extensive research has been conducted on cell cycle checkpoints and their regulators in mammalian cells in response to high-dose ionizing radiation. However, it is unclear how low-dose ionizing radiation (LDIR) regulates the cell cycle progression. In this review, Drs. Khan and Wang aim at summarizing the recent advances related to the role of LDIR

(200 mGy or lower) in cell cycle regulation. This review also discusses how different cell cycle regulators are modulated by LDIR exposure in normal, cancerous, and stem cells. Moreover, the authors propose a new conceptual mechanism to explain how LDIR differentially regulates the nucleocytoplasmic shuttling of p21^{Waf1}, a key cell cycle regulator and transcriptional target of p53 [11].

Cell cycle control is vital for cell proliferation in all eukaryotic organisms. Through more than one hundred years of efforts by scientists, we now have a much clearer picture of cell cycle progression and its regulation. The research articles and reviews published in this Special Issue reflect the current and future efforts made by scientists to better understand the progression and regulation of the cell cycle, and how they may serve as targets for novel cancer therapies.

Funding: This research received no external funding.

Conflicts of Interest: The authors declare no conflict of interest.

References

1. Moudříková, Š.; Ivanov, I.N.; Vítová, M.; Nedbal, L.; Zachleder, V.; Mojzeš, P.; Bišová, K. Comparing biochemical and raman microscopy analyses of starch, lipids, polyphosphate, and guanine pools during the cell cycle of desmodesmus quadricauda. *Cells* **2021**, *10*, 62. [CrossRef] [PubMed]
2. Di Bari, M.; Tombolillo, V.; Alessandrini, F.; Guerriero, C.; Fiore, M.; Asteriti, I.A.; Castigli, E.; Sciaccaluga, M.; Guarguaglini, G.; Degrassi, F.; et al. M2 muscarinic receptor activation impairs mitotic progression and bipolar mitotic spindle formation in human glioblastoma cell lines. *Cells* **2021**, *10*, 1727. [CrossRef] [PubMed]
3. Vannini, M.; Mingione, V.R.; Meyer, A.; Sniffen, C.; Whalen, J.; Seshan, A. A novel hyperactive nud1 mitotic exit network scaffold causes spindle position checkpoint bypass in budding yeast. *Cells* **2022**, *11*, 46. [CrossRef] [PubMed]
4. Limatola, N.; Chun, J.T.; Cherraben, S.; Schmitt, J.-L.; Lehn, J.-M.; Santella, L. Effects of dithiothreitol on fertilization and early development in sea urchin. *Cells* **2021**, *10*, 3573. [CrossRef] [PubMed]
5. Evans, T.; Rosenthal, E.T.; Youngblom, J.; Distel, D.; Hunt, T. Cyclin: A protein specified by maternal mrna in sea urchin eggs that is destroyed at each cleavage division. *Cell* **1983**, *33*, 389–396. [CrossRef]
6. Martí-Clúa, J. Incorporation of 5-bromo-2′-deoxyuridine into DNA and proliferative behavior of cerebellar neuroblasts: All that glitters is not gold. *Cells* **2021**, *10*, 1453. [CrossRef] [PubMed]
7. Assis, L.H.C.; Andrade-Silva, D.; Shiburah, M.E.; de Oliveira, B.C.D.; Paiva, S.C.; Abuchery, B.E.; Ferri, Y.G.; Fontes, V.S.; de Oliveira, L.S.; da Silva, M.S.; et al. Cell cycle, telomeres, and telomerase in *leishmania* spp.: What do we know so far? *Cells* **2021**, *10*, 3195. [CrossRef] [PubMed]
8. Wang, Z. Regulation of cell cycle progression by growth factor-induced cell signaling. *Cells* **2021**, *10*, 3327. [CrossRef] [PubMed]
9. Tokunaga, Y.; Otsuyama, K.-I.; Hayashida, N. Cell cycle regulation by heat shock transcription factors. *Cells* **2022**, *11*, 203. [CrossRef] [PubMed]
10. Zheng, Z.-L. Cyclin-dependent kinases and ctd phosphatases in cell cycle transcriptional control: Conservation across eukaryotic kingdoms and uniqueness to plants. *Cells* **2022**, *11*, 279. [CrossRef] [PubMed]
11. Khan, M.G.M.; Wang, Y. Advances in the current understanding of how low-dose radiation affects the cell cycle. *Cells* **2022**, *11*, 356. [CrossRef] [PubMed]

Review

Advances in the Current Understanding of How Low-Dose Radiation Affects the Cell Cycle

Md Gulam Musawwir Khan [1] **and Yi Wang** [1,2,*]

[1] Radiobiology and Health, Canadian Nuclear Laboratories (CNL), Chalk River, ON K0J 1J0, Canada; Md.Gulam.Musawwir.Khan@USherbrooke.ca

[2] Department of Biochemistry, Microbiology and Immunology, Faculty of Medicine, University of Ottawa, Ottawa, ON K1H 8M5, Canada

* Correspondence: yi.wang@cnl.ca

Abstract: Cells exposed to ionizing radiation undergo a series of complex responses, including DNA damage, reproductive cell death, and altered proliferation states, which are all linked to cell cycle dynamics. For many years, a great deal of research has been conducted on cell cycle checkpoints and their regulators in mammalian cells in response to high-dose exposures to ionizing radiation. However, it is unclear how low-dose ionizing radiation (LDIR) regulates the cell cycle progression. A growing body of evidence demonstrates that LDIR may have profound effects on cellular functions. In this review, we summarize the current understanding of how LDIR (of up to 200 mGy) regulates the cell cycle and cell-cycle-associated proteins in various cellular settings. In light of current findings, we also illustrate the conceptual function and possible dichotomous role of p21^{Waf1}, a transcriptional target of p53, in response to LDIR.

Keywords: LDIR; cell cycle; hormesis; cancer; p21^{Waf1} (CDKN1A)

Citation: Khan, M.G.M.; Wang, Y. Advances in the Current Understanding of How Low-Dose Radiation Affects the Cell Cycle. *Cells* **2022**, *11*, 356. https://doi.org/10.3390/cells11030356

Academic Editor: Michal Goldberg

Received: 30 November 2021
Accepted: 20 January 2022
Published: 21 January 2022

Publisher's Note: MDPI stays neutral with regard to jurisdictional claims in published maps and institutional affiliations.

1. Introduction

It is generally accepted that ionizing radiation is harmful to living organisms, including humans, particularly at high doses [1]. Exposure to ionizing radiation can lead to two broad categories of adverse health effects: Deterministic effects that occur above a threshold dose and whose severity is dose related (e.g., skin reaction), and stochastic effects that have neither a threshold dose nor their severity is dose related (e.g., cancer). However, the probability of incidence of either effect increases with dose [1].

The use of ionizing radiation in the treatment of cancer began shortly after the discovery of X-rays in 1895. In 1896, Emil Grubbe used X-rays to treat a recurrent carcinoma of the breast [2]. Currently, controlled utilization of HDIR is a standard option in treating 20–60% of all new cancer cases [3,4]. However, in the process, normal tissue toxicity, followed by the emergence of second cancers, may arise due to the genotoxic properties of HDIR. Permanent changes in the coding sequence of essential genes may lead to a cascade of events associated with the neoplastic transformation of normal tissue that is unavoidably exposed.

HDIR has well-documented and evident impacts on biological processes, molecular pathways, and cellular functions, but the effect of LDIR on human health remains unclear. Human beings are inevitably exposed throughout their lives to low doses of radiation from natural sources (such as cosmic rays and radon gas) as well as human-made sources. A clear understanding of how LDIR impacts cellular processes and molecular mechanisms is becoming increasingly important, as LDIR is frequently used in research, industrial products, security, and modern medicine. Particularly, per-capita medical radiation exposure has been on the rise at an alarming rate over the last few decades, to the point where it is now roughly equivalent to natural background radiation exposure [5,6].

A growing body of evidence supports that LDIR exposure results in distinct molecular-, cellular-, and tissue-level responses when compared with those observed after HDIR exposure [7]. In contrast to HDIR, which causes numerous alterations to macromolecules, including DNA/RNA damage, robust modulation of cell signaling pathways, and degenerative/carcinogenic effects, LDIR may promote hormesis. Radiation-induced homeostasis, often termed hormesis, is a theoretical concept that suggests that exposure to LDIR stimulates beneficial pathways. Several reports indicate that the hormetic process boosts cell survival and growth, improves immune functions, and enhances cytogenetic protection [8,9]. Most population-based epidemiological studies do not show a threshold for cancer incidence; however, at LDIR doses below 100 mGy, there is uncertainty as to whether significant increases in cancer incidence occur in humans. Intriguingly, accumulating experimental evidence indicates that the linear no-threshold (LNT) hypothesis cannot be supported by the biological findings following LDIR exposure, demonstrating that LDIR exposure rather reduces the risk of spontaneous cancer [10–12]. Even though the phenomenon is not universal, multiple cellular and molecular data support the notion that LDIR-mediated adaptive responses induce hormesis and cell proliferation in normal cells, but not in cancer cells [13–15]. A half-century-old study first described how normal cells respond differently to LDIR exposure than cancerous cells of the same species; J.B. Little showed in 1968 that normal cells exhibited a transient G1 arrest after 100 mGy, implying that delaying in the cell cycle following LDIR exposure commences DNA synthesis machinery for subsequent cell proliferation [16]. Another study demonstrated that the downregulation of several cell-cycle-regulated genes occurs in normal human fibroblasts even at 100 mGy exposure of X-rays and cesium-137 γ-rays, which occurs in a p53-dependent manner and requires its effector p21$^{\mathrm{Waf1}}$ [17]. Coincidently, exposure of the same cells to doses $\leq$100 mGy of γ-rays upregulates the level of a protein (TCTP) involved in DNA repair [18].

The cell cycle pathway is one of the pivotal pathways that has been intricately connected with the cellular responses to radiation for many years. HDIR negatively impacts the progression of the cell cycle, and irreparable DNA damage caused by radiation causes the cycle to stall [19–22]. Even though there have been recent breakthroughs in understanding the molecular mechanisms of the cellular responses to LDIR, its role in the cell cycle arrests/enhanced progression remains unclear in different cellular contexts. There is increasing evidence that LDIR activates multiple signaling pathways to promote cell proliferation (e.g., activation of the Raf, AKT pathways) [23,24]. Through an intricate communication between DNA damage and cell cycle checkpoints, LDIR can confer radioprotection to normal cells as part of the adaptive response [25]. Several cell cycle regulators have been implicated in adaptive response following LDIR exposure via distinct mechanisms; for instance, in human keratinocytes, LDIR triggers cyclin D1 accumulation in the cytoplasm and regulates apoptosis [26]. The cell cycle arrest caused by LDR exposure was observed in human lymphoblast cells as an adaptive response that requires wild-type p53 [27]. The expression of p21$^{\mathrm{Waf1}}$, a cyclin-dependent kinase inhibitor that is regulated by p53, has also been reported to increase in human U397 cells and normal breast fibroblasts following LDIR exposure [28,29]. Another study confirmed transient and permanent G1 cell cycle arrest in human fibroblast populations after LDIR exposure (alpha particles) at a dose of 10 mGy, which was accompanied by increased p53 and p21$^{\mathrm{Waf1}}$ expression [30].

Increasing numbers of studies have been conducted to understand how LDIR induces hormesis, adaptive responses to subsequent challenge exposures, radioresistance, bystander effects, and genomic instability in various cellular and radiation exposure contexts. Our goal is to summarize recent advances related to the role of LDIR in cell cycle regulation. This review will also discuss how different cell cycle regulators are modulated by LDIR exposure in normal, cancerous, and stem cells. One of the major problems in LDIR research is that arbitrary exposure ranges are used on different cells and in different studies, making it challenging to extrapolate consistent outcomes. Therefore, this review has considered exposure of up to 200 mGy as LDIR to analyze cell cycle effects. Furthermore, a new conceptual mechanism is proposed to explain how LDIR differentially regulates the nucleocytoplasmic shuttling of p21, a key cell cycle regulator, when compared with HDIR.

2. Molecular Mechanisms of Low-Dose Ionizing Radiation

2.1. Controversy over the Linear No-Threshold Model

The linear no-threshold (LNT) model is commonly used to estimate the cancer risk assessment caused by ionizing radiation exposure and is based on extensive studies of the Japanese atomic bomb survivors. According to the LNT model, the radiation-induced radiological risk for cancer follows a linear relationship with no threshold between absorbed radiation dose and the incidence of cancer [31]. Several, mostly epidemiological, studies using HDIR and LDIR (below 100 mGy) have endorsed the LNT model to predict the risk of cancer or other radiological diseases [32–34]. An increasing number of experimental and epidemiological studies highlight inherent uncertainty in the LNT model, particularly at lower dose points, which raises controversy regarding the validity of this model [11,35,36]. In fact, nonlinear patterns are evident at low levels of radiation, demonstrating deterministic beneficial effects via activation of protective mechanisms that defend against disease, also known as hormesis [37–40].

2.2. Radiation Hormesis

Despite extensive studies on the chemical hormesis model, a specific radiation hormesis dose–response model is necessary for a better understanding of the cellular response to LDIR. Growing scientific evidence suggests that radiation hormesis caused by LDIR confers beneficial effects that outweigh any harmful, thereby reducing the risk of cancer [10,41,42]. Experimental evidence suggests that hormesis induced by LDIR provides a protective effect by increasing the proliferation of normal cells and stem cells, subsequent to a transient cell cycle arrest [10,36–38]. Many signaling pathways are activated in LDIR-induced hormesis, including Raf, AKT, ERK MAPK, and Wnt, which promote cell proliferation [14,23,43,44]. LDIR induces hormesis by triggering the DNA-damage repair mechanism and an augmented antioxidant response against reactive oxygen species (ROS), which protects chromosomes from mutations, potentially preventing neoplastic transformation [9,12,45,46]. Experimental evidence demonstrates that LDIR induced hormesis enhances innate immunity (through an increased abundance of dendritic cells, natural killer cells, and macrophage cells) and adaptive immunity (through CD4+ T cells, CD8+ T cells, and regulatory T cells), which helps to combat cancer [47–50]. Furthermore, LDIR-induced hormesis was also associated with significant modulations of cytokines or chemokines production; for example, an increase in stimulatory cytokines and a decrease in immuno-suppressive cytokines promote the proliferation of immune cells and confers anticancer immunity [49]. Figure 1 summarizes our current knowledge of LDIR-induced hormesis. Even so, how the cell cycle modulation in the context of LDIR confers hormesis remains unclear and, therefore, needs to be further investigated.

2.3. Biological Responses Associated with Low-Dose Ionizing Radiation

To better comprehend the potential beneficial or harmful impacts of LDIR on human health, radiation researchers have been increasingly focusing on several radiation responses such as adaptive responses, bystander effects, hypersensitivity, radioresistance, and genomic instability [51].

In adaptive response, a low priming radiation dose can stimulate intrinsic stress response mechanisms that allow cells to protect themselves from subsequent higher doses of radiation. In adaptive response, various cellular events occur, including activation of multiple signaling pathways, augmented DNA-damage response, increased antioxidant function, enhanced antiapoptotic function, modulation of mitochondrial function, and cell cycle regulation [52,53]. Experimental evidence confirmed that a number of cell cycle regulators, including CDK2, cyclin E, and p53, play a crucial role in mediating the radio-adaptive response [54].

Nonirradiated or nontargeted cells may develop biological effects due to the signal transmitted from adjacent radiation-exposed cells, which is called a bystander response. Multiple signaling pathways (NF-κB/NADPH oxidase/TGFβ pathways), gap junctional intercellular communication, cytokine release, ROS/RNS or nitric oxide (NO) production, and oxidized cell-free DNA play significant roles in the bystander responses [52,55]. Different cellular contexts (e.g., state of the cellular redox environment), radiation type (sparsely or densely ionizing), and levels of absorbed radiation determine the consequences of a bystander response—namely, whether it is beneficial or detrimental [51]. While most bystander effects resulting in adverse outcomes have been studied mainly with low fluences of densely ionizing radiations (α particles, high atomic number (Z) and high energy (E) HZE particles, as well as exposure to high doses of sparsely ionizing radiations (X-rays, γ-rays), beneficial bystander effects have been observed in several studies with low-dose exposures to sparsely ionizing radiations, with to the use of LDIR in anticancer treatments [55–57]. LDIR-mediated bystander effects on the regulation of cell cycle in nontargeted cells would be an engaging research topic in the future. Preliminary evidence shows the relevance of this topic to both radiotherapy and radiation risk assessment [58].

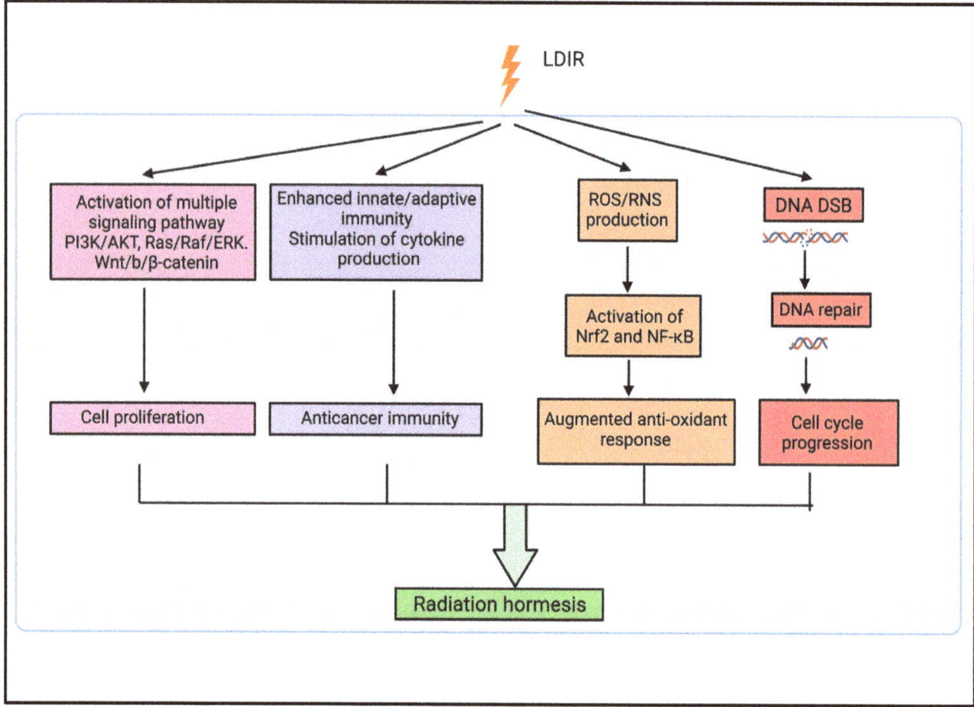

Figure 1. Multiple mechanisms through which LDIR induces hormesis. (RNS, reactive nitrogen species; DSB, double-stranded break).

Low-dose hyper-radiosensitivity (HRS) and increased radioresistance (IRR) are two phenomena that are associated with LDIR-induced biological responses. The in vitro experiments show that even very small doses of radiation (<100 mGy) can trigger HRS, but as the dose increases (>300 mGy), most cells will begin to display radioresistance until the dose reaches 1 Gy [59]; this phenomenon inherent to the T-N-PR model of radiation response described by John Calkins in 1973 to help understand how high resistance of living organisms is achieved in an environment where they are erratically or chronically exposed to injurious levels of irradiation [60]. Low-dose HRS is often linked to the impairment of DNA repair mechanisms, NO-mediated cell death, and premitotic cell cycle checkpoints; in particular, cells in the G2 phase of the cell cycle exhibit a stronger HRS effect [51,61,62]. Radiation exposure can activate multiple signaling pathways (JAK2/STAT3/AKT/ERK/JNK) and transcription factors (i.e., p53), which can modulate the radiosensitivity of cancer cells [63,64]. Since the description of the HRS phenomenon, there has been a growing interest in fractionated exposure to multiple LDIR and in low-dose-rate radiation due to its stimulation of system responses that lead to the more effective killing of radioresistant cancer cells [65–67]. The association between the radiation-induced arrest of G2 phase cells entering mitotic phase prematurely and low-dose HRS has been confirmed by experimental evidence [61]; however, the regulation of cell cycle in the context of HRS, IRR, and low-dose-rate radiation is poorly understood and requires further investigation at different cellular settings.

Recent studies have brought considerable attention to the effects of LDIR on genomic stability. In vitro LDIR exposure leads to chromosomal aberrations of lymphocytes in the human peripheral blood [68]. Another recent study demonstrated an interesting connection between the cell cycle and genomic instability caused by LDIR. The fractionated LDIR exposure can lead to persistent ROS accumulation in mitochondria and can disrupt the AKT/cyclin D1 signaling pathway. Subsequently, the nuclear accumulation of cyclin D1 can result in cell cycle retardation and genomic instability [69,70]. Based on experimental and epidemiological evidence, a recent metanalytic study attempted to estimate the lowest radiation dose that would lead to molecular changes; in this study, chromosomal aberration in cells began in a range of 1 mGy to 500 mGy and in animal models between 50 mGy and 100 mGy; chromosomal aberration has also been observed in children shortly after computed tomography scans with LDIR exposures less than 200 mGy [71].

3. Cell Cycle and Radiation

The cell cycle is a controlled process involving a complex network of regulatory mechanisms with appropriate checkpoints that contributes to cell growth, proliferation, and reproduction. There are four primary phases of the cell cycle: G1 (preparatory phase for division), S (chromosome replication), G2 (preparatory phase for mitosis), and M (mitosis, when chromosomes are distributed to two progeny cells) [72]. The cell cycle process is governed by a number of regulatory proteins, ensuring unidirectional and synchronized progression, which includes cyclin family proteins, cyclin-dependent kinases (CDKs), retinoblastoma protein, transcription factors (e.g., E2F), CDK inhibitors (e.g., $p16^{INK4}$ and $p21^{Waf1}$), CDC25 isoforms, p53 family proteins, and MDM2 [73]. It has been well established that radiation (regardless of type) disrupts the regular course of the cell cycle in normal cells, causing affected cells to stop at a checkpoint during the cell cycle. Radiation-induced DNA damage triggers the activation of G1/S, G2/M, and intra-S cell cycle checkpoints, consequently slowing the progress of radiation-exposed cells in the cell cycle [74]. Radiation-induced DNA damage is sensed by ATM/ATR kinases, whose downstream action initiates the DNA damage response and cell cycle arrest by activating the p53 pathway and its target proteins (i.e., $p21^{Waf1}$) [75,76]. Contrary to what is observed in normal cells, extensive in vitro work by Nagasawa et al. showed in various cancer cell types that exposure to radiation exposure does not induce G1 arrests in tumor cells regardless of the presence or absence of functional p53 [77]. In contrast, other studies showed that such arrests exist and occur in a P53-dependent mechanism, particularly in myeloid malignancies (e.g., lymphoma and myeloblastoma) in a p53-dependent mechanism [78–81]. Clearly, additional research is required to elucidate the role of the microenvironment in which the experiments are performed. Notably, research that addresses the role of cancer-associated fibroblasts (CAFs) in the induction of G1 arrests in cancer cells will be of particular interest. In this context, recent studies have shown that the presence of CAFs alongside cancer cells greatly contributes to the radioresponse of cancer cells [82].

Exposure of normal cells to radiation results in interruption of the G1/S transition, thereby halting further advancement into the S-phase progression, which allows more time to repair DNA damage prior to DNA replication. The arrest is often transient but can become permanent following exposure to HDIR. When double-stranded DNA breaks occur, a G2/M arrest occurs, which prevents cells from entering the M (mitosis) phase, and this cell cycle arrest in G2 allows the coordinated repair of the damage [83]. In severe cases of radiation, the recovery process can be delayed, and sometimes, irreparable DNA damage can lead to mitotic catastrophes that cause cell death [84]. It has been extensively discussed how HDIR modulates the cell cycle, but too little attention has been devoted to how LDIR affects the cell cycle, as discussed in the next section.

4. Effect of Low-Dose Ionizing Radiation on the Cell Cycle

Exposure of normal tissue to HDIR (>0.5 Gy) can induce persistent perturbations in molecular and cellular functions, which can lead to adverse effects on health. By contrast, the effects of LDIR have not been thoroughly explored in the context of fundamental cellular pathways/signaling, biological process, and molecular functions. Especially, the impact of LDIR on the cell cycle remains ambiguous, and it needs to be examined thoroughly in both cancer cells and noncancerous cells. Here, we summarize the evidence on how LDIR (<0.2 Gy) can regulate the cell cycle in different cell types, ranging from cancerous cells to normal cells and stem cells, and evaluate the clinical potential of these observations.

4.1. Cancer Cells

One of the earliest studies that examined the cell cycle response) to LDIR exposure in a human myeloid tumor cell line (ML-1 was published in 2002 and a microarray profile of cDNA was used to analyze global transcriptional response in this study [85]. This study revealed that LDIR exposure to 20 mGy and 50 mGy of γ-rays resulted in the overexpression of *CDKN1A* (encoding p21$^{\text{Waf1}}$) and *GADD45A* (encoding GADD45 alpha protein) genes without any detectable increase in apoptosis and showed that such an LDIR dose range is sufficient to induce a transient cell cycle arrest at the S phase. A study conducted concurrently in the same ML-1 cells also noted that low doses of radiation also resulted in rapid induction of *CDKN1A* and *GADD45A* mRNA [86]. It is noteworthy that cell cycle inhibitor p21$^{\text{Waf1}}$, a major transcriptional target of p53, commonly inhibits cyclin/CDK complexes [87,88]. In addition, *GADD45A*, a stress response gene, is also known to be involved in the regulation of the cell cycle [89].

The effect of LDIR on modulating the cell cycle in different cancer cells has been poorly studied compared to the effect of high doses. In a papillary thyroid carcinoma model, the effects of low-dose X-ray irradiation were examined with wild-type p53 (TPC-1) and mutated p53 (BCPAP) cells [90]. TPC-1 cells treated with an LDIR dose of 62.5 mGy of X-ray showed significant decreases in the fraction of cells in the S phase of the cell cycle, along with a concomitant upregulation of p16; however, no changes in cell cycle distribution were observed in BCPAP cells in response to LDIR.

Human prostate cells respond to LDIR via activation of the ATM/p53/p21 axis [91]. Prostate cancer cells PC-3 lacking functional p53 were observed to exhibit a significant S and G2/M phase arrest following 75 mGy exposure, whereas normal prostate cells RWPE-1 did not show detectable changes in cell cycle distribution. In the absence of functioning p53, the ATM/p21 pathway is activated in a p53-independent manner, providing an insight into radiotherapy treatment of prostate cancer [91].

Contrary to the findings in prostate cancer, p53-mutated breast cancer cell MDA-MB-231 responded differently upon LDIR exposure. A dose of 150 mGy significantly increased the growth of MDA-MB-231 cells and accelerated their entry into the S phase of the cell cycle, whereas the growth of Hs578Bst cells (harboring wild-type p53) remained unaltered [28]. Accordingly, LDIR exposure activates cyclin-dependent kinases with increased CDK4, CDK6, and cyclin D1 expression, along with a decreased expression of p21$^{\text{Waf1}}$ [28].

The role of p53 in DNA damage-induced arrest in G2 is unclear. In DT40 B-lymphoma cells that lack functional p53, LDIR (100 mGy) induced a Chk2-dependent G2 cell cycle arrest that prevented mitotic entry with damaged DNA [92]. Therefore, LDIR-induced cell cycle regulation is strongly influenced by the status of the p53 gene in cancer cells. Contrary to this notion, the capacity of p53 to mediate a radiation-induced G1 arrest in several human tumor cell lines is disputed by several lines of evidence [77,93]. A previous study investigated the association between low-dose hyper-radiosensitivity and activation of a novel arrest checkpoint in the G2 phase [94]. An asynchronous population of Chinese hamster V79 lung fibroblasts and human T98G glioblastoma cells exhibited significant low-dose hyper-radiosensitivity at doses <200 mGy and <300 mGy, respectively, and both displayed delayed mitosis of damaged G2-phase cells in response to LDIR. However, this study does not demonstrate the variability in LDIR thresholds for controlling progression within the cell cycle in different cells, nor does it examine the modulation of cell cycle regulators in the context of LDIR [94].

4.2. Normal Cells

LDIR was exploited in another microarray-based gene expression study to determine the biological effects of 100 mGy of X-rays on normal human lymphoblastoid cells (AHH-1) [95]. The results of this study confirm that LDIR exposure substantially modulates a range of signature genes, such as *GADD45A* and *CDKN2A*. Intriguingly, The *CDKN2A* gene encodes two well-known tumor suppressor proteins p16(INK4a) and the p14(ARF) proteins that are capable of controlling the cell cycle [96]. A recent review revealed that p53 is crucial in LDIR-induced hormesis, adaptive response, radioresistance, and genomic stability [52]. Nevertheless, it remains unclear how LDIR exposure modulates the dynamics of p53 transcription in different cellular contexts. There is increasing evidence that LDIR triggers the expression of CDKN1A [97], which is one of the major transcriptional targets of the p53 gene. Furthermore, p21^{Waf1} can also be regulated independently of p53 through other transcription factors as well as kinases [98].

Another study found that 100 mGy X-ray exposure stimulated cyclin D1 expression in HK-18 human keratinocytes, which is another important regulator of the cell cycle upon DNA damage [26]. However, that study found no detectable difference in cell cycle arrest between cells exposed to LDIR and those subsequently exposed to high dose radiation. Therefore, cyclin D1 does not affect cell cycle regulation in human keratinocytes following LDIR exposure; however, it might play a role in LDIR-induced adaptive radioresistance. A recent review discussed the roles of cyclin D1/CDK4 and cyclin B1/CDK1 in LDIR-induced adaptive responses as well as in the modulation of the mitochondrial signaling network in response to LDIR-induced DNA damage to coordinate cellular responses [70].

As conventional two-dimensional monolayer cell culture models do not replicate the complexity of the human tissue, researchers have expanded their study to three-dimensional (3D) tissue models to gain more insights into LDIR response. In a 3D skin model exposed to 0.1 Gy of LDIR or 1 Gy of HDIR (X-rays) and with subsequent post-radiation harvesting, RNA samples were analyzed for the global transcriptional response using microarrays [99]. As expected HDIR modulates a larger number of genes compared to LDIR over the course of the study (24 h). Intriguingly, global gene expression analysis revealed that LDIR exposure triggered a greater number of differentially expressed genes within the first 3 h. In response to LDIR, genes regulating cell cycle distribution displayed a different dynamic when compared with high doses. After 3 h post-irradiation, LDIR-exposed cells showed prolonged G1/S checkpoint arrest, supported by upregulation of p21^{Waf1}, Rb, and p130. At this point, LDIR significantly modulated eight cell-cycle-related pathways, whereas HDIR modulated only two. The findings of this study suggested that the initial response within 3 h of LDIR exposure was associated with tissue protection and enhanced survival against stress [99].

A recent in vitro study indicates that LDIR exposure of 100 mGy to primary keratinocytes and U937 (lymphoma) cell lines causes cell cycle arrest and impairs protein synthesis; however, the phase of the cell cycle in which the arrest occurs is not discussed [29].

4.3. Stem Cells

Hormesis effects of LDIR on rat mesenchymal stem cells (MSCs) have also been assessed in a recent study through the investigation of proliferation and signaling pathway (MEK/ERK) activation [14]. Cell proliferation was significantly augmented in MSCs exposed to LDIR at doses of 50 mGy and 75 mGy, and there was a significant shift from the G1 phase to the S phase in response to 75 mGy X-rays. A panel of human embryonic stem cells (hESCs) was also studied to identify the radiobiological effects of LDIR on human cells [97]. *CDKN1A* gene expression was elevated significantly in only one of the hESCs, and it occurred only at 2 h of LDIR exposure (50 mGy). However, the gene expression profiles for the cell-cycle-related genes did not show any linear dose–response relationship, even at the lowest doses of ionizing radiation exposure. Human MSCs also undergo a brief cell cycle arrest at the G1 phase after LDIR exposure (100 mGy) as part of the adaptive responses [100].

Another study examined the effect of LDIR on neural stem/progenitor cells (NSPCs), which were isolated from the subventricular zone of newborn mouse pups and further analyzed for proliferation, self-renewal, and differentiation [22]. There was no apparent increase in the fraction of dying cells shortly after exposure to 100 mGy of cesium-137 γ-rays, and the number of single cells that formed neurospheres was not significantly different from the control. Upon differentiation, irradiated neural precursors did not differ from those that were not irradiated in their ability to generate neurons, astrocytes, and oligodendrocytes. By contrast, the progression of NSPCs through the cell cycle decreased dramatically after exposure to higher doses of radiation.

5. LDIR Might Regulate Nucleocytoplasmic Localization of p21$^{\text{Waf1}}$

Carcinogenesis or neoplasia is characterized by a break in the control of the cell cycle [101]. Recently, p21$^{\text{Waf1}}$ has gained attention for its crucial role in many fundamental biological processes beyond cell cycle regulation including DNA replication and repair, gene transcription, apoptosis, and cell motility [98,102–104]. In contrast to its role as an inhibitor of the cell cycle, p21$^{\text{Waf1}}$ expression is found to be correlated positively with tumor aggressiveness in a variety of cancers [98]. Moreover, p53-independent regulation of p21$^{\text{Waf1}}$ and its ability to nucleo-cytoplasmic shuttling are other aspects of its dichotomous oncogenic nature. The nuclear p21$^{\text{Waf1}}$ favors its cell cycle inhibitory function, whereas the cytoplasmic p21$^{\text{Waf1}}$ expression is oncogenic [105,106]. When aberrant signaling is triggered by extrinsic factors (including radiation), p21$^{\text{Waf1}}$ might be phosphorylated and phosphorylation causes its localization to the cytoplasm, where it exerts its antiapoptotic or oncogenic properties [98].

The radiation level might have a major impact on the nucleocytoplasmic localization of p21$^{\text{Waf1}}$, although it is unclear how this occurs. It is likely that in normal cells HDIR might induce several kinases that phosphorylate p21$^{\text{Waf1}}$ and cause its localization to the cytoplasm, where it might promote oncogenesis through various mechanisms (e.g., regulation of the nuclear assembly of the cyclin D1.CDK4 complex or conferring protection against apoptosis) [107,108]. Conversely, LDIR might inhibit phosphorylation of p21$^{\text{Waf1}}$ (or promote dephosphorylation of p21$^{\text{Waf1}}$), thereby exhibiting its nuclear cell cycle inhibitory or tumor suppressor action, which is beneficial in controlling oncogenesis (Figure 2). To confirm this hypothetical concept, a more systematic and extensive investigation must be conducted to ascertain the LDIR-mediated regulation of p21$^{\text{Waf1}}$ in different cellular contexts.

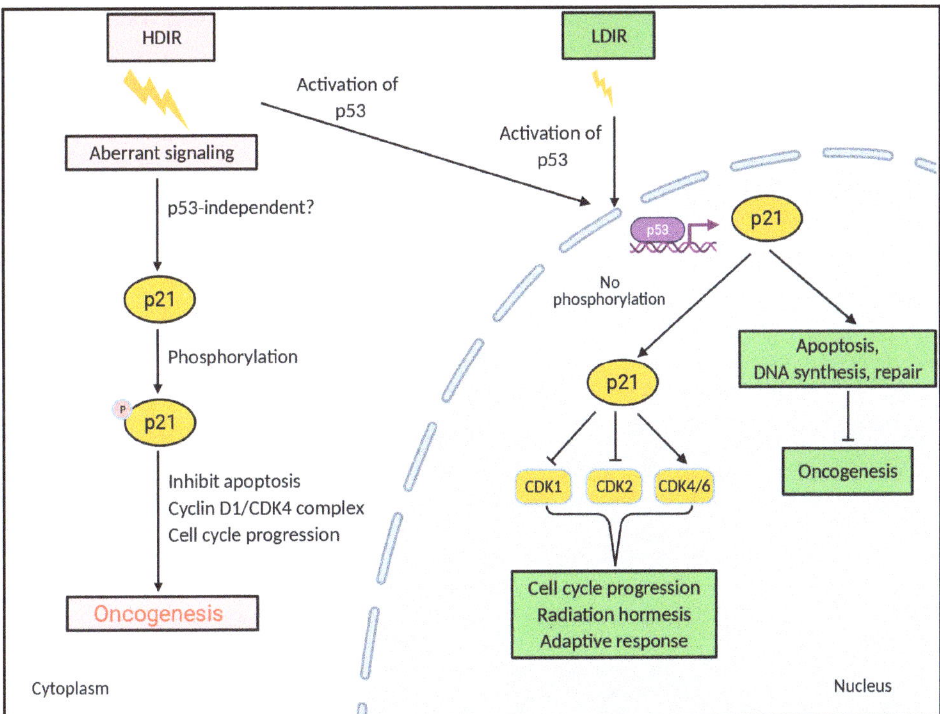

Figure 2. Conceptual diagram showing that radiation dose might differentially regulate the nucleocytoplasmic shuttling of p21$^{\text{Waf1}}$. The nuclear function of p21$^{\text{Waf1}}$ is predominantly cell cycle inhibitory and tumor suppressive, whereas cytoplasmic p21$^{\text{Waf1}}$ can exert antiapoptotic functions. Aberrant cell signaling induced by HDIR might cause phosphorylation of p21$^{\text{Waf1}}$ leading to its cytoplasmic localization. Similar to LDIR, HDIR also induces p53-dependent transcription of p21$^{\text{Waf1}}$ and p21$^{\text{Waf1}}$ nuclear functions.

6. Future Directions

To assess the risk and benefits of LDIR in humans, ongoing studies are investigating molecular, cellular, and tissue responses in different genetic and environmental backgrounds. It is known that LDIR can modulate many cell cycle regulators in normal, tumor, and stem cells, which might have different implications in a particular context. The coordination of cell cycle activities in response to low doses of radiation must be extensively studied to learn how cells cope with genotoxic stress; such studies could lead to the development of cell-cycle-targeted therapies for many diseases, including cancer. It would also be interesting to compare the transition through cell cycle phases after LDIR exposure with the determination of genes associated with early cell cycle response.

LDIR can induce a number of cell cycle regulators, including p53/p21$^{\text{Waf1}}$, which are known to be involved in several cell-intrinsic functions (such as apoptosis and senescence) besides cell cycle inhibition. LDIR-exposed nontransformed cells should also be examined in terms of cell cycle regulation and the key players involved. This would be a promising innovative solution for eliminating precancerous cells from the population through the LDIR-stimulated bystander effects. A recent study demonstrated that LDIR exposure on HPKs propagates effects that induce cell cycle arrest and protein synthesis repression in U937 cells; this bystander effect induced by LDIR in U937 cells is associated with increased expression of p21$^{\text{Waf1}}$ via TGFβ signaling and activation of p38 MAPK activation [29]. It is interesting to note that p53 plays an important role in communicating a stress response to neighboring cells, in addition to coordinating intrinsic and extrinsic cell signals [109]. Due to its key role in controlling the cell cycle, additional research on the role of the tumor suppressor p53 in studies of bystander effects in normal and cancer cells is warranted.

Regulation of the cell cycle is connected to many important biological processes, metabolic pathways, various stress responses, DNA damage responses, DNA repair, etc. The differential regulation of cell cycle mechanisms in response to LDIR and HDIR might help us determine LDIR's beneficial or detrimental effects. LDIR-induced regulation of cell cycle and radioresistance in stem cells are the next frontiers to be explored, which can lead to advances in regenerative medicine and tissue engineering.

The cancer cell cycle study is a complex area to explore due to its heterogeneous mutational nature and the multitude of deregulated signaling pathways behind it. Due to the heterogeneous mutational landscape prevailing in cancers, LDIR may affect cell cycle distribution and multiple cellular processes differently in different cancers. Nevertheless, the advent of modern genomics, transcriptomics, and proteomics technology can provide a deeper understanding of the global expression and modulation of genes and proteins associated with the cell cycle in cancer cells maintained in vitro or in vivo response of tumors to LDIR. Such research would provide us with a more comprehensive picture of the regime of cell cycle regulation in cancer cells and would illustrate the effects of nearly all mutations implicated in cancer following LDIR exposure.

Pulse low-dose-rate irradiation (PLDR) has gained increasing acceptance in recent years as the treatment of choice for recurrent cancers and radiation-resistant cancers [110,111]. As PLDR takes advantage of the phenomenon of HRS, it is less toxic to normal cells. The effects of PLDR on the cell cycle have not been well explored. In a recent study, PLDR has been shown to be effective against radioresistant head and neck squamous cell carcinoma (HNSCC), which showed more low-dose HRS than an isogenic population of non-radioresistant parental HNSCC cells; as a result of PLDR-mediated G2/M arrest, the cell cycle resumed 24 h after irradiation on parental cells but not on radioresistant cells [112]. Therefore, it is vital to map out the landscape of cell cycle regulators in the context of radioresistant and radiosensitive cells as well as normal cells.

Recently, several new tools for cell cycle analysis, such as the fluorescence ubiquitination cell-cycle indicator (FUCCI) system, chromobodies, and Cycletest reagent, have been developed. For example, the FUCCI system is a technology that uses the cell cycle's phase-specific expression of proteins and their degradation by the ubiquitin–proteasome degradation system to color-coded phases of the cell cycle in real time. The new tools, combined with time-lapse imaging, will allow us to explore the spatiotemporal patterns of cell cycle dynamics after exposure to LDIR.

Author Contributions: M.G.M.K. prepared the original draft; Y.W. reviewed and edited the manuscript; Y.W. acquired the funding. All authors have read and agreed to the published version of the manuscript.

Funding: Funding and support for the research was provided by a Federal Nuclear Science and Technology Work Plan of the Atomic Energy of Canada Limited (AECL) (FST-51320.50.19.07) and CANDU Owners Group (COG) (COG-73600).

Institutional Review Board Statement: Not applicable.

Informed Consent Statement: Not applicable.

Acknowledgments: Funding and support for the research were provided by a Federal Nuclear Science and Technology Work Plan of the AECL and COG. The authors are especially grateful to Edouard Azzam and Helena Rummens for the inspiring review idea and suggestions.

Conflicts of Interest: The authors declare no conflict of interest.

References

1. Hall, E.J.; Giaccia, A. *Radiobiology for the Radiologist*; Elsevier Inc.: Amsterdam, The Netherlands, 2012.
2. Grubbe, E.H. The Origin and Birth of X-ray Therapy. *Urol. Cutan. Rev.* **1947**, *51*, 375–379.

3. Shah, D.J.; Sachs, R.K.; Wilson, D.J. Radiation-Induced Cancer: A Modern View. *Br. J. Radiol.* **2012**, *85*, e1166–e1173. [CrossRef]
4. Baskar, R.; Lee, K.A.; Yeo, R.; Yeoh, K.-W. Cancer and Radiation Therapy: Current Advances and Future Directions. *Int. J. Med. Sci.* **2012**, *9*, 193–199. [CrossRef]
5. Mettler, F.A.; Thomadsen, B.R.; Bhargavan, M.; Gilley, D.B.; Gray, J.E.; Lipoti, J.A.; McCrohan, J.; Yoshizumi, T.T.; Mahesh, M. Medical Radiation Exposure in the U.S. in 2006: Preliminary Results. *Health Phys.* **2008**, *95*, 502–507. [CrossRef]
6. Einstein, A.J. Medical Radiation Exposure to the U.S. Population: The Turning Tide. *Radiology* **2020**, *295*, 428–429. [CrossRef] [PubMed]
7. Sokolov, M.; Neumann, R. Global Gene Expression Alterations as a Crucial Constituent of Human Cell Response to Low Doses of Ionizing Radiation Exposure. *Int. J. Mol. Sci.* **2015**, *17*, 55. [CrossRef] [PubMed]
8. Jia, C.; Wang, Q.; Yao, X.; Yang, J. The Role of DNA Damage Induced by Low/High Dose Ionizing Radiation in Cell Carcinogenesis. *Explor. Res. Hypothesis Med.* **2021**, *6*, 177–184. [CrossRef]
9. Feinendegen, L.E.; Pollycove, M.; Sondhaus, C.A. Responses to Low Doses of Ionizing Radiation in Biological Systems. *Nonlinearity Biol. Toxicol. Med.* **2004**, *2*, 154014204905074. [CrossRef] [PubMed]
10. Tubiana, M. Dose–Effect Relationship and Estimation of the Carcinogenic Effects of Low Doses of Ionizing Radiation: The Joint Report of the Académie Des Sciences (Paris) and of the Académie Nationale de Médecine. *Int. J. Radiat. Oncol.* **2005**, *63*, 317–319. [CrossRef] [PubMed]
11. Tubiana, M.; Feinendegen, L.E.; Yang, C.; Kaminski, J.M. The Linear No-Threshold Relationship Is Inconsistent with Radiation Biologic and Experimental Data. *Radiology* **2009**, *251*, 13–22. [CrossRef]
12. Azzam, E.I.; de Toledo, S.M.; Raaphorst, G.P.; Mitchel, R.E. Low-Dose Ionizing Radiation Decreases the Frequency of Neoplastic Transformation to a Level below the Spontaneous Rate in C3H 10T1/2 Cells. *Radiat. Res.* **1996**, *146*, 369–373. [CrossRef]
13. Li, W.; Wang, G.; Cui, J.; Xue, L.; Cai, L. Low-Dose Radiation (LDR) Induces Hematopoietic Hormesis: LDR-Induced Mobilization of Hematopoietic Progenitor Cells into Peripheral Blood Circulation. *Exp. Hematol.* **2004**, *32*, 1088–1096. [CrossRef]
14. Liang, X.; So, Y.H.; Cui, J.; Ma, K.; Xu, X.; Zhao, Y.; Cai, L.; Li, W. The Low-Dose Ionizing Radiation Stimulates Cell Proliferation via Activation of the MAPK/ERK Pathway in Rat Cultured Mesenchymal Stem Cells. *J. Radiat. Res.* **2011**, *52*, 380–386. [CrossRef] [PubMed]
15. Jiang, H.; Xu, Y.; Li, W.; Ma, K.; Cai, L.; Wang, G. Low-Dose Radiation Does Not Induce Proliferation in Tumor Cells in Vitro and in Vivo. *Radiat. Res.* **2008**, *170*, 477–487. [CrossRef] [PubMed]
16. Little, J.B. Delayed Initiation of DNA Synthesis in Irradiated Human Diploid Cells. *Nature* **1968**, *218*, 1064–1065. [CrossRef] [PubMed]
17. De Toledo, S.M.; Azzam, E.I.; Keng, P.; Laffrenier, S.; Little, J.B. Regulation by Ionizing Radiation of CDC2, Cyclin A, Cyclin B, Thymidine Kinase, Topoisomerase IIalpha, and RAD51 Expression in Normal Human Diploid Fibroblasts Is Dependent on P53/P21Waf1. *Cell Growth Differ. Mol. Biol. J. Am. Assoc. Cancer Res.* **1998**, *9*, 887–896.
18. Zhang, J.; de Toledo, S.M.; Pandey, B.N.; Guo, G.; Pain, D.; Li, H.; Azzam, E.I. Role of the Translationally Controlled Tumor Protein in DNA Damage Sensing and Repair. *Proc. Natl. Acad. Sci. USA* **2012**, *109*, E926–E933. [CrossRef]
19. Wilson, G.D. Radiation and the Cell Cycle, Revisited. *Cancer Metastasis Rev.* **2004**, *23*, 209–225. [CrossRef]
20. Lonati, L.; Barbieri, S.; Guardamagna, I.; Ottolenghi, A.; Baiocco, G. Radiation-Induced Cell Cycle Perturbations: A Computational Tool Validated with Flow-Cytometry Data. *Sci. Rep.* **2021**, *11*, 925. [CrossRef]
21. Iliakis, G.; Wang, Y.; Guan, J.; Wang, H. DNA Damage Checkpoint Control in Cells Exposed to Ionizing Radiation. *Oncogene* **2003**, *22*, 5834–5847. [CrossRef]
22. Chen, H.; Goodus, M.T.; de Toledo, S.M.; Azzam, E.I.; Levison, S.W.; Souayah, N. Ionizing Radiation Perturbs Cell Cycle Progression of Neural Precursors in the Subventricular Zone Without Affecting Their Long-Term Self-Renewal. *ASN Neuro* **2015**, *7*, 1–16. [CrossRef] [PubMed]
23. Kim, C.S.; Kim, J.K.; Nam, S.Y.; Yang, K.H.; Jeong, M.; Kim, H.S.; Kim, C.S.; Jin, Y.-W.; Kim, J. Low-Dose Radiation Stimulates the Proliferation of Normal Human Lung Fibroblasts via a Transient Activation of Raf and Akt. *Mol. Cells* **2007**, *24*, 424–430. [PubMed]
24. Yang, G.; Li, W.; Jiang, H.; Liang, X.; Zhao, Y.; Yu, D.; Zhou, L.; Wang, G.; Tian, H.; Han, F.; et al. Low-Dose Radiation May Be a Novel Approach to Enhance the Effectiveness of Cancer Therapeutics. *Int. J. Cancer* **2016**, *139*, 2157–2168. [CrossRef] [PubMed]
25. Olivieri, G.; Bodycote, J.; Wolff, S. Adaptive Response of Human Lymphocytes to Low Concentrations of Radioactive Thymidine. *Science* **1984**, *223*, 594–597. [CrossRef] [PubMed]
26. Ahmed, K.M.; Fan, M.; Nantajit, D.; Cao, N.; Li, J.J. Cyclin D1 in Low-Dose Radiation-Induced Adaptive Resistance. *Oncogene* **2008**, *27*, 6738–6748. [CrossRef]
27. Hendrikse, S.; Hunter, A.J.; Keraan, M.; Blekkenhorst, G.H. Effects of Low Dose Irradiation on TK6 and U937 Cells: Induction of P53 and Its Role in Cell-Cycle Delay and the Adaptive Response. *Int. J. Radiat. Biol.* **2000**, *76*, 11–21. [CrossRef]
28. Li, S.-J.; Liang, X.-Y.; Li, H.-J.; Li, W.; Zhou, L.; Chen, H.-Q.; Ye, S.-G.; Yu, D.-H.; Cui, J.-W. Low-Dose Irradiation Promotes Proliferation of the Human Breast Cancer MDA-MB-231 Cells through Accumulation of Mutant P53. *Int. J. Oncol.* **2017**, *50*, 290–296. [CrossRef] [PubMed]
29. Sekihara, K.; Saitoh, K.; Yang, H.; Kawashima, H.; Kazuno, S.; Kikkawa, M.; Arai, H.; Miida, T.; Hayashi, N.; Sasai, K.; et al. Low-Dose Ionizing Radiation Exposure Represses the Cell Cycle and Protein Synthesis Pathways in in Vitro Human Primary Keratinocytes and U937 Cell Lines. *PLoS ONE* **2018**, *13*, e0199117. [CrossRef]

30. Azzam, E.I.; de Toledo, S.M.; Waker, A.J.; Little, J.B. High and Low Fluences of Alpha-Particles Induce a G1 Checkpoint in Human Diploid Fibroblasts. *Cancer Res.* **2000**, *60*, 2623–2631.
31. Lindell, B.; Sowby, D. The 1958 UNSCEAR Report. *J. Radiol. Prot.* **2008**, *28*, 277–282. [CrossRef]
32. Zielinski, J.; Garner, M.; Band, P.; Krewski, D.; Shilnikova, N.; Jiang, H.; Ashmore, P.; Sont, W.; Fair, M.; Letourneau, E.; et al. Health Outcomes of Low-Dose Ionizing Radiation Exposure among Medical Workers: A Cohort Study of the Canadian National Dose Registry of Radiation Workers. *Int. J. Occup. Med. Environ. Health* **2009**, *22*, 149–156. [CrossRef]
33. Hazelton, W.D.; Moolgavkar, S.H.; Curtis, S.B.; Zielinski, J.M.; Patrick Ashmore, J.; Krewski, D. Biologically Based Analysis of Lung Cancer Incidence in a Large Canadian Occupational Cohort with Low-Dose Ionizing Radiation Exposure, and Comparison with Japanese Atomic Bomb Survivors. *J. Toxicol. Environ. Health Part A* **2006**, *69*, 1013–1038. [CrossRef] [PubMed]
34. Zablotska, L.B.; Little, M.P.; Cornett, R.J. Potential Increased Risk of Ischemic Heart Disease Mortality With Significant Dose Fractionation in the Canadian Fluoroscopy Cohort Study. *Am. J. Epidemiol.* **2014**, *179*, 120–131. [CrossRef] [PubMed]
35. Sanders, C.L. (Ed.) *Radiation Hormesis and the Linear-No-Threshold Assumption*; Springer: Berlin/Heidelberg, Geramny, 2010; ISBN 978-3-642-03719-1.
36. Golden, R.; Bus, J.; Calabrese, E. An Examination of the Linear No-Threshold Hypothesis of Cancer Risk Assessment: Introduction to a Series of Reviews Documenting the Lack of Biological Plausibility of LNT. *Chem. Biol. Interact.* **2019**, *301*, 2–5. [CrossRef]
37. Luckey, T.D. Radiation Hormesis: The Good, the Bad, and the Ugly. *Dose-Response* **2006**, *4*, 169–190. [CrossRef] [PubMed]
38. Lehrer, S.; Rosenzweig, K.E. Lung Cancer Hormesis in High Impact States Where Nuclear Testing Occurred. *Clin. Lung Cancer* **2015**, *16*, 152–155. [CrossRef]
39. Dobrzyński, L.; Fornalski, K.W.; Feinendegen, L.E. Cancer Mortality Among People Living in Areas With Various Levels of Natural Background Radiation. *Dose-Response* **2015**, *13*, 155932581559239. [CrossRef]
40. Tharmalingam, S.; Sreetharan, S.; Brooks, A.L.; Boreham, D.R. Re-Evaluation of the Linear No-Threshold (LNT) Model Using New Paradigms and Modern Molecular Studies. *Chem. Biol. Interact.* **2019**, *301*, 54–67. [CrossRef]
41. Kant, K.; Chauhan, R.P.; Sharma, G.S.; Chakarvarti, S.K. Hormesis in Humans Exposed to Low-Level Ionising Radiation. *Int. J. Low Radiat.* **2003**, *1*, 76. [CrossRef]
42. Mitchel, R.E.J. Low Doses of Radiation Are Protective in Vitro and in Vivo: Evolutionary Origins. *Dose-Response* **2006**, *4*, 75–90. [CrossRef]
43. Kim, C.S.; Kim, J.-M.; Nam, S.Y.; Yang, K.H.; Jeong, M.; Kim, H.S.; Lim, Y.-K.; Kim, C.S.; Jin, Y.-W.; Kim, J. Low-Dose of Ionizing Radiation Enhances Cell Proliferation via Transient ERK1/2 and P38 Activation in Normal Human Lung Fibroblasts. *J. Radiat. Res.* **2007**, *48*, 407–415. [CrossRef]
44. Wei, L.-C.; Ding, Y.-X.; Liu, Y.-H.; Duan, L.; Bai, Y.; Shi, M.; Chen, L.-W. Low-Dose Radiation Stimulates Wnt/β-Catenin Signaling, Neural Stem Cell Proliferation and Neurogenesis of the Mouse Hippocampus in Vitro and in Vivo. *Curr. Alzheimer Res.* **2012**, *9*, 278–289. [CrossRef]
45. Rothkamm, K.; Lobrich, M. Evidence for a Lack of DNA Double-Strand Break Repair in Human Cells Exposed to Very Low x-Ray Doses. *Proc. Natl. Acad. Sci. USA* **2003**, *100*, 5057–5062. [CrossRef]
46. Scott, B.R. Low-Dose Radiation-Induced Protective Process and Implications for Risk Assessment, Cancer Prevention, and Cancer Therapy. *Dose-Response* **2007**, *5*, 131–149. [CrossRef]
47. Pandey, R.; Shankar, B.S.; Sharma, D.; Sainis, K.B. Low Dose Radiation Induced Immunomodulation: Effect on Macrophages and CD8 + T Cells. *Int. J. Radiat. Biol.* **2005**, *81*, 801–812. [CrossRef]
48. Shigematsu, A.; Adachi, Y.; Koike-Kiriyama, N.; Suzuki, Y.; Iwasaki, M.; Koike, Y.; Nakano, K.; Mukaide, H.; Imamura, M.; Ikehara, S. Effects of Low-Dose Irradiation on Enhancement of Immunity by Dendritic Cells. *J. Radiat. Res.* **2007**, *48*, 51–55. [CrossRef]
49. Yang, G.; Kong, Q.; Wang, G.; Jin, H.; Zhou, L.; Yu, D.; Niu, C.; Han, W.; Li, W.; Cui, J. Low-Dose Ionizing Radiation Induces Direct Activation of Natural Killer Cells and Provides a Novel Approach for Adoptive Cellular Immunotherapy. *Cancer Biother. Radiopharm.* **2014**, *29*, 428–434. [CrossRef] [PubMed]
50. Wang, B.; Li, B.; Dai, Z.; Ren, S.; Bai, M.; Wang, Z.; Li, Z.; Lin, S.; Wang, Z.; Huang, N.; et al. Low-Dose Splenic Radiation Inhibits Liver Tumor Development of Rats through Functional Changes in CD4$^+$CD25$^+$Treg Cells. *Int. J. Biochem. Cell Biol.* **2014**, *55*, 98–108. [CrossRef]
51. Lau, Y.S.; Chew, M.T.; Alqahtani, A.; Jones, B.; Hill, M.A.; Nisbet, A.; Bradley, D.A. Low Dose Ionising Radiation-Induced Hormesis: Therapeutic Implications to Human Health. *Appl. Sci.* **2021**, *11*, 8909. [CrossRef]
52. Tang, F.R.; Loke, W.K. Molecular Mechanisms of Low Dose Ionizing Radiation-Induced Hormesis, Adaptive Responses, Radioresistance, Bystander Effects, and Genomic Instability. *Int. J. Radiat. Biol.* **2015**, *91*, 13–27. [CrossRef] [PubMed]
53. Guéguen, Y.; Bontemps, A.; Ebrahimian, T.G. Adaptive Responses to Low Doses of Radiation or Chemicals: Their Cellular and Molecular Mechanisms. *Cell. Mol. Life Sci.* **2019**, *76*, 1255–1273. [CrossRef] [PubMed]
54. Saini, D.; Shelke, S.; Mani Vannan, A.; Toprani, S.; Jain, V.; Das, B.; Seshadri, M. Transcription Profile of DNA Damage Response Genes at G0 Lymphocytes Exposed to Gamma Radiation. *Mol. Cell. Biochem.* **2012**, *364*, 271–281. [CrossRef] [PubMed]
55. Portess, D.I.; Bauer, G.; Hill, M.A.; O'Neill, P. Low-Dose Irradiation of Nontransformed Cells Stimulates the Selective Removal of Precancerous Cells via Intercellular Induction of Apoptosis. *Cancer Res.* **2007**, *67*, 1246–1253. [CrossRef]

56. Ermakov, A.V.; Konkova, M.S.; Kostyuk, S.V.; Egolina, N.A.; Efremova, L.V.; Veiko, N.N. Oxidative Stress as a Significant Factor for Development of an Adaptive Response in Irradiated and Nonirradiated Human Lymphocytes after Inducing the Bystander Effect by Low-Dose X-Radiation. *Mutat. Res./Fundam. Mol. Mech. Mutagenesis* **2009**, *669*, 155–161. [CrossRef] [PubMed]

57. Abdelrazzak, A.B.; Stevens, D.L.; Bauer, G.; O'Neill, P.; Hill, M.A. The Role of Radiation Quality in the Stimulation of Intercellular Induction of Apoptosis in Transformed Cells at Very Low Doses. *Radiat. Res.* **2011**, *176*, 346–355. [CrossRef] [PubMed]

58. Leung, C.N.; Canter, B.S.; Rajon, D.; Bäck, T.A.; Fritton, J.C.; Azzam, E.I.; Howell, R.W. Dose-Dependent Growth Delay of Breast Cancer Xenografts in the Bone Marrow of Mice Treated with 223Ra: The Role of Bystander Effects and Their Potential for Therapy. *J. Nucl. Med.* **2020**, *61*, 89–95. [CrossRef]

59. Joiner, M.C.; Marples, B.; Lambin, P.; Short, S.C.; Turesson, I. Low-Dose Hypersensitivity: Current Status and Possible Mechanisms. *Int. J. Radiat. Oncol.* **2001**, *49*, 379–389. [CrossRef]

60. Calkins, J. The T-N-PR Model of Radiation Response. *J. Theor. Biol.* **1973**, *39*, 609–622. [CrossRef]

61. Marples, B. Is Low-Dose Hyper-Radiosensitivity a Measure of G2-Phase Cell Radiosensitivity? *Cancer Metastasis Rev.* **2004**, *23*, 197–207. [CrossRef]

62. Marples, B.; Wouters, B.G.; Collis, S.J.; Chalmers, A.J.; Joiner, M.C. Low-Dose Hyper-Radiosensitivity: A Consequence of Ineffective Cell Cycle Arrest of Radiation-Damaged G2 -Phase Cells. *Radiat. Res.* **2004**, *161*, 247–255. [CrossRef]

63. Shimura, T. Targeting the AKT/Cyclin D1 Pathway to Overcome Intrinsic and Acquired Radioresistance of Tumors for Effective Radiotherapy. *Int. J. Radiat. Biol.* **2017**, *93*, 381–385. [CrossRef] [PubMed]

64. Park, S.-Y.; Lee, C.-J.; Choi, J.-H.; Kim, J.-H.; Kim, J.-W.; Kim, J.-Y.; Nam, J.-S. The JAK2/STAT3/CCND2 Axis Promotes Colorectal Cancer Stem Cell Persistence and Radioresistance. *J. Exp. Clin. Cancer Res.* **2019**, *38*, 399. [CrossRef] [PubMed]

65. Thomas, C.; Fertil, B.; Foray, N. Very Low-Dose Hyper-Radiosensitivity: Impact for Radiotherapy of Micrometastases. *Cancer Radiother. J. Soc. Fr. Radiother. Oncol.* **2007**, *11*, 260–265. [CrossRef]

66. Amundson, S.A.; Lee, R.A.; Koch-Paiz, C.A.; Bittner, M.L.; Meltzer, P.; Trent, J.M.; Fornace, A.J. Differential Responses of Stress Genes to Low Dose-Rate Gamma Irradiation. *Mol. Cancer Res. MCR* **2003**, *1*, 445–452. [PubMed]

67. Short, C.; Kelly, J.; Mayes, C.R.; Woodcock, M.; Joiner, M.C. Low-Dose Hypersensitivity after Fractionated Low-Dose Irradiation in Vitro. *Int. J. Radiat. Biol.* **2001**, *77*, 655–664. [CrossRef] [PubMed]

68. Tewari, S.; Khan, K.; Husain, N.; Rastogi, M.; Mishra, S.P.; Srivastav, A.K. Peripheral Blood Lymphocytes as In Vitro Model to Evaluate Genomic Instability Caused by Low Dose Radiation. *Asian Pac. J. Cancer Prev.* **2016**, *17*, 1773–1777. [CrossRef] [PubMed]

69. Shimura, T.; Sasatani, M.; Kamiya, K.; Kawai, H.; Inaba, Y.; Kunugita, N. Mitochondrial Reactive Oxygen Species Perturb AKT/Cyclin D1 Cell Cycle Signaling via Oxidative Inactivation of PP2A in Lowdose Irradiated Human Fibroblasts. *Oncotarget* **2016**, *7*, 3559–3570. [CrossRef]

70. Alexandrou, A.T.; Li, J.J. Cell Cycle Regulators Guide Mitochondrial Activity in Radiation-Induced Adaptive Response. *Antioxid. Redox Signal.* **2014**, *20*, 1463–1480. [CrossRef]

71. Shimura, N.; Kojima, S. The Lowest Radiation Dose Having Molecular Changes in the Living Body. *Dose-Response* **2018**, *16*, 1–17. [CrossRef]

72. Harashima, H.; Dissmeyer, N.; Schnittger, A. Cell Cycle Control across the Eukaryotic Kingdom. *Trends Cell Biol.* **2013**, *23*, 345–356. [CrossRef]

73. Barnum, K.J.; O'Connell, M.J. Cell Cycle Regulation by Checkpoints. In *Cell Cycle Control*; Humana Press: New York, NY, USA, 2014; pp. 29–40.

74. Sia, J.; Szmyd, R.; Hau, E.; Gee, H.E. Molecular Mechanisms of Radiation-Induced Cancer Cell Death: A Primer. *Front. Cell Dev. Biol.* **2020**, *8*, 41. [CrossRef]

75. Niida, H.; Nakanishi, M. DNA Damage Checkpoints in Mammals. *Mutagenesis* **2006**, *21*, 3–9. [CrossRef]

76. Maier, P.; Hartmann, L.; Wenz, F.; Herskind, C. Cellular Pathways in Response to Ionizing Radiation and Their Targetability for Tumor Radiosensitization. *Int. J. Mol. Sci.* **2016**, *17*, 102. [CrossRef]

77. Nagasawa, H.; Li, C.Y.; Maki, C.G.; Imrich, A.C.; Little, J.B. Relationship between Radiation-Induced G1 Phase Arrest and P53 Function in Human Tumor Cells. *Cancer Res.* **1995**, *55*, 1842–1846. [PubMed]

78. El-Deiry, W.S.; Harper, J.W.; O'Connor, P.M.; Velculescu, V.E.; Canman, C.E.; Jackman, J.; Pietenpol, J.A.; Burrell, M.; Hill, D.E.; Wang, Y. WAF1/CIP1 Is Induced in P53-Mediated G1 Arrest and Apoptosis. *Cancer Res.* **1994**, *54*, 1169–1174. [PubMed]

79. Fei, P.; El-Deiry, W.S. P53 and Radiation Responses. *Oncogene* **2003**, *22*, 5774–5783. [CrossRef] [PubMed]

80. Kastan, M.B.; Radin, A.I.; Kuerbitz, S.J.; Onyekwere, O.; Wolkow, C.A.; Civin, C.I.; Stone, K.D.; Woo, T.; Ravindranath, Y.; Craig, R.W. Levels of P53 Protein Increase with Maturation in Human Hematopoietic Cells. *Cancer Res.* **1991**, *51*, 4279–4286. [PubMed]

81. Kastan, M.B.; Onyekwere, O.; Sidransky, D.; Vogelstein, B.; Craig, R.W. Participation of P53 Protein in the Cellular Response to DNA Damage. *Cancer Res.* **1991**, *51*, 6304–6311. [CrossRef] [PubMed]

82. Domogauer, J.D.; de Toledo, S.M.; Howell, R.W.; Azzam, E.I. Acquired Radioresistance in Cancer Associated Fibroblasts Is Concomitant with Enhanced Antioxidant Potential and DNA Repair Capacity. *Cell Commun. Signal.* **2021**, *19*, 30. [CrossRef]

83. Wang, B. Analyzing Cell Cycle Checkpoints in Response to Ionizing Radiation in Mammalian Cells. In *Cell Cycle Control*; Humana Press: New York, NY, USA, 2014; pp. 313–320.

84. Huang, R.-X.; Zhou, P.-K. DNA Damage Response Signaling Pathways and Targets for Radiotherapy Sensitization in Cancer. *Signal Transduct. Target. Ther.* **2020**, *5*, 60. [CrossRef] [PubMed]

85. Fornace, A.J.; Amundson, S.A.; Do, K.T.; Meltzer, P.; Trent, J.; Bittner, M. Stress-Gene Induction by Low-Dose Gamma Irradiation. *Mil. Med.* **2002**, *167*, 13.

86. Daino, K.; Ichimura, S.; Nenoi, M. Early Induction of CDKN1A (P21) and GADD45 MRNA by a Low Dose of Ionizing Radiation Is Due to Their Dose-Dependent Post-Transcriptional Regulation. *Radiat. Res.* **2002**, *157*, 478–482. [CrossRef]

87. Xiong, Y.; Hannon, G.J.; Zhang, H.; Casso, D.; Kobayashi, R.; Beach, D. P21 Is a Universal Inhibitor of Cyclin Kinases. *Nature* **1993**, *366*, 701–704. [CrossRef]

88. El-Deiry, W.S.; Tokino, T.; Velculescu, V.E.; Levy, D.B.; Parsons, R.; Trent, J.M.; Lin, D.; Mercer, W.E.; Kinzler, K.W.; Vogelstein, B. WAF1, a Potential Mediator of P53 Tumor Suppression. *Cell* **1993**, *75*, 817–825. [CrossRef]

89. Wang, X.W.; Zhan, Q.; Coursen, J.D.; Khan, M.A.; Kontny, H.U.; Yu, L.; Hollander, M.C.; O'Connor, P.M.; Fornace, A.J.; Harris, C.C. GADD45 Induction of a G2/M Cell Cycle Checkpoint. *Proc. Natl. Acad. Sci. USA* **1999**, *96*, 3706–3711. [CrossRef]

90. Abou-El-Ardat, K. Response to Low-Dose X-Irradiation Is P53-Dependent in a Papillary Thyroid Carcinoma Model System. *Int. J. Oncol.* **2011**, *39*, 1429–1441. [CrossRef] [PubMed]

91. Li, S.-J.; Liang, X.-Y.; Li, H.-J.; Yang, G.-Z.; Li, W.; Li, Z.; Zhou, L.; Wen, X.; Yu, D.-H.; Cui, J.-W. Low-Dose Irradiation Inhibits Proliferation of the P53null Type Human Prostate Cancer Cells through the ATM/P21 Pathway. *Int. J. Mol. Med.* **2017**, *41*, 548–554. [CrossRef] [PubMed]

92. Rainey, M.D.; Black, E.J.; Zachos, G.; Gillespie, D.A.F. Chk2 Is Required for Optimal Mitotic Delay in Response to Irradiation-Induced DNA Damage Incurred in G2 Phase. *Oncogene* **2008**, *27*, 896–906. [CrossRef] [PubMed]

93. Li, C.Y.; Nagasawa, H.; Dahlberg, W.K.; Little, J.B. Diminished Capacity for P53 in Mediating a Radiation-Induced G1 Arrest in Established Human Tumor Cell Lines. *Oncogene* **1995**, *11*, 1885–1892. [PubMed]

94. Marples, B.; Wouters, B.G.; Joiner, M.C. An Association between the Radiation-Induced Arrest of G2-Phase Cells and Low-Dose Hyper-Radiosensitivity: A Plausible Underlying Mechanism? *Radiat. Res.* **2003**, *160*, 38–45. [CrossRef] [PubMed]

95. Li, J.; Hou, D.; Ma, Y.; Feng, L.; Li, W.; Liu, W.; Qiao, J.; Jia, X.; Li, K. Effect of 0.1 Gy X-Ray Irradiation on Gene Expression Profiles in Normal Human Lymphoblastoid Cells. *Zhonghua Lao Dong Wei Sheng Zhi Ye Bing Za Zhi = Zhonghua Laodong Weisheng Zhiyebing Zazhi = Chin. J. Ind. Hyg. Occup. Dis.* **2013**, *31*, 749–752.

96. Chin, L.; Pomerantz, J.; DePinho, R.A. The INK4a/ARF Tumor Suppressor: One Gene–Two Products–Two Pathways. *Trends Biochem. Sci.* **1998**, *23*, 291–296. [CrossRef]

97. Sokolov, M.; Neumann, R. Effects of Low Doses of Ionizing Radiation Exposures on Stress-Responsive Gene Expression in Human Embryonic Stem Cells. *Int. J. Mol. Sci.* **2014**, *15*, 588–604. [CrossRef]

98. Abbas, T.; Dutta, A. P21 in Cancer: Intricate Networks and Multiple Activities. *Nat. Rev. Cancer* **2009**, *9*, 400–414. [CrossRef]

99. Yunis, R.; Albrecht, H.; Kalanetra, K.M.; WU, S.; Rocke, D.M. Genomic Characterization of a Three-Dimensional Skin Model Following Exposure to Ionizing Radiation. *J. Radiat. Res.* **2012**, *53*, 860–875. [CrossRef] [PubMed]

100. Sergeeva, V.A.; Ershova, E.S.; Veiko, N.N.; Malinovskaya, E.M.; Kalyanov, A.A.; Kameneva, L.V.; Stukalov, S.V.; Dolgikh, O.A.; Konkova, M.S.; Ermakov, A.V.; et al. Low-Dose Ionizing Radiation Affects Mesenchymal Stem Cells via Extracellular Oxidized Cell-Free DNA: A Possible Mediator of Bystander Effect and Adaptive Response. *Oxidat. Med. Cell. Longev.* **2017**, *2017*, 9515809. [CrossRef] [PubMed]

101. Hanahan, D.; Weinberg, R.A. Hallmarks of Cancer: The next Generation. *Cell* **2011**, *144*, 646–674. [CrossRef]

102. Cazzalini, O.; Scovassi, A.I.; Savio, M.; Stivala, L.A.; Prosperi, E. Multiple Roles of the Cell Cycle Inhibitor P21(CDKN1A) in the DNA Damage Response. *Mutat. Res.* **2010**, *704*, 12–20. [CrossRef] [PubMed]

103. Dotto, G.P. P21(WAF1/Cip1): More than a Break to the Cell Cycle? *Biochim. Biophys. Acta* **2000**, *1471*, M43–M56. [PubMed]

104. Gartel, A.L.; Tyner, A.L. The Role of the Cyclin-Dependent Kinase Inhibitor P21 in Apoptosis. *Mol. Cancer Ther.* **2002**, *1*, 639–649.

105. Romanov, V.S.; Rudolph, K.L. P21 Shapes Cancer Evolution. *Nat. Cell Biol.* **2016**, *18*, 722–724. [CrossRef] [PubMed]

106. Romanov, V.S.; Pospelov, V.A.; Pospelova, T.V. Cyclin-Dependent Kinase Inhibitor P21(Waf1): Contemporary View on Its Role in Senescence and Oncogenesis. *Biochemistry* **2012**, *77*, 575–584. [CrossRef] [PubMed]

107. Alt, J.R.; Gladden, A.B.; Diehl, J.A. P21Cip1 Promotes Cyclin D1 Nuclear Accumulation via Direct Inhibition of Nuclear Export. *J. Biol. Chem.* **2002**, *277*, 8517–8523. [CrossRef]

108. Wang, Y.A.; Elson, A.; Leder, P. Loss of P21 Increases Sensitivity to Ionizing Radiation and Delays the Onset of Lymphoma in Atm-Deficient Mice. *Proc. Natl. Acad. Sci. USA* **1997**, *94*, 14590–14595. [CrossRef] [PubMed]

109. Jaiswal, H.; Lindqvist, A. Bystander Communication and Cell Cycle Decisions after DNA Damage. *Front. Genet.* **2015**, *6*, 63. [CrossRef] [PubMed]

110. Ma, C.-M.C.; Luxton, G.; Orton, C.G. Pulsed Reduced Dose Rate Radiation Therapy Is Likely to Become the Treatment Modality of Choice for Recurrent Cancers. *Med. Phys.* **2011**, *38*, 4909–4911. [CrossRef]

111. Zhang, P.; Wang, B.; Chen, X.; Cvetkovic, D.; Chen, L.; Lang, J.; Ma, C.-M. Local Tumor Control and Normal Tissue Toxicity of Pulsed Low-Dose Rate Radiotherapy for Recurrent Lung Cancer. *Dose-Response* **2015**, *13*, 155932581558850. [CrossRef]

112. Todorovic, V.; Prevc, A.; Zakelj, M.N.; Savarin, M.; Bucek, S.; Groselj, B.; Strojan, P.; Cemazar, M.; Sersa, G. Pulsed Low Dose-Rate Irradiation Response in Isogenic HNSCC Cell Lines with Different Radiosensitivity. *Radiol. Oncol.* **2020**, *54*, 168–179. [CrossRef]

Review

Cyclin-Dependent Kinases and CTD Phosphatases in Cell Cycle Transcriptional Control: Conservation across Eukaryotic Kingdoms and Uniqueness to Plants

Zhi-Liang Zheng [1,2]

[1] Department of Biological Sciences, Lehman College, City University of New York, Bronx, NY 10468, USA; zhiliang.zheng@lehman.cuny.edu
[2] Biology PhD Program, Graduate Center, City University of New York, New York, NY 10016, USA

Abstract: Cell cycle control is vital for cell proliferation in all eukaryotic organisms. The entire cell cycle can be conceptually separated into four distinct phases, Gap 1 (G1), DNA synthesis (S), G2, and mitosis (M), which progress sequentially. The precise control of transcription, in particular, at the G1 to S and G2 to M transitions, is crucial for the synthesis of many phase-specific proteins, to ensure orderly progression throughout the cell cycle. This mini-review highlights highly conserved transcriptional regulators that are shared in budding yeast (*Saccharomyces cerevisiae*), *Arabidopsis thaliana* model plant, and humans, which have been separated for more than a billion years of evolution. These include structurally and/or functionally conserved regulators cyclin-dependent kinases (CDKs), RNA polymerase II C-terminal domain (CTD) phosphatases, and the classical versus shortcut models of Pol II transcriptional control. A few of CDKs and CTD phosphatases counteract to control the Pol II CTD Ser phosphorylation codes and are considered critical regulators of Pol II transcriptional process from initiation to elongation and termination. The functions of plant-unique CDKs and CTD phosphatases in relation to cell division are also briefly summarized. Future studies towards testing a cooperative transcriptional mechanism, which is proposed here and involves sequence-specific transcription factors and the shortcut model of Pol II CTD code modulation, across the three eukaryotic kingdoms will reveal how individual organisms achieve the most productive, large-scale transcription of phase-specific genes required for orderly progression throughout the entire cell cycle.

Keywords: CDK; CTD phosphatase; RNA polymerase II; CTD code; transcription; cell cycle

Citation: Zheng, Z.-L. Cyclin-Dependent Kinases and CTD Phosphatases in Cell Cycle Transcriptional Control: Conservation across Eukaryotic Kingdoms and Uniqueness to Plants. *Cells* **2022**, *11*, 279. https://doi.org/10.3390/cells11020279

Academic Editor: Zhixiang Wang

Received: 1 December 2021
Accepted: 10 January 2022
Published: 14 January 2022

Publisher's Note: MDPI stays neutral with regard to jurisdictional claims in published maps and institutional affiliations.

1. Introduction

The control of cell cycle is a vital process for cell proliferation in all eukaryotic organisms. The cell cycle from unicellular organisms, such as budding yeast (*Saccharomyces cerevisiae*), to multicellular organisms including plants and humans has shared some common features. These include four phases that progress from Gap 1 (G1) to DNA synthesis (S) and then to G2 and mitosis (M), completing a cell cycle. In the cell cycle, the G1 to S and G2 to M transitions are considered two key control points, with the former making an irreversible decision to enter the cycle, and the latter ensuring completion of faithful DNA synthesis before distributing genomes into new daughter cells [1]. In addition, the cyclin and cyclin-dependent kinases (CDKs) act as a core of the regulatory network that governs the progression of each phase in order [2,3]. Moreover, the targets of CDK-cyclin complex and the downstream transcriptional events are also evolutionarily conserved [1,4]. Among those, transcription factors and RNA polymerase II (Pol II) have been intensively studied. A few of the CDKs are involved in phosphorylation of Ser residues at the Pol II C-terminal domain (CTD), which is opposed by several CTD phosphatases. On the one hand, these evolutionarily conserved features have allowed scientists to harness the power of genetic and biochemical studies in model organisms including *S. cerevisiae* and the *Arabidopsis thaliana* model plant. On the other hand, yeasts, plants, and humans have been separated for more than a billion years of evolution. Therefore, humans and plants also have evolved unique mechanisms to enable them to adapt to various internal cues and external stimuli during growth and development.

A comparison of cell cycle durations at each phase between *S. cerevisiae*, *A. thaliana*, and humans (*Homo sapiens*) reveals their similarities and differences (Figure 1). For budding yeast with a size of approximately 10 μm, a typical cell cycle is approximately 80–90 min. G2 and M phases largely overlap, and thus are conceptually merged. Although the G1 phase can be longer in relatively smaller, daughter cells (43 min) than in larger, mother cells (21 min), durations of S (26–31 min) and G2/M (28–30 min) phases do not have a big difference and are size-independent [5,6]. A typical human cell in culture, with a size of around 100 μm, has a cell cycle of approximately 24 h, with 11, 8, 4, and 1 h for G1, S, G2, and M phases, respectively [7]. Similar to human cells, the cell cycle lengths of plant cells, with a size range of 10–100 μm, can also vary, depending on the cell types. Recent studies of Arabidopsis root cells found that their cell cycles are in the range of approximately 9–10 h, which is similar in tomato, tobacco, and other plant species with bigger genomes than Arabidopsis [8]. While humans have a relatively longer G1 phase (46% of the total cell cycle duration) than budding yeast (23% of the cycle duration), the absolute time for DNA duplication in yeast is understandably shorter than in humans, given that the yeast genome (12 Mb) is considerably smaller than that in humans (3 Gb). However, relative to the total cell cycle duration, they both have a long S phase (i.e., approximately 1/3 of the total cycle duration). In contrast, plants have a very short S phase of approximately 20–30 min, which corresponds to approximately only 3–5% of the total cycle duration, although plants and humans have similar genome sizes, with 125 Mb in Arabidopsis and 4.5 Gb in tobacco. The reason for this difference in S phase lengths is not clear. Plants differ from humans and other animals in that plant cells have a cell wall that limits their mobility. Therefore, these sessile organisms have to develop unique mechanisms in order to respond and adapt to dynamic environmental conditions.

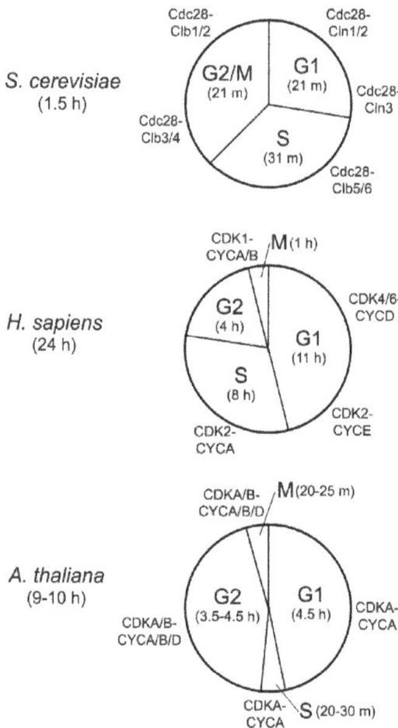

Figure 1. Varying lengths of cell cycle phases and the key CDK-cyclin pairs involved in each phase for budding yeast (mother cells), Arabidopsis plant (root cells), and human (cell culture). Of note, G2 and M phases in budding yeast largely overlap, and thus are conceptually merged. m stands for minute.

In this mini-review, I will first summarize the conserved features in cell cycle transcriptional control across the three eukaryotic kingdoms, Fungi, Plantae, and Animalia, Fungi, by highlighting some of the most important findings from studies of budding yeast, *Arabidopsis* plants, and human. These include CDKs, CTD phosphatases, and Pol II transcription regulated by these protein kinases and phosphatases. These CDKs and CTD phosphatases have conserved domains across the three kingdoms, but some of them also evolve with unique structure and function in each of the kingdoms. The unique aspects and potential challenges of transcriptional control in plant cell cycle will be discussed. For an in-depth review of recent updates in plant cell cycle control, readers are referred to several most recent, comprehensive reviews by other authors [9–11].

2. Cyclin-Dependent Kinases in Yeast, Human, and Arabidopsis

CDKs represent a subgroup of the CMGC Ser/Thr kinases first discovered from the budding yeast model system for their important role in cell cycle control [3]. Their catalytic function requires an association with multiple cyclins, which act as a regulatory subunit. It seems that the same CDK associated with different cyclins act at distinct phases of cell cycle (Figure 1). Subsequent studies in budding yeast and other eukaryotes revealed that more CDKs are involved in gene transcription and they are bound by a single cyclin. Therefore, there are two types of CDKs, cell cycle-related and transcription-related CDKs. As presented in Figure 2, among a total of six CDKs in budding yeast, two (Cdc28 and Pho85) belong to cell cycle-related CDKs, while four other CDKs (Kin28, Srb10, Bur1, and Ctk1) are grouped into transcription-related CDKs [3].

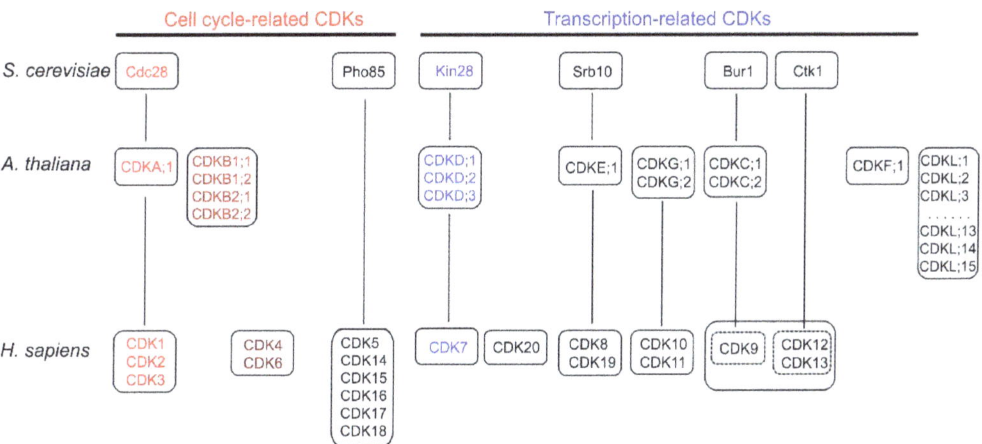

Figure 2. Groups of CDKs from yeast (*S. cerevisiae*), Arabidopsis (*A. thaliana*), and human (*H. sapiens*). Note that Arabidopsis contains 15 CDK-like genes (CDKL1 to CDKL15). The vertical line indicates an orthologous relationship. CDKs in color represent the two major regulators that are involved in cell cycle and transcriptional control.

Human and Arabidopsis plant contain a larger number of CDKs than budding yeast [3,9]. In humans, a total of 20 CDKs are divided into eight groups, with three groups (CDK1/2/3, CDK4/6, and CDK5/14/15/16) recognized as cell cycle-related CDKs and five other groups as transcription-related CDKs (Figure 2). CDK9 (the Bur1 ortholog) and CDK12/13 (orthologs of Ctk1) are regarded as the eighth group represented by CDK9. Furthermore, humans have three groups of CDKs (CDK4/6, CDK10/11, and CDK20) that are not orthologous to any yeast CDKs. Compared to slightly more than 6000 protein-coding genes in budding yeast and approximately 20,000 protein-coding genes in human, it seems that the number of human CDKs expanded in a similar scale as the whole set of protein-coding genes.

However, Arabidopsis genome has a significantly larger expansion of CDKs than human, with a total of 30 CDK and CDK-like (CDKL) genes [9,12,13], given only a slightly larger number of protein-coding genes in Arabidopsis (about 25,000) than human (approximately 20,000). These Arabidopsis CDK and CDKL genes can be separated into eight groups (Figure 2), and four of them are orthologous to both yeast and human, including CDKA;1, CDKD;1/2/3, CDKF;1, and CDKC1/2. It seems that the orthologs of yeast Cdc28 in human (CDK1/2/3) and Arabidopsis (CDKA;1) have played a similar and important role in controlling different phases of cell cycle (Figure 1). Note that the Arabidopsis CDKB group and the human CDK4 group are not orthologous to each other and to budding yeast, but they have been found to regulate the cell cycle, indicating the expansion of animal-specific and plant-specific CDKs in cell cycle control. While Arabidopsis does not have yeast Pho85 and Ctk1 orthologs, unlike human, CDKF;1 is unique to plants. In addition, Arabidopsis has 15 CDKL genes, which are not the orthologs of any of yeast and human CDKs.

3. Pol II CTD Phosphatases in Yeast, Human, and Arabidopsis

Protein phosphatases involved in dephosphorylation of Ser at 2, 5, and 7 positions in the highly conserved heptad peptide ($Y_1S_2P_3T_4S_5P_6S_7$) repeat of the CTD of the largest subunit of Pol II have been identified in yeast, human, and Arabidopsis (Figure 3), and some of these phosphatases have been functionally characterized. The phosphorylated forms of Ser2, Ser5, and Ser7 are designated Ser2P, Ser5P, and Ser7P, respectively, in this review. Yeast contains four known CTD phosphatases (Fcp1, Rtr1, Ssu72, and Cdc14) and two probable phosphatases (PSR1 and PSR2) that share a sequence homology with another type of CTD phosphatases initially classified as small CTD phosphatases (SCP). Human has one ortholog for each of the three classical CTD phosphatases, FCP1 (renamed as CTDP1), RPAP2 (claimed to be orthologous to yeast Rtr1), and SSU72, but two orthologs for yeast Cdc14, CDC14A, and CDC14B. In addition, human contains four SCP members, renamed as CTD small phosphatase 1 (CTDSP1), CTDSP2, CTDSP-like (CTDSPL), and CTDSPL2.

Arabidopsis has the largest number of CTD phosphatases among these three eukaryotes. In the Arabidopsis genome, there are five CTD phosphatase-like (CPL) genes, CPL1 to CPL5 [14,15], and one ortholog for Ssu72 and RPAP2, respectively, designated SSU72 [16] and RIMA [17]. Surprisingly, Arabidopsis contains a large number of SCP1-like small phosphatases, designated SSP. Initially, a total of 18 SSP genes (SSP1–SSP18) were proposed [18]. However, the original SSP7 (At3g19600) was renamed as CPL5 [15], and a new member, SSP4B, was added [19]. An extensive genome sequence search revealed that the original SSP15 (At3g15330) is a pseudogene and that At3g19590 is also closely related to all other SSP proteins. In addition, the updated Arabidopsis genome does not have SSP8, which was originally proposed as a fused gene spanning the CPL5 (At3g19600) locus. In order not to cause any confusion in gene name designation for SSP [18], At3g19590 is proposed to encode SSP7, and SSP16 and SSP17 are kept without change. Therefore, there are a total of 16 SSP genes in Arabidopsis, *SSP1–SSP3, SSP4, SSP4B, SSP5–SSP7, SSP9, SSP11–SSP17* (Figure 3). Of note, only SSP4, SSP4B, and SSP5 have been shown to exhibit a CTD phosphatase activity [19], while 13 other SSP proteins are annotated as the haloacid dehalogenase-like hydrolase (HAD) superfamily proteins, with SSP7–SSP17 having a gene ontology (GO) term of dephosphorylation of Pol II CTD. Given that Fcp1, the founding member of FCP/SCP phosphatases, has a biochemical mechanism more closely resembling the HAD superfamily proteins [20], it is conceivable that the 13 other SSP proteins might also possess CTD phosphatase activity. Overall, relative to the total number of protein-coding genes in the yeast, Arabidopsis, and human genomes, CTD phosphatases in Arabidopsis have expanded considerably more than that in humans, a situation similar to CDKs. It remains to be determined whether this unequal expansion of

CDK and CTD phosphatase genes in Arabidopsis is relevant to the unique aspects of plant growth and development.

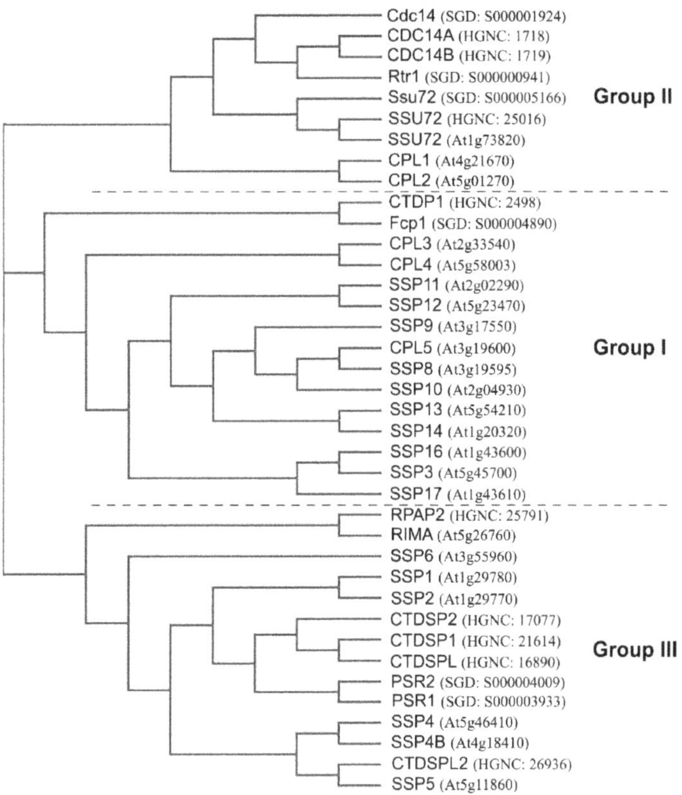

Figure 3. CTD phosphatase protein tree analysis. Full-length protein sequences of CTD phosphatases from yeast, Arabidopsis, and humans were used in the tree analysis. CTD phosphatases are separated into three groups (I, II and III). The gene ID for each protein was given in parenthesis for yeast (SGD), Arabidopsis (At), and humans (HGNC).

The protein tree analysis of these CTD phosphatases and their closely related phosphatases indicates that they are classified into three groups (Figure 3). Group I contains Fcp1, its orthologs in human (CTDP1) and Arabidopsis (CPL3/4), and Arabidopsis CPL5 and other SSP proteins. Fcp1 and its orthologs possess an N-terminal FCP1 homology (FCPH) domain, a breast cancer protein-related C-terminal domain (BRCT), and a C-terminal region involved in the interaction with RAP74, a component of general transcription factor TFIIF [20]. The FCPH domain is important for phosphatase activity, while the BRCT domain is involved in protein-protein interaction. Group II contains CPL1/2, SSU72, Cdc14, and Rtr1. This group of CTD phosphatases acts to dephosphorylate Ser5P (although some of them also impact Ser2P dephosphorylation when mutated) and are structurally more diverse, with Rtr1 considered as an atypical CTD phosphatase (due to its weak in vitro CTD phosphatase activity and the lack of apparent CTD phosphatase domain). Indeed, the functional ortholog of yeast Rtr1 in human, RPAP2, is placed in Group III. SSU72 and Cdc14 in yeast and human are Ser5 phosphatases, but they can also dephosphorylate Ser2P (Cdc14) or Ser7P (SSU72). However, the CTD phosphatase activity of Arabidopsis SSU72 has not been reported, and there is no Arabidopsis ortholog for Cdc14. Arabidopsis CPL1 and CPL2 are unique in that in addition to a CTD phosphatase domain, they also contain one (for CPL2) or two (for CPL1) double-stranded RNA binding motifs [21]. CPL1/2 possess Ser5P-specific phosphatase activity in vitro [22], but our in vivo studies found that the Ser2P level is elevated in both single gene mutants and the double mutants [23]. Group III includes human RPAP2 and its Arabidopsis ortholog RIMA, the biochemically validated small CTD phosphatases in human (CTDSP1/2/L) and Arabidopsis (SSP4/4B/5), and all other putative SSP proteins in Arabidopsis (SSP1/2/5/6) and yeast (PSR1/1). RPAP2 and CTDSP1/2/L in human and SSP5 in Arabidopsis have been demonstrated to dephosphorylate Ser5P specifically. In addition, Arabidopsis SSP4 and SSP4B can dephosphorylate Ser2P [19]. Overall, the phosphatases for the critical Ser residues in the Pol II CTD have been identified, with diverse structural and functional conservation, as well as divergence.

4. Control of Transcription during the Cell Cycle

4.1. Importance of Precise, Global Transcriptional Control in Cell Cycle

Progression through the cell cycle requires synthesis of more than 1000 cell cycle-dependent or cell cycle-related proteins [1,24]. This process is regulated at the transcriptional level primarily during the S and M phases. During the S phase, a group of genes encoding proteins important for DNA replication and DNA repair are transcribed, while another group of genes that encode proteins involved in mitosis and cytokinesis are expressed during the M phase [24]. Then, these cell cycle phase-specific proteins are subjected to proteasome-mediated degradation after completion of the S phase and the M phases, respectively, to ensure precisely ordered progression through the cell cycle and eventual exit from the cycle until mitotic cues are perceived to initiate another round of cell division.

Transcription of all protein-coding and many non-coding RNA genes requires the function of RNA polymerase II (Pol II) together with key transcription factors. While Pol II binds to the core promoter (which is relatively conserved among numerous genes) and thus functions in basal-level, global transcription, sequence-specific transcription factors control the transcription of genes with the corresponding enhancer sequence in the right tissues or cells and at the right time. Intensive studies of transcriptional control during G1-S and G2-M transitions have led to a consensus that a combination of several groups of transcription factors or complexes dictate the phase-dependent expression of genes [1]. These transcription factors include the RB pocket protein family, the E2F small transcription factor family, and MuvB complexes [1]. The details of the dynamic control are described in a comprehensive review [1]. In brief, upon activation, E2F transcription factors recognize the E2F binding elements present in G1-S transition- or S phase-related genes to turn on the expression of these target genes. For transcription of the G2-M transition or M phase-specific genes, the activating B-MYB and FOXM1 transcription factors, respectively, form the complexes with MuvB, which are then recruited to the CHR promoter elements of G2/M phase-related genes. In other phases, when those genes do not need to be transcribed, distinct repressor complexes (such as DREAM) are recruited to the E2F or CHR promoter elements. Through this mechanism, the precise timing of cell cycle-dependent transcription can be achieved to ensure the orderly progression of various phases in the cell cycle.

4.2. Pol II CTD Phosphorylation Is Controlled by CDKs and CTD Phosphatases

Pol II is a large multi-unit protein complex, with its largest subunit RPB1 as the core of Pol II transcriptional machinery. The CTD of RPB1 contains various numbers of highly conserved heptad peptide ($Y_1S_2P_3T_4S_5P_6S_7$) repeat, ranging from 26 in budding yeast and 29 in fission yeast to 34 in Arabidopsis and 52 in human [20,25–30]. Each of the seven amino acids in the repeat may undergo different modifications (e.g., S/T/Y phosphorylation, S glycosylation, and P isomerization), and each repeat may have a different posttranslational modification pattern [31]. Therefore, the CTD potentially exhibits a large and complex pattern collectively called the CTD code [20,25,26,28–31]. The CTD code, in particular the levels of Ser2P, Ser5P, and Ser7P during three stages of transcription, is critical for transcriptional control. In general, before transcription starts, Ser2, Ser5, and Ser7 are all unphosphorylated, and the initiation of transcription requires Ser5P. However, the

Ser5P level declines when gene transcription enters the elongation stage. In the meantime, Ser2P and Ser7P levels increase during productive elongation, and they all decrease at the termination stage, in order that the Pol II CTD enters another transcriptional cycle.

The dynamic CTD phosphorylation pattern during transcription is tightly regulated by various CDKs and CTD phosphatases [20,25–30]. In vitro and in vivo biochemical studies, together with genetic evidence, have shown that most of CDKs and CTD phosphatases have targeted two or three Ser positions [20,26,27,32]. The result is summarized in Table 1. Note that overall, orthologs of CDKs and CTD phosphatases in yeast and human have almost identical Ser-specificity. However, Arabidopsis orthologs of CDKs and CTD phosphatases have slightly different Ser specificity, except for CDKA1, SSU72, and RIMA, whose CTD Ser phosphorylation or dephosphorylation activity has not been reported yet. For example, CDKD;1/2/3 also phosphorylate Ser2, while yeast and human counterparts do not. In contrast, while Arabidopsis CDKC;1/2 are specific to Ser2P, their yeast and human orthologs phosphorylate Ser at all three positions (although they predominantly phosphorylate Ser2).

Table 1. CDKs and CTD phosphatases with known CTD Ser activities in yeast, Arabidopsis, and human. * Denotes the CTD phosphatases not yet studied regarding their Ser dephosphorylation activity.

	S. cerevisiae	*H. sapiens*	*A. thaliana*	Ser-Specificity
CDKs	Kin28	CDK7		Ser5, Ser7
			CDKD;1, CDKD;2, CDKD;3	Ser2, Ser5, Ser7
	Bur1	CDK9		Ser2, Ser5, Ser7
			CDKC;1, CDKC;2	Ser2
	Ctk1	CDK12, CDK13		Ser2, Ser5, Ser7
			CDKF;1	Ser7
	Cdc28		CDKA;1 *	Ser5
		CDK1		Ser2, Ser5
CTD Phosphatases	Fcp1	CTDP1	CPL3 *, CPL4	Ser2, Ser5
	Ssu72	SSU72	SSU72 *	Ser5, Ser7
	Rtr1	RPAP2	RIMA *	Ser5
	Cdc14	CDC14A, CDC14B		Ser2, Ser5
		CTDSP1, CTDSP2, CTDSPL	SSP5	Ser5
			SSP4, SSP4B	Ser2, Ser5
			CPL5	Ser2
			CPL1, CPL2	Ser2, Ser5

4.3. Transcriptional Control: Global vs. Centralized?

Given the large-scale gene expression needed to fulfil the distinct tasks of the S and M phases, two conflicting models have been proposed to explain the precise timing of phase-specific transcriptional control: The centralized, autonomous CDK-APC/C oscillator vs. the global transcriptional oscillator [33]. In the CDK-APC/C model, CDKs act to oppose the anaphase-promoting complex/cyclosome (APC/C, which possesses E3 ubiquitin ligase activity to degrade cyclin) and trigger phase-specific events, including phosphorylation of transcription factors. In turn, this precisely times the transcription of many phase-specific genes. The global transcriptional oscillator model [34] was proposed based on the findings on the yeast transcription factor network. In this model, transcription factors transcribed during one cell-cycle phase can bind the promoters of the next set of transcription factors that control phase-specific transcription.

The strongest evidence supporting the centralized CDK-APC/C oscillator model came from a single-cell study that observed time-series transcriptome changes during different phases of the budding yeast cell cycle in the B-cyclins (*CLN2*, *CLB2*, *Swi5*) mutants with "on" and "off" switches for controlling these individual cyclins [35]. The resulting transcription data were inconsistent with the global transcriptional oscillator model, and thus it was proposed that the CDK-APC/C oscillator predominantly entrains periodic cell cycle transcription. However, not all phase-specific genes were studied, and some of those genes under study did not exhibit a consistent transcriptional pattern.

In an effort to address which of the two models more likely operates in periodic control of transcription during the cell cycle, Cho et al. [33] analyzed transcriptome data using the yeast mutants depleted of B-cyclins and the *cdc14* and *cdc15* mutants as well. They found that a large subset of the cell cycle transcriptional program continued to oscillate in those yeast mutants arrested with constitutive Clb-CDK activity, which is inconsistent with the APC/C oscillator model. However, CDKs are required to maintain amplitudes of global transcriptional oscillations [35]. To reconcile these findings, Cho et al. [33] proposed an integrated CDK-APC/C and transcription factor network model. In this refined model, a global transcription oscillator drives periodic transcription, but CDKs are highly interconnected with transcription factors and contribute to robust, high-amplitude oscillations.

While this integrated model explaining how the CDK-APC/C oscillator and transcription factor network work together is attractive, it remains unknown whether Pol II itself is actively involved or simply serves as a machinery for basal transcription during precisely timed progression in the cell cycle. A recent study used the single-cell and single-molecule mRNA fluorescence in situ hybridization (smFISH) approach to count the number of mRNA molecules per cell in each phase of budding yeast cell cycle [36]. Their result surprisingly showed that all of the three main G1-S transition genes tested (*SIC1*, *CLN2*, and *CLB5*) had basal expression throughout the cell cycle. In contrast to the findings obtained using cell population, this single cell-based result indicates that these genes are not simply turned on or off completely, but instead they are expressed at high or low levels. While the biological relevance of this contrasting expression pattern needs to be further investigated, it is important to distinguish the possibilities whether this ubiquitous basal expression is under the control of the integrated CDK-APC/C and transcription factor network model or simply a reflection of Pol II basal activity.

5. Functional Conservation of CDK and CTD Phosphatases

In this section, three aspects of functional conservation for CDK and CTD phosphatases are discussed: Conserved CDKs in cell cycle and transcription, common substrates RB and E2F, and pathways leading to Pol II CTD Ser phosphorylation. Other aspects of functional conservation, such as the involvement of CDK activators (CAK) and inhibitors (CKI, Kip1, and Sic1) and the APC/C in regulation of CDK activities, can be found in other excellent comprehensive reviews [1,3,9].

5.1. CDK-Cyclin in Cell Cycle and Transcription

Ser/Thr kinase activity of CDKs is dependent on their regulatory subunit cyclin. There are more cyclin genes than CDKs, with 22 in budding yeast, 29 in human, and a significantly expanded number (at least 50) in Arabidopsis [3,9,37]. Therefore, one would expect that the large number of CDK-cyclin combinations would enable complex multicellular organisms, such as Arabidopsis and human, to undergo a wide range of growth and developmental behaviors in response to dynamic cues. However, the basic function of CDK-cyclin in cell cycle control is conserved although with certain degree of diversity in the regulatory patterns, which are contributed from other CDKs and their cyclin subunits [4]. As depicted in Figure 1, a single CDK (Cdc28) in yeast is sufficient to drive the progression of each phase during the cell cycle, with distinct cyclins at each phase. Cdc28 contains a conserved PSTAIRE motif in the cyclin binding domain. In the human cell cycle, almost all Cdc28 orthologs (CDK1/2/3) that also contain the PSTAIRE motif are involved, with CDK1 alone sufficient to drive the entire cell cycle and distinct CDK-cyclin pairs that predominantly function at different phases, such as CDK2-CYCE at late G1, CDK2/CYCA at S, and CDK1-CYCA/B at late G2 and M phases. In addition, other types of cell cycle-related CDKs, CDK4/6-CYCD, are also important at the early G1. However, only CDK1 seems to be essential, since the knockout of the CDK1 ortholog in mouse caused lethality, while *cdk2/4/6* triple knockout mice were still viable [38,39]. In Arabidopsis, CDKA1;1, the ortholog of yeast Cdc28 and human CDK1, also contains the PSTAIRE motif and has kinase activity peaked at the G1-S and G2-M transitions [9], suggesting its critical role in controlling the entire cell cycle (Figure 1). However, the *cdka;1* null mutant was still viable although the mutant was severely impacted, and the *cdka;1 cdkb1;1 cdkb1;2* triple mutant caused the cell cycle arrest [40]. Indeed, CDKBs and their cyclins are involved together with CDKA;1 in the control of the cell cycle (Figure 1). CDKBs have two subgroups, each with two members (Figure 2), and they all contain altered PSTAIRE motifs (PPTALRE in CDKB1 and P[S/P]TTLRE in CDKB2 subgroups). Due to this structural difference and the observation that Arabidopsis CDKBs could not complement yeast *cdc28* or *cdc2* mutants [40], CDKBs are considered a plant-unique group. Taken together, it seems that although human and Arabidopsis have evolved with an expansion of several cell cycle-related CDKs compared to yeast, the canonical PSTAIRE motif-containing CDKs (Cdc28, CDK1, and CDKA;1) have a conserved function in controlling the entire cell cycle. Nevertheless, the role of cell cycle CDKs, Cdc28, and CDK1, has been expanded to transcriptional control in yeast and human. Human CDK1 (previously called CDC2) can phosphorylate Pol II CTD Ser2 and Ser5 in vitro [41], and yeast Cdc28 only phosphorylates Ser5 [42]. This activity is believed to stimulate the Pol II basal transcriptional machinery to boost transcription of a subset of housekeeping genes upon entrance into the cell cycle [43]. However, it remains unknown whether this dual role in cell cycle and transcription is also conserved in Arabidopsis CDKA;1.

Another functionally well conserved CDK is CDK7, which is mainly involved in Pol II CTD phosphorylation and thus transcriptional control during the cell cycle. As discussed above, human CDK7 and its yeast ortholog Kin28 are believed to phosphorylate Ser5 and Ser7, but not Ser2 based on in vitro biochemical studies [44], while Arabidopsis counterparts CDKD1;1/1;2/1;3 also phosphorylate Ser2P, in addition to Ser5 and Ser7 [45]. However, using a potent and specific CDK7 inhibitor THZ1, we found a dramatic decrease of Ser2P, Ser5P, and Ser7P in human cells [46]. Despite this functional divergence, human CDK7 has emerged as a critical target in containing uncontrolled cell division in various tumor types [47,48]. CDK7 is a member of the general transcription factor TFIIH complex composed of 10 subunits. Phosphorylation by CDK7 of Ser5 at the hypophosphorylated Pol II CTD leads to transcriptional initiation and clearance from the promoter [48]. In addition, CDK7 also phosphorylates CDK9, which then becomes active to phosphorylate Ser2 of Pol II CTD, enabling productive transcription. Therefore, CDK7 has been considered as a key transcriptional CDK in Pol II control of transcriptional cycle, although CDK7 is completely dispensable for global transcription [48,49]. Consistent with the critical importance of CDK7 in human cell cycle transcription, inactivation of mouse *CDK7* [49] and yeast *Kin28* [50] led to cell cycle arrest, and the Arabidopsis triple mutant *cdkd;1 cdkd;2 cdkd;3* exhibited severely impacted plant growth [45]. Furthermore, studies in yeast have shown that Cdc28 cooperates with Kin28 to achieve full Ser5P in the Pol II CTD. Therefore, Kin28-mediated CTD Ser5P serves as a priming site for recruitment of Cdc28 to Pol II, linking the two most important CDKs, Cdc28/CDK1 and Kin28/CDK7, which are commonly perceived as cell cycle-related and transcription-related CDKs, respectively, to achieve the precise control of productive transcription during progression through the whole cell cycle [42,43].

5.2. Substrates RB and E2F

Accumulating evidence suggests that CDK control of the G1-S transition is more conserved in yeast, human, and Arabidopsis than in the G2-M transition. The G1-S phase transition in human cell cycle is mainly controlled by CDK2-CYCE and CDK2-CYCA (Figure 2) that regulate two opposing transcriptional regulators, RB and E2F. RB was identified as a tumor suppressor gene from a retina cancer called retinoblastoma, while E2F is a small transcription factor family. When RB binds to E2F1–3, the RB-E2F complex is formed via the dimerization partner DP, which inhibits the E2F1–3 activity, and thus transcription is repressed. At the late G1 phase, CDK2 phosphorylates RB, which becomes inactive but releases E2F1–3 [1,51]. Ultimately, genes required for DNA synthesis and DNA repair are transcribed. At the late M phase, RB is dephosphorylated, and thus binds to and inhibits E2F, which then inhibits transcription until a new round of cell division is executed. Homologs of RB and E2F have been identified in yeast and Arabidopsis: Whi5 (functional homolog of RB with no sequence homology) and SBF, respectively, in yeast [51], and RBR and E2F in Arabidopsis [52,53]. Arabidopsis RBR contains several domains similar to RB, and it has been shown to be the CDKA;1 target in the G1-S transition [40]. In Arabidopsis, there are at least six E2F genes (*E2F1–3* and *E2FA-C*) and several E2F-like or atypical E2F genes [52]. Despite some divergence between Arabidopsis and yeast or human, it seems that the double-negative regulatory feedback loops between CDK and RB/Whi5/RBR are conserved in these three eukaryotes [4]. Therefore, the findings that the substrates (Whi5/RB/RBR) of cdc28/CDK1/CDKA;1 and the associated transcription factors (SBF/E2F) are also functionally conserved suggest that these three canonical CDKs containing the PSTAIRE motif act as universal regulators of the cell cycle with a conserved biochemical mechanism.

5.3. Substrate Pol II CTD and Its Upstream Regulatory Pathways: Classical vs. Shortcut?

The highly conserved heptad peptide repeats in the CTD of Pol II are dynamically regulated by several CDKs and CTD phosphatases in response to mitogenic signals. Overall, CDKs and CTD phosphatases in yeast, human, and Arabidopsis have similar CTD Ser specificity, although some of the orthologs have more or less Ser specificity (Table 1). How these kinases and phosphatases are controlled by upstream signals in gene expression regulation have received increasing attention. Accumulating evidence suggests the existence of two models of transcriptional control (Figure 4) [54]. In the *"classical"* model of transcriptional control, which is frequently described in molecular genetics or cell biology textbooks, extracellular proliferation cues first activate intracellular signaling switches, such as the well-studied Ras and Rho families of small GTPases, which in turn activate the MAP kinase cascade. Subsequently, phosphorylated MAPKs phosphorylate various sequence-specific transcription factors, which then become active and bind to the gene-specific enhancer and consequently, recruit Pol II, by interacting with the general transcription factor TFIIH complex, to the core promoter of those genes to be transcribed [54]. Moreover, this interaction stimulates Ser5P in the CTD of RPB1 via activation of CDK7 present in the general transcription factor TFIIH complex and/or the mediator complex [55–57]. Ultimately, transcription is initiated. Since many components and steps are indirectly involved in the Pol II CTD Ser5P, this classical model is also called the indirect model and has been considered an intracellular signaling paradigm.

The other model, called a *shortcut* model, depicts the direct modulation of Pol II CTD Ser5P and Ser2P status by Ras GTPase-exerted PKA signaling to the mediator component Srb9 in yeast [58] or by Rho GTPase-mediated degradation of CTD phosphatases in yeast and Arabidopsis [23]. Importantly, we have found that the Rho signaling shortcut to Pol II CTD Ser2P and Ser5P was controlled by proteasome-mediated degradation of CPL1 and CPL2 in Arabidopsis or Fcp1 in yeast [23]. Furthermore, Rho family GTPases (Cdc42 and Rac1) in human cells also seem to suppress CTD phosphatases in a GTPase-specific manner: Suppression of RPAP2 by Cdc42 signaling but not Rac1, and suppression of CTDP1 (FCP1) by Rac1 signaling but not Cdc42 [46]. This strongly suggests that the shortcut model of Pol II transcription is conserved from yeast to Arabidopsis and human. In addition to the control of CTD phosphatases, CDKs (for example, CDK7 and CDK13) are also activated by Rho signaling, although these two CDKs do not exhibit any specificity for Rac1 and Cdc42, as the knockdown of both GTPases by RNA interference reduced the levels of these two CDKs.

What is the implication for the existence of both classical and shortcut models in Ras or Rho GTPase signaling to Pol II transcription across three eukaryotic kingdoms? Here, a cooperative control hypothesis is proposed (Figure 4). Since Pol II CTD can be directly targeted by signaling pathways in the shortcut model, rather than via the MAP kinase cascade in the classical model, the shortcut model has the advantage of rapidly bringing up large-scale gene expression changes in response to urgent growth or proliferation cues. Yet, the spatial and temporal control of transcription for those cell cycle-related genes depends on those sequence-specific transcription factors. Therefore, in response to a signal for cell division, a cell can activate Rho or probably other signaling molecules as well. In addition, a cell can use the classical model to promote the precise binding of sequence-specific transcription factor to the enhancer sequence of the cell cycle-related genes required at each phase, and in the meantime, it also can use the shortcut model to quickly modulate the Pol II CTD phosphorylation code. Therefore, this cooperative mechanism enables a cell to quickly achieve the most precise and productive control of large-scale transcription. This mechanism may be essential for cellular organisms to efficiently complete a cell cycle and determine whether additional rounds of cell division are needed when facing dynamic internal cues and external stimuli.

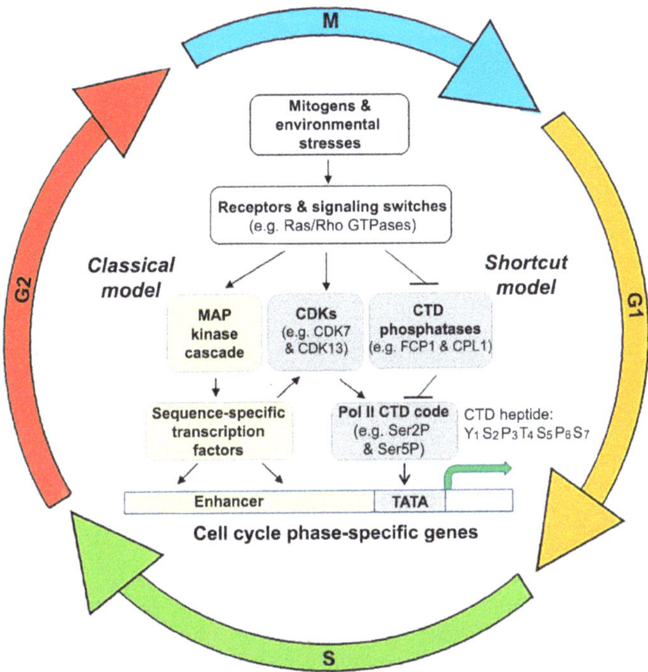

Figure 4. A proposed cooperative mechanism of Pol II transcriptional control via integration of the classical and shortcut models. Spatial and temporal transcription control of cell cycle phase-specific genes is conferred by sequence-specific transcription factors, which are activated by the classical model and consequently bind to the enhancer of those cell cycle genes. Direct modulation of the Pol II CTD code via regulating the abundance or activity of CDKs and CTD phosphatases allows Pol II, which recognizes the core promoter including the TATA box, to undergo productive transcription. Together, this cooperative mechanism by integrating the two intracellular signaling models enables a cell to quickly achieve the most precise and productive control of large-scale transcription critical for completing each phase of the cell cycle.

6. Functions of CDKs and CTD Phosphatases Unique to Arabidopsis

As discussed above, Arabidopsis plants have expanded the families of CDKs and CTD Ser phosphatases dramatically compared to yeast and human. Moreover, although the Arabidopsis orthologs of CDKs and CTD phosphatases have conserved functions as in yeast and human cell cycle control, they also exhibit some diversification in performing their biochemical functions or participating in cell cycle progression. Therefore, in order to understand why plants evolved with many CDKs and CTD phosphatases, it is important to address the key question: What are the functions for those plant-specific CDKs and CTD phosphatases in relation to cell cycle control?

6.1. Plant-Specific CDKs

Since the identification of two CDK genes (*cdc2a* and *cdc2b* now called *CDKA;1* and *CDKB1;1*, respectively) from Arabidopsis 30 years ago [59,60], functions of many CDKs and their cyclin partners have been reported. Overall, we have more knowledge of CDKs than CTD phosphatases regarding cell cycle control. For the functional details of individual CDKs, including those plant-unique CDKs, readers are referred to prior reviews [27,61,62] regarding the cell cycle or transcriptional control or the three most recent, comprehensive reviews [9–11] regarding other biological processes, such as growth and development, hormone response, and nutrient or biotic/abiotic stress response. Here, only the plant-specific CDKF, CDKG, and CDKL genes in Arabidopsis are summarized in relation to their cell cycle transcriptional control.

CDKF;1 is a plant-unique CDK activating kinase (previously called CAK1) that can phosphorylate two other CAKs now designated as CDKD;2 and CDKD;3, but not CDKD;1 [63]. Although when it was first identified and characterized, CDKF;1 was believed not to phosphorylate Pol II CTD based on the fractionation assay [64], a subsequent in vitro study demonstrated that it specifically phosphorylates Ser7 [45]. However, the *cdkf;1* knockout mutant also had lower levels of Ser2P and Ser5P than the wildtype during later stages of seedling development, but not in 7-day-old young seedlings [45]. The *CDKF;1* transcript level gradually increased during seedling development, suggesting that CDKF;1 is developmentally regulated. However, the alteration of Ser2P could not be explained simply by the loss of function in CDKD group kinases, which phosphorylate all three Ser resides in the Pol II CTD, given that the Ser2P level was even lower than in the *cdkd;1 cdkd;2 cdkd;3* triple mutant [45]. Moreover, genetic evidence suggests that CDKF;1 and CDKDs have slightly different functions, in which CDKF;1 plays a distinct role, mainly in post-embryonic development, while CDKD;1 and CDKD;3 function as CAKs in the control of mitosis [65]. Therefore, it was proposed that CDKF;1 is also required for regulating CDKD-independent Ser2 kinase activity [27,45]. In addition to phosphorylating CDKDs, CDKF;1 was also found to phosphorylate and activate CDKA;1 in Arabidopsis root protoplasts [66]. However, genetic evidence shows that CDKF;1 is dispensable for CDKA;1 activation [67]. Therefore, CDKF;1 is suggested to play a more important role in CDK phosphorylation than in CTD phosphorylation. Consistent with its role as CAK for both cell cycle-related CDKA;1 and transcription-related CDKDs and as a CTD kinase, knockout of *CDKF;1* led to the formation of curling serrated leaves, arrested root growth, and severe dwarfism, which were caused by the decreased cell number and cell size [45,67]. Therefore, genetic and biochemical evidence support the fact that CDKF;1 is a major regulator of cell proliferation, although it remains unknown whether CDKF;1 acts at a specific cell cycle phase or throughout the cell cycle. The fact that *CDKF;1* expression did not seem to considerably change during the cell cycle progression indicates that it is likely regulated at the translational or kinase activity level during the cell cycle progression [13].

CDKG;2 has been found to regulate RNA splicing [68–70], similar to its counterparts in human (CDK10/11). However, CDKG;2 acts in different biological processes than CDKG;1. *CDKG;1* is essential for synapsis and recombination during male meiosis [71,72], while *CDKG;2* is involved in flowering and response to heat and salt stresses [68,69,73]. Interestingly, the *CDKG;1* transcript level decreased from G1/G0 to S and G2, although *CDKG;2* expression stayed almost the same during cell cycle progression [13]. Therefore, it will be important to determine whether the CDKG;1-controlled meiosis is related to the transcriptional control of cell cycle-related genes.

The role of 15 CDKL genes has not been extensively studied. Using synchronized Arabidopsis cell cultures to survey expression profiles of Arabidopsis core cell cycle regulators [13], it was found that several *CKL* (now called *CDKL*) genes exhibited cell cycle phase-correlated expression patterns. For example, when re-entering the cell cycle, *CDKL;3* had a gradual increase from G0/G1 to S, but then decreased at G2, while *CDKL;5* and *CDKL;6* expression decreased overall from G0/G1 to S and G2. *CDKL;3* was also identified from a mutant impaired in beta-aminobutyric acid (BABA)-induced sterility (*ibs1*), but how its kinase activity is involved in cell cycle progression and whether this role is related to priming for defense gene expression remain to be determined [74,75]. In addition, two *CDKL* genes are specifically (*CDKL;1*) [76] or preferentially (*CDKL;15*) [13] expressed in flowers, but their expression did not change dramatically during cell cycle progression [13].

6.2. Plant-Specific CTD Phosphatases

Among the three groups of CTD phosphatases, several members of Group I (CPL3, CPL4, CPL5) and Group III (SSP4, SSP4B, SSP5, and RIMA) have been functionally characterized [14,15,17,19,77–81]. These include Pol II CTD Ser dephosphorylation activity (except for CPL3 and RIMA), which is similar to their orthologs or closest homologs in yeast and human. When these genes are inactivated or overexpressed, they exhibited phenotypic alterations in hormone, nutrient, biotic, and abiotic stress responses. However, none of the mutants or transgenic lines are characterized regarding their cell division phenotype. Group II contains an SSU72 ortholog, which is shown to act in flowering time control [16], although whether it possesses CTD phosphatase activity remains unknown. Therefore, CPL1 and CPL2, which belong to a unique subgroup within Group II (Figure 3) due to the presence of RNA binding motifs not found in any other CTD phosphatases, represent unique plant-specific CTD phosphatases with a likely involvement in cell division [23], as discussed in detail below.

CPL1 was first identified from a genetic screen as allelic mutants showing high expression of a presumably stress responsive *RD29A-Luciferase (LUC)* reporter gene [21,82], and *CPL2* was then found based on the sequence homology with CPL1. CPL1 was found to dephosphorylate Ser5 specifically in vitro [22], but the loss-of-function and gain-of-function of *CPL1* alleles led to a consistent increase or decrease of both Ser5P and Ser2P [23]. Therefore, it remains to be clarified whether the observed Ser2P impact is due to an indirect effect caused by genetic perturbation of *CPL1* or whether the lack of Ser2 dephosphorylation by CPL1 is due to the lack of a critical cellular factor in the in vitro biochemical assay. Surprisingly, *CPL1* was frequently identified from various mutant screens, including from mutants showing increased expression of reporter genes, such as silenced *miRNA-LUC* [83], salt-inducible *SOT12-LUC* [84], cold-inducible *CBF2:LUC* [85], and disease responsive *GSTF8:LUC* [86], or from an enhancer of CA-rop2 (constitutively active form of ROP2) in cell shape [23]. Together with other phenotypes in the *cpl1* mutants, such as iron deficiency response [87] and floral transition [88], these lines of genetic evidence demonstrate CPL1 as a critical and global regulator in growth, development, and stress responses. Consistent with its role in transcriptional control, expression of many genes is affected in the *cpl1* mutants [23,83,86,87]. Of note, CPL1 is frequently regarded as a negative regulator of transcription due to its dephosphorylation impact on Pol II CTD Ser2 and Ser5. However, the aforementioned transcriptomic studies revealed that a similar set of genes are upregulated and downregulated. Therefore, the differing roles of the plant-unique CPL1 and probably CPL2 in transcriptional control may be context-dependent or due to multiple regulators that are influenced by CPL1 and CPL2.

As only our recent study observed the cell number increase and the cell size decrease during cotyledon development in the *cpl1 CA-rop2* background or the *cpl1 cpl2* double mutants [23], there is no considerable knowledge regarding how CPL1/2 control the cell cycle. Based on the biochemical and genetic evidence, we only know to date that they are inhibited by the signaling of Arabidopsis ROP family GTPases via a proteasome-mediated degradation mechanism. Moreover, a major growth hormone auxin, which has been shown to activate ROP GTPases [89,90], also stimulates Pol II CTD Ser2P and Ser5P in a ROP2/4-dependent manner [23]. Therefore, auxin-exerted gene expression probably involves the shortcut model described above to achieve the rapid and large-scale transcriptome changes needed for cell growth. As ROP GTPases act as a universal signaling switch for multiple hormones or developmental processes and various stress responses, it remains to be determined how the ROP2/4-CPL1/2 signaling shortcut is involved in many different processes. The finding that yeast Cdc42 GTPase signaling also promotes Fcp1 degradation [23] and the observation that human Rac1 and Cdc42 activity differentially inhibits CTDP1 (FCP1) and RPAP2 [46] may provide clues to this signaling specificity. For example, the shortcut model may involve different members of Rho GTPases and target any of three groups of CTD phosphatases (and CDKs, as well). Together with sequence-specific transcription factors, these complex interplays will enable plants to precisely time the expression of genes required for cell division and other biological processes.

7. Future Perspectives

Studies at the genetic, biochemical, and system levels from the simple unicellular yeast model to complex eukaryotes such as Arabidopsis model plants and humans have started to reveal the mechanisms for the precise control of phase-related transcription during the cell cycle progression. One question that remains to be answered is: Why do Arabidopsis plants with a considerably smaller genome size than humans evolve with substantially more CDKs and CTD phosphatases? A few possibilities have been proposed to explain the uniqueness of plant structure and function, including a bigger demand for these sessile organisms to respond and adapt to dynamic environmental challenges, and consequently with a high degree of developmental plasticity. However, molecular details are needed to answer this question and ultimately will contribute to our mechanistic understanding of convergence and divergency in the transcriptional control that makes an organism decide to enter the cell cycle or not and if that is the case, complete the entire cell cycle without delay and error. Therefore, many prior functional studies of genes, which have been emphasized at the organismal level, need to be assessed with regards to cell cycle progression. Moreover, given the consensus that individual cells frequently deviate from the population of cells, the use or development of single-cell technologies that minimally disturb the physiological state of cells, coupled with single-molecule techniques to count the individual cell cycle-related mRNA per cell, will be critical. Another challenge lies in the functional redundancy of many cell cycle-related genes in Arabidopsis and human (and the rodent or primate animal models, as well), such as several CDKs, cyclins or CTD phosphatases even within a subgroup. While some specific CDKs have been shown to regulate distinct cell cycle phases, it will be interesting to determine whether CTD phosphatases act in the same fashion during the cell cycle progression. Thus, functional redundancy and specification of these cell cycle transcriptional regulators need to be dissected using CRISPR/Cas9-based multiplex genetic manipulation. With these advanced single-cell and genome editing tools, the cooperative transcriptional mechanism, which is proposed here and involves sequence-specific transcription factors as well as the shortcut model of Pol II CTD code modulation (via CDKs and CTD phosphatases), can be tested and refined. Ultimately, a complete regulatory network can be assembled, which governs how individual organisms quickly achieve the most precise and productive, large-scale transcription of phase-specific genes required for orderly progression throughout the entire cell cycle.

Funding: The work described in this paper was partly supported by a grant from the US National Science Foundation (IOS-1121551 to Z.-L.Z.) and by intramural grants from CUNY Bridge Fund (80233-0315 to Z.-L.Z.), PSC-CUNY Research Award (61302-0049 and 68389-0046 to Z.-L.Z.), and the Shuster Faculty Fellowship Fund (71731-0001 to Z.-L.Z.).

Institutional Review Board Statement: Not applicable.

Informed Consent Statement: Not applicable.

Conflicts of Interest: The authors declare no conflict of interest.

References

1. Fischer, M.; Muller, G.A. Cell cycle transcription control: DREAM/MuvB and RB-E2F complexes. *Crit. Rev. Biochem. Mol. Biol.* **2017**, *52*, 638–662. [CrossRef] [PubMed]
2. Sanso, M.; Fisher, R.P. Pause, play, repeat: CDKs push RNAP II's buttons. *Transcription* **2013**, *4*, 146–152. [CrossRef] [PubMed]
3. Malumbres, M. Cyclin-dependent kinases. *Genome Biol.* **2014**, *15*, 122. [CrossRef] [PubMed]
4. Harashima, H.; Dissmeyer, N.; Schnittger, A. Cell cycle control across the eukaryotic kingdom. *Trends Cell Biol.* **2013**, *23*, 345–356. [CrossRef] [PubMed]
5. Brewer, B.J.; Chlebowicz-Sledziewska, E.; Fangman, W.L. Cell cycle phases in the unequal mother/daughter cell cycles of Saccharomyces cerevisiae. *Mol. Cell. Biol.* **1984**, *4*, 2529–2531. [CrossRef]
6. Turner, J.J.; Ewald, J.C.; Skotheim, J.M. Cell size control in yeast. *Curr. Biol.* **2012**, *22*, R350–R359. [CrossRef]
7. Cooper, G.M. The Eukaryotic Cell Cycle. Available online: https://www.ncbi.nlm.nih.gov/books/NBK9876/ (accessed on 1 October 2021).
8. Pasternak, T.; Kircher, S.; Palme, K. Estimation of cell cycle kinetics in higher plant root meristem links organ position with cellular fate and chromatin structure. *bioRxiv* **2021**. [CrossRef]
9. Shimotohno, A.; Aki, S.S.; Takahashi, N.; Umeda, M. Regulation of the Plant Cell Cycle in Response to Hormones and the Environment. *Annu. Rev. Plant Biol.* **2021**, *72*, 273–296. [CrossRef] [PubMed]
10. Carneiro, A.K.; Montessoro, P.D.F.; Fusaro, A.F.; Araujo, B.G.; Hemerly, A.S. Plant CDKs-Driving the Cell Cycle through Climate Change. *Plants* **2021**, *10*, 1804. [CrossRef] [PubMed]
11. Qi, F.; Zhang, F. Cell Cycle Regulation in the Plant Response to Stress. *Front. Plant Sci.* **2020**, *10*, 1765. [CrossRef]
12. Vandepoele, K.; Raes, J.; De Veylder, L.; Rouze, P.; Rombauts, S.; Inze, D. Genome-wide analysis of core cell cycle genes in Arabidopsis. *Plant Cell* **2002**, *14*, 903–916. [CrossRef]
13. Menges, M.; de Jager, S.M.; Gruissem, W.; Murray, J.A. Global analysis of the core cell cycle regulators of Arabidopsis identifies novel genes, reveals multiple and highly specific profiles of expression and provides a coherent model for plant cell cycle control. *Plant J. Cell Mol. Biol.* **2005**, *41*, 546–566. [CrossRef]
14. Bang, W.; Kim, S.; Ueda, A.; Vikram, M.; Yun, D.; Bressan, R.A.; Hasegawa, P.M.; Bahk, J.; Koiwa, H. Arabidopsis carboxyl-terminal domain phosphatase-like isoforms share common catalytic and interaction domains but have distinct in planta functions. *Plant Physiol.* **2006**, *142*, 586–594. [CrossRef]
15. Jin, Y.M.; Jung, J.; Jeon, H.; Won, S.Y.; Feng, Y.; Kang, J.S.; Lee, S.Y.; Cheong, J.J.; Koiwa, H.; Kim, M. AtCPL5, a novel Ser-2-specific RNA polymerase II C-terminal domain phosphatase, positively regulates ABA and drought responses in Arabidopsis. *New Phytol.* **2011**, *190*, 57–74. [CrossRef]
16. Tian, Y.; Zheng, H.; Zhang, F.; Wang, S.; Ji, X.; Xu, C.; He, Y.; Ding, Y. PRC2 recruitment and H3K27me3 deposition at FLC require FCA binding of COOLAIR. *Sci. Adv.* **2019**, *5*, eaau7246. [CrossRef]
17. Munoz, A.; Mangano, S.; Gonzalez-Garcia, M.P.; Contreras, R.; Sauer, M.; De Rybel, B.; Weijers, D.; Sanchez-Serrano, J.J.; Sanmartin, M.; Rojo, E. RIMA-Dependent Nuclear Accumulation of IYO Triggers Auxin-Irreversible Cell Differentiation in Arabidopsis. *Plant Cell* **2017**, *29*, 575–588. [CrossRef] [PubMed]
18. Koiwa, H. Phosphorylation of RNA polymerase II C-terminal domain and plant osmotic-stress responses. In *Abiotic Stress Tolerance in Plants: Toward the Improvement of Global Environment and Food*; Rai, A.K., Takabe, T., Eds.; Springer: Dordrecht, The Netherlands, 2006; pp. 47–57.
19. Feng, Y.; Kang, J.S.; Kim, S.; Yun, D.J.; Lee, S.Y.; Bahk, J.D.; Koiwa, H. Arabidopsis SCP1-like small phosphatases differentially dephosphorylate RNA polymerase II C-terminal domain. *Biochem. Biophys. Res. Commun.* **2010**, *397*, 355–360. [CrossRef] [PubMed]
20. Mayfield, J.E.; Burkholder, N.T.; Zhang, Y.J. Dephosphorylating eukaryotic RNA polymerase II. *Biochim. Biophys. Acta* **2016**, *1864*, 372–387. [CrossRef] [PubMed]
21. Koiwa, H.; Barb, A.W.; Xiong, L.; Li, F.; McCully, M.G.; Lee, B.H.; Sokolchik, I.; Zhu, J.; Gong, Z.; Reddy, M.; et al. C-terminal domain phosphatase-like family members (AtCPLs) differentially regulate Arabidopsis thaliana abiotic stress signaling, growth, and development. *Proc. Natl. Acad. Sci. USA* **2002**, *99*, 10893–10898. [CrossRef]
22. Koiwa, H.; Hausmann, S.; Bang, W.Y.; Ueda, A.; Kondo, N.; Hiraguri, A.; Fukuhara, T.; Bahk, J.D.; Yun, D.J.; Bressan, R.A.; et al. Arabidopsis C-terminal domain phosphatase-like 1 and 2 are essential Ser-5-specific C-terminal domain phosphatases. *Proc. Natl. Acad. Sci. USA* **2004**, *101*, 14539–14544. [CrossRef]
23. Zhang, B.; Yang, G.; Chen, Y.; Zhao, Y.; Gao, P.; Liu, B.; Wang, H.; Zheng, Z.L. C-terminal domain (CTD) phosphatase links Rho GTPase signaling to Pol II CTD phosphorylation in Arabidopsis and yeast. *Proc. Natl. Acad. Sci. USA* **2016**, *113*, E8197–E8206. [CrossRef] [PubMed]
24. Delgado-Roman, I.; Munoz-Centeno, M.C. Coupling Between Cell Cycle Progression and the Nuclear RNA Polymerases System. *Front. Mol. Biosci.* **2021**, *8*, 691636. [CrossRef] [PubMed]
25. Hsin, J.P.; Manley, J.L. The RNA polymerase II CTD coordinates transcription and RNA processing. *Genes Dev.* **2012**, *26*, 2119–2137. [CrossRef] [PubMed]
26. Egloff, S.; Dienstbier, M.; Murphy, S. Updating the RNA polymerase CTD code: Adding gene-specific layers. *Trends Genet.* **2012**, *28*, 333–341. [CrossRef]

27. Hajheidari, M.; Koncz, C.; Eick, D. Emerging roles for RNA polymerase II CTD in Arabidopsis. *Trends Plant Sci.* **2013**, *18*, 633–643. [CrossRef] [PubMed]

28. Aristizabal, M.J.; Kobor, M.S. A single flexible RNAPII-CTD integrates many different transcriptional programs. *Transcription* **2016**, *7*, 50–56. [CrossRef]

29. Zaborowska, J.; Egloff, S.; Murphy, S. The pol II CTD: New twists in the tail. *Nat. Struct. Mol. Biol.* **2016**, *23*, 771–777. [CrossRef]

30. Harlen, K.M.; Churchman, L.S. The code and beyond: Transcription regulation by the RNA polymerase II carboxy-terminal domain. *Nat. Rev. Mol. Cell Biol.* **2017**, *18*, 263–273. [CrossRef]

31. Schwer, B.; Sanchez, A.M.; Shuman, S. Punctuation and syntax of the RNA polymerase II CTD code in fission yeast. *Proc. Natl. Acad. Sci. USA* **2012**, *109*, 18024–18029. [CrossRef]

32. Jeronimo, C.; Collin, P.; Robert, F. The RNA Polymerase II CTD: The Increasing Complexity of a Low-Complexity Protein Domain. *J. Mol. Biol.* **2016**, *428*, 2607–2622. [CrossRef]

33. Cho, C.Y.; Motta, F.C.; Kelliher, C.M.; Deckard, A.; Haase, S.B. Reconciling conflicting models for global control of cell-cycle transcription. *Cell Cycle* **2017**, *16*, 1965–1978. [CrossRef]

34. Simon, I.; Barnett, J.; Hannett, N.; Harbison, C.T.; Rinaldi, N.J.; Volkert, T.L.; Wyrick, J.J.; Zeitlinger, J.; Gifford, D.K.; Jaakkola, T.S.; et al. Serial regulation of transcriptional regulators in the yeast cell cycle. *Cell* **2001**, *106*, 697–708. [CrossRef]

35. Rahi, S.J.; Pecani, K.; Ondracka, A.; Oikonomou, C.; Cross, F.R. The CDK-APC/C Oscillator Predominantly Entrains Periodic Cell-Cycle Transcription. *Cell* **2016**, *165*, 475–487. [CrossRef] [PubMed]

36. Amoussouvi, A.; Teufel, L.; Reis, M.; Seeger, M.; Schlichting, J.K.; Schreiber, G.; Herrmann, A.; Klipp, E. Transcriptional timing and noise of yeast cell cycle regulators-a single cell and single molecule approach. *NPJ Syst. Biol. Appl.* **2018**, *4*, 17. [CrossRef] [PubMed]

37. Andrews, B.; Measday, V. The cyclin family of budding yeast: Abundant use of a good idea. *Trends Genet.* **1998**, *14*, 66–72. [CrossRef]

38. Santamaria, D.; Barriere, C.; Cerqueira, A.; Hunt, S.; Tardy, C.; Newton, K.; Caceres, J.F.; Dubus, P.; Malumbres, M.; Barbacid, M. Cdk1 is sufficient to drive the mammalian cell cycle. *Nature* **2007**, *448*, 811–815. [CrossRef]

39. Satyanarayana, A.; Berthet, C.; Lopez-Molina, J.; Coppola, V.; Tessarollo, L.; Kaldis, P. Genetic substitution of Cdk1 by Cdk2 leads to embryonic lethality and loss of meiotic function of Cdk2. *Development* **2008**, *135*, 3389–3400. [CrossRef]

40. Nowack, M.K.; Harashima, H.; Dissmeyer, N.; Zhao, X.; Bouyer, D.; Weimer, A.K.; De Winter, F.; Yang, F.; Schnittger, A. Genetic framework of cyclin-dependent kinase function in Arabidopsis. *Dev. Cell* **2012**, *22*, 1030–1040. [CrossRef]

41. Zhang, J.; Corden, J.L. Identification of phosphorylation sites in the repetitive carboxyl-terminal domain of the mouse RNA polymerase II largest subunit. *J. Biol. Chem.* **1991**, *266*, 2290–2296. [CrossRef]

42. Chymkowitch, P.; Eldholm, V.; Lorenz, S.; Zimmermann, C.; Lindvall, J.M.; Bjoras, M.; Meza-Zepeda, L.A.; Enserink, J.M. Cdc28 kinase activity regulates the basal transcription machinery at a subset of genes. *Proc. Natl. Acad. Sci. USA* **2012**, *109*, 10450–10455. [CrossRef]

43. Chymkowitch, P.; Enserink, J.M. The cell cycle rallies the transcription cycle: Cdc28/Cdk1 is a cell cycle-regulated transcriptional CDK. *Transcription* **2013**, *4*, 3–6. [CrossRef] [PubMed]

44. Larochelle, S.; Amat, R.; Glover-Cutter, K.; Sanso, M.; Zhang, C.; Allen, J.J.; Shokat, K.M.; Bentley, D.L.; Fisher, R.P. Cyclin-dependent kinase control of the initiation-to-elongation switch of RNA polymerase II. *Nat. Struct. Mol. Biol.* **2012**, *19*, 1108–1115. [CrossRef]

45. Hajheidari, M.; Farrona, S.; Huettel, B.; Koncz, Z.; Koncz, C. CDKF;1 and CDKD protein kinases regulate phosphorylation of serine residues in the C-terminal domain of Arabidopsis RNA polymerase II. *Plant Cell* **2012**, *24*, 1626–1642. [CrossRef] [PubMed]

46. Zhang, B.; Zhong, X.; Sauane, M.; Zhao, Y.; Zheng, Z.L. Modulation of the Pol II CTD Phosphorylation Code by Rac1 and Cdc42 Small GTPases in Cultured Human Cancer Cells and Its Implication for Developing a Synthetic-Lethal Cancer Therapy. *Cells* **2020**, *9*, 621. [CrossRef] [PubMed]

47. Kwiatkowski, N.; Zhang, T.; Rahl, P.B.; Abraham, B.J.; Reddy, J.; Ficarro, S.B.; Dastur, A.; Amzallag, A.; Ramaswamy, S.; Tesar, B.; et al. Targeting transcription regulation in cancer with a covalent CDK7 inhibitor. *Nature* **2014**, *511*, 616–620. [CrossRef]

48. Fisher, R.P. Cdk7: A kinase at the core of transcription and in the crosshairs of cancer drug discovery. *Transcription* **2019**, *10*, 47–56. [CrossRef]

49. Ganuza, M.; Saiz-Ladera, C.; Canamero, M.; Gomez, G.; Schneider, R.; Blasco, M.A.; Pisano, D.; Paramio, J.M.; Santamaria, D.; Barbacid, M. Genetic inactivation of Cdk7 leads to cell cycle arrest and induces premature aging due to adult stem cell exhaustion. *EMBO J.* **2012**, *31*, 2498–2510. [CrossRef]

50. Simon, M.; Seraphin, B.; Faye, G. KIN28, a yeast split gene coding for a putative protein kinase homologous to CDC28. *EMBO J.* **1986**, *5*, 2697–2701. [CrossRef]

51. Cooper, K. Rb, whi it's not just for metazoans anymore. *Oncogene* **2006**, *25*, 5228–5232. [CrossRef]

52. Mariconti, L.; Pellegrini, B.; Cantoni, R.; Stevens, R.; Bergounioux, C.; Cella, R.; Albani, D. The E2F family of transcription factors from Arabidopsis thaliana. Novel and conserved components of the retinoblastoma/E2F pathway in plants. *J. Biol. Chem.* **2002**, *277*, 9911–9919. [CrossRef]

53. Weimer, A.K.; Nowack, M.K.; Bouyer, D.; Zhao, X.; Harashima, H.; Naseer, S.; De Winter, F.; Dissmeyer, N.; Geldner, N.; Schnittger, A. Retinoblastoma related1 regulates asymmetric cell divisions in Arabidopsis. *Plant Cell* **2012**, *24*, 4083–4095. [PubMed]

54. Zheng, Z.L. Ras and Rho GTPase regulation of Pol II transcription: A shortcut model revisited. *Transcription* **2017**, *8*, 268–274. [CrossRef]
55. Luse, D.S. The RNA polymerase II preinitiation complex: Through what pathway is the complex assembled? *Transcription* **2013**, *5*, e27050. [CrossRef] [PubMed]
56. Wong, K.H.; Jin, Y.; Struhl, K. TFIIH phosphorylation of the Pol II CTD stimulates mediator dissociation from the preinitiation complex and promoter escape. *Mol. Cell* **2014**, *54*, 601–612. [CrossRef]
57. Allen, B.L.; Taatjes, D.J. The Mediator complex: A central integrator of transcription. *Nat. Rev. Mol. Cell Biol.* **2015**, *16*, 155–166. [CrossRef]
58. Chang, Y.W.; Howard, S.C.; Herman, P.K. The Ras/PKA signaling pathway directly targets the Srb9 protein, a component of the general RNA polymerase II transcription apparatus. *Mol. Cell* **2004**, *15*, 107–116. [CrossRef] [PubMed]
59. Ferreira, P.C.; Hemerly, A.S.; Villarroel, R.; Van Montagu, M.; Inze, D. The Arabidopsis functional homolog of the p34cdc2 protein kinase. *Plant Cell* **1991**, *3*, 531–540. [CrossRef]
60. Hirayama, T.; Imajuku, Y.; Anai, T.; Matsui, M.; Oka, A. Identification of two cell-cycle-controlling cdc2 gene homologs in Arabidopsis thaliana. *Gene* **1991**, *105*, 159–165. [CrossRef]
61. Mironov, V.V.; De Veylder, L.; Van Montagu, M.; Inze, D. Cyclin-dependent kinases and cell division in plants-the nexus. *Plant Cell* **1999**, *11*, 509–522. [CrossRef]
62. Umeda, M.; Shimotohno, A.; Yamaguchi, M. Control of cell division and transcription by cyclin-dependent kinase-activating kinases in plants. *Plant Cell Physiol.* **2005**, *46*, 1437–1442. [CrossRef]
63. Shimotohno, A.; Umeda-Hara, C.; Bisova, K.; Uchimiya, H.; Umeda, M. The plant-specific kinase CDKF;1 is involved in activating phosphorylation of cyclin-dependent kinase-activating kinases in Arabidopsis. *Plant Cell* **2004**, *16*, 2954–2966. [CrossRef] [PubMed]
64. Umeda, M.; Bhalerao, R.P.; Schell, J.; Uchimiya, H.; Koncz, C. A distinct cyclin-dependent kinase-activating kinase of Arabidopsis thaliana. *Proc. Natl. Acad. Sci. USA* **1998**, *95*, 5021–5026. [CrossRef]
65. Takatsuka, H.; Umeda-Hara, C.; Umeda, M. Cyclin-dependent kinase-activating kinases CDKD;1 and CDKD;3 are essential for preserving mitotic activity in Arabidopsis thaliana. *Plant J. Cell Mol. Biol.* **2015**, *82*, 1004–1017. [CrossRef] [PubMed]
66. Shimotohno, A.; Ohno, R.; Bisova, K.; Sakaguchi, N.; Huang, J.; Koncz, C.; Uchimiya, H.; Umeda, M. Diverse phosphoregulatory mechanisms controlling cyclin-dependent kinase-activating kinases in Arabidopsis. *Plant J. Cell Mol. Biol.* **2006**, *47*, 701–710. [CrossRef] [PubMed]
67. Takatsuka, H.; Ohno, R.; Umeda, M. The Arabidopsis cyclin-dependent kinase-activating kinase CDKF;1 is a major regulator of cell proliferation and cell expansion but is dispensable for CDKA activation. *Plant J. Cell Mol. Biol.* **2009**, *59*, 475–487. [CrossRef]
68. Cavallari, N.; Nibau, C.; Fuchs, A.; Dadarou, D.; Barta, A.; Doonan, J.H. The cyclin-dependent kinase G group defines a thermo-sensitive alternative splicing circuit modulating the expression of Arabidopsis ATU2AF65A. *Plant J. Cell Mol. Biol.* **2018**, *94*, 1010–1022. [CrossRef]
69. Nibau, C.; Gallemi, M.; Dadarou, D.; Doonan, J.H.; Cavallari, N. Thermo-Sensitive Alternative Splicing of FLOWERING LOCUS M Is Modulated by Cyclin-Dependent Kinase G2. *Front. Plant Sci.* **2019**, *10*, 1680. [CrossRef]
70. Kanno, T.; Venhuizen, P.; Wu, M.T.; Chiou, P.; Chang, C.L.; Kalyna, M.; Matzke, A.J.M.; Matzke, M. A Collection of Pre-mRNA Splicing Mutants in Arabidopsis thaliana. *Genes Genomes Genet.* **2020**, *10*, 1983–1996. [CrossRef]
71. Zheng, T.; Nibau, C.; Phillips, D.W.; Jenkins, G.; Armstrong, S.J.; Doonan, J.H. CDKG1 protein kinase is essential for synapsis and male meiosis at high ambient temperature in Arabidopsis thaliana. *Proc. Natl. Acad. Sci. USA* **2014**, *111*, 2182–2187. [CrossRef]
72. Nibau, C.; Dadarou, D.; Kargios, N.; Mallioura, A.; Fernandez-Fuentes, N.; Cavallari, N.; Doonan, J.H. A Functional Kinase Is Necessary for Cyclin-Dependent Kinase G1 (CDKG1) to Maintain Fertility at High Ambient Temperature in Arabidopsis. *Front. Plant Sci.* **2020**, *11*, 586870. [CrossRef]
73. Ma, X.; Qiao, Z.; Chen, D.; Yang, W.; Zhou, R.; Zhang, W.; Wang, M. CYCLIN-DEPENDENT KINASE G2 regulates salinity stress response and salt mediated flowering in Arabidopsis thaliana. *Plant Mol. Biol.* **2015**, *88*, 287–299. [CrossRef]
74. Ton, J.; Jakab, G.; Toquin, V.; Flors, V.; Iavicoli, A.; Maeder, M.N.; Metraux, J.P.; Mauch-Mani, B. Dissecting the beta-aminobutyric acid-induced priming phenomenon in Arabidopsis. *Plant Cell* **2005**, *17*, 987–999. [CrossRef] [PubMed]
75. Slaughter, A.; Daniel, X.; Flors, V.; Luna, E.; Hohn, B.; Mauch-Mani, B. Descendants of primed Arabidopsis plants exhibit resistance to biotic stress. *Plant Physiol.* **2012**, *158*, 835–843. [CrossRef] [PubMed]
76. Lessard, P.; Bouly, J.P.; Jouannic, S.; Kreis, M.; Thomas, M. Identification of cdc2cAt: A new cyclin-dependent kinase expressed in Arabidopsis thaliana flowers. *Biochim. Biophys. Acta (BBA)-Gene Struct. Expr.* **1999**, *1445*, 351–358. [CrossRef]
77. Li, F.; Cheng, C.; Cui, F.; de Oliveira, M.V.; Yu, X.; Meng, X.; Intorne, A.C.; Babilonia, K.; Li, M.; Li, B.; et al. Modulation of RNA polymerase II phosphorylation downstream of pathogen perception orchestrates plant immunity. *Cell Host Microbe* **2014**, *16*, 748–758. [CrossRef]
78. Fukudome, A.; Aksoy, E.; Wu, X.; Kumar, K.; Jeong, I.S.; May, K.; Russell, W.K.; Koiwa, H. Arabidopsis CPL4 is an essential C-terminal domain phosphatase that suppresses xenobiotic stress responses. *Plant J. Cell Mol. Biol.* **2014**, *80*, 27–39. [CrossRef] [PubMed]
79. Fukudome, A.; Sun, D.; Zhang, X.; Koiwa, H. Salt Stress and CTD PHOSPHATASE-LIKE4 Mediate the Switch between Production of Small Nuclear RNAs and mRNAs. *Plant Cell* **2017**, *29*, 3214–3233. [CrossRef]

80. Fukudome, A.; Goldman, J.S.; Finlayson, S.A.; Koiwa, H. Silencing Arabidopsis CARBOXYL-TERMINAL DOMAIN PHOSPHATASE-LIKE 4 induces cytokinin-oversensitive de novo shoot organogenesis. *Plant J. Cell Mol. Biol.* **2018**, *94*, 799–812. [CrossRef]
81. Li, T.; Natran, A.; Chen, Y.; Vercruysse, J.; Wang, K.; Gonzalez, N.; Dubois, M.; Inze, D. A genetics screen highlights emerging roles for CPL3, RST1 and URT1 in RNA metabolism and silencing. *Nat. Plants* **2019**, *5*, 539–550. [CrossRef]
82. Xiong, L.; Lee, H.; Ishitani, M.; Tanaka, Y.; Stevenson, B.; Koiwa, H.; Bressan, R.A.; Hasegawa, P.M.; Zhu, J.K. Repression of stress-responsive genes by FIERY2, a novel transcriptional regulator in Arabidopsis. *Proc. Natl. Acad. Sci. USA* **2002**, *99*, 10899–10904. [CrossRef]
83. Manavella, P.A.; Hagmann, J.; Ott, F.; Laubinger, S.; Franz, M.; Macek, B.; Weigel, D. Fast-forward genetics identifies plant CPL phosphatases as regulators of miRNA processing factor HYL1. *Cell* **2012**, *151*, 859–870. [CrossRef] [PubMed]
84. Jiang, J.; Wang, B.; Shen, Y.; Wang, H.; Feng, Q.; Shi, H. The arabidopsis RNA binding protein with K homology motifs, SHINY1, interacts with the C-terminal domain phosphatase-like 1 (CPL1) to repress stress-inducible gene expression. *PLoS Genet.* **2013**, *9*, e1003625. [CrossRef] [PubMed]
85. Guan, Q.; Yue, X.; Zeng, H.; Zhu, J. The protein phosphatase RCF2 and its interacting partner NAC019 are critical for heat stress-responsive gene regulation and thermotolerance in Arabidopsis. *Plant Cell* **2014**, *26*, 438–453. [CrossRef]
86. Thatcher, L.F.; Foley, R.; Casarotto, H.J.; Gao, L.L.; Kamphuis, L.G.; Melser, S.; Singh, K.B. The Arabidopsis RNA Polymerase II Carboxyl Terminal Domain (CTD) Phosphatase-Like1 (CPL1) is a biotic stress susceptibility gene. *Sci. Rep.* **2018**, *8*, 13454. [CrossRef]
87. Aksoy, E.; Jeong, I.S.; Koiwa, H. Loss of function of Arabidopsis C-terminal domain phosphatase-like1 activates iron deficiency responses at the transcriptional level. *Plant Physiol.* **2013**, *161*, 330–345. [CrossRef]
88. Yuan, C.; Xu, J.; Chen, Q.; Liu, Q.; Hu, Y.; Jin, Y.; Qin, C. C-terminal domain phosphatase-like 1 (CPL1) is involved in floral transition in Arabidopsis. *BMC Genom.* **2021**, *22*, 642. [CrossRef]
89. Tao, L.Z.; Cheung, A.Y.; Wu, H.M. Plant Rac-like GTPases are activated by auxin and mediate auxin-responsive gene expression. *Plant Cell* **2002**, *14*, 2745–2760. [CrossRef] [PubMed]
90. Xu, T.; Wen, M.; Nagawa, S.; Fu, Y.; Chen, J.G.; Wu, M.J.; Perrot-Rechenmann, C.; Friml, J.; Jones, A.M.; Yang, Z. Cell surface- and rho GTPase-based auxin signaling controls cellular interdigitation in Arabidopsis. *Cell* **2010**, *143*, 99–110. [CrossRef] [PubMed]

Review

Cell Cycle Regulation by Heat Shock Transcription Factors

Yasuko Tokunaga [1], Ken-Ichiro Otsuyama [2] and Naoki Hayashida [1,*]

1   Division of Molecular Gerontology and Anti-Ageing Medicine, Department of Biochemistry and Molecular Biology, Graduate School of Medicine, Yamaguchi University, Ube 7558505, Japan; yako1229@yamaguchi-u.ac.jp
2   Department of Laboratory Science, Graduate School of Medicine, Yamaguchi University, Ube 7558505, Japan; otsuyama@yamaguchi-u.ac.jp
*   Correspondence: hayasida@yamaguchi-u.ac.jp; Tel.: +81-836-22-2359

Abstract: Cell division and cell cycle mechanism has been studied for 70 years. This research has revealed that the cell cycle is regulated by many factors, including cyclins and cyclin-dependent kinases (CDKs). Heat shock transcription factors (HSFs) have been noted as critical proteins for cell survival against various stresses; however, recent studies suggest that HSFs also have important roles in cell cycle regulation-independent cell-protective functions. During cell cycle progression, HSF1, and HSF2 bind to condensed chromatin to provide immediate precise gene expression after cell division. This review focuses on the function of these HSFs in cell cycle progression, cell cycle arrest, gene bookmarking, mitosis and meiosis.

Keywords: cell cycle; HSF1; HSF2; cell cycle arrest; APC/C complex

1. Introduction

The mitosis phenomenon was discovered over one hundred years ago, and many scientists have performed experiments to make novel discoveries. In the early 20th century, germ cells were the predominant focus of studies and the change in cell morphology during mitosis was widely known among researchers [1,2]. Important studies were performed all over the world and many essential findings were reported [3–8]. Changes in cytoplasmic structures were also revealed in studies that took place over a short period, and structural changes of mitochondria, Golgi substance, new cell walls, mitotic spindles, and centromere were identified [9–14]. Classification into Gap1 (G_1), synthesis (S), Gap2 (G_2), and mitosis (M) phases was suggested following the observation of cytoplasmic structures [15–24].

In the early 1970s, Leland Hartwell and his colleagues discovered several genes responsible for the eukaryotic cell cycle in budding yeast (*Saccharomyces cerevisiae*) (reviewed in [25]). In 1970, they discovered three genes: *cdc-1*, *cdc-2*, and *cdc-3* (*cdc* stands for "cell division cycle") [26]. Several years later, it was revealed that *cdc-1* encodes metallophosphodiesterase, *cdc-2* encodes the DNA polymerase delta catalytic subunit and *cdc-3* encodes septin family members [25]. Hartwell and his colleagues examined approximately 1500 temperature-sensitive mutants and isolated 148 mutants. They characterized these yeast mutants and succeeded in determining the locations of 14 genes on the yeast genetic map [27]. Subsequently, they found that four genes, including *cdc-4*, *cdc-7*, and *cdc-28*, are required for the initiation of yeast DNA synthesis and discovered that the *cdc-28* gene is required for DNA synthesis and budding [28]. In the *cdc-28* mutant, cells arrested as unbudded cells, and the cell cycle was also arrested before the initiation of DNA synthesis. Due to the importance of *cdc-28* in cell division, *cdc-28* became the first cloned cell cycle gene [29].

Discoveries by Hartwell and his colleagues contributed to cell division and cell cycle research and the progression of other researchers' studies. Their research became a pioneering study.

Citation: Tokunaga, Y.; Otsuyama, K.-I.; Hayashida, N. Cell Cycle Regulation by Heat Shock Transcription Factors. *Cells* **2022**, *11*, 203. https://doi.org/10.3390/cells11020203

Academic Editor: Zhixiang Wang

Received: 16 November 2021
Accepted: 4 January 2022
Published: 8 January 2022

Publisher's Note: MDPI stays neutral with regard to jurisdictional claims in published maps and institutional affiliations.

Paul Nurse and his colleagues also made valuable discoveries. They used temperature-sensitive mutants of the fission yeast *Schizosaccharomyces pombe*, different from the budding yeast *Saccharomyces cerevisiae* that Hartwell used, and found that their mutants have genetic mutations involved in cell size control over DNA synthesis and a second control acting on nuclear division [30]. Moreover, they discovered that *cdc-2* gene product activity is important for determining when mitosis takes place and is required for starting and controlling mitosis. They also successfully made a genetic map of the *cdc-2* locus and its isolation [31,32]. The understanding of *cdc-2* in the cell cycle was advanced by Nurse and his colleagues.

While Leland Hartwell and Paul Nurse made many discoveries using yeast, Tim Hunt examined protein synthesis using fertilized sea urchin eggs. Hunt and his colleagues analyzed the pattern of protein synthesis before and after fertilization using a two-dimensional gel separation technique and found that the pattern of expressed proteins was changed [33]. They first noticed that one of these proteins is destroyed every time the cells divide. Then, they performed protein analysis experiments using ^{35}S-methionine-added egg suspensions and discovered that some proteins start to be synthesized after fertilization and are destroyed at certain points in the cell division cycle. They examined this protein in detail and proposed to call these proteins "cyclins" [33]. Importantly, they also showed data indicating the existence of multiple cyclins and described two cyclins, cyclin A and cyclin B, discovered from the clam *Spisula solidissima* in this paper.

After the discovery of cyclins, molecular cloning was succeeded very soon. Ruderman and his colleagues cloned cyclin A from Xenopus oocytes [34]. Hunt's group also reported the cloning of the cyclin from sea urchin eggs [35].

Leland Hartwell, Paul Nurse, and Tim Hunt were awarded a Nobel Prize in 2001, and the importance of cell cycle regulation has been more widely recognized among researchers in various fields. It is notable that the essential genes in the cell cycle were discovered through experiments using temperature-sensitive mutant yeast cells. Furthermore, the most famous transcription factor against temperature stress is heat shock transcription factor/heat shock factor (HSF).

HSF was discovered as a transcription factor essential for the heat shock response. The heat shock response is a cellular protective mechanism against heat stress that was discovered by Ritossa in 1962. While he was keeping larvae of *Drosophila melanogaster* at 25 degrees, he observed that the salivary gland of *Drosophila* was puffing when under additional heat stress at 37 degrees, which is a sign of enlarged chromosomal formation and indicates that some specific mRNA synthesis was accelerated [36–38]. Subsequently, Tissières and his colleagues showed that accelerated mRNA synthesis occurs in the heat shock protein (HSP) genes [39,40].

Five years before the discovery of HSF, it was reported that there are some proteins that specifically bind to the heat shock element (HSE) sequence commonly found in the HSP promoter following in vitro and in vivo experiments performed by Wu's group [41,42]. Several research groups discovered HSF protein from *Saccharomyces cerevisiae*, *Drosophila melanogaster*, and HeLa cells in 1987 [43–46]. In 1991, two HSFs, HSF1 and HSF2, were cloned in humans and mice [47–49]. HSF1 is activated by heat stress, but HSF2 was found to be activated first by hemin [50]. As two HSFs were cloned at the same time, the previously discovered HSF was called HSF1, and the other HSF was called HSF2 (Figure 1).

As HSF1 was discovered to be a transcription factor protecting cell survival through the heat shock response, the HSF1–HSP pathway was exclusively studied in the early period. However, as already described, the essential genes in the cell cycle were discovered in experiments using temperature-sensitive mutant yeast cells. Interestingly, other experiments using temperature-sensitive mutant yeast cells opened the door to the study of the cell cycle and HSFs.

Significance in Section 1:

- The essential genes in the cell cycle were discovered through experiments using temperature-sensitive mutant yeast cells.

- HSFs were discovered as proteins specifically binding to HSE sequence commonly found in the *HSP* promoter.

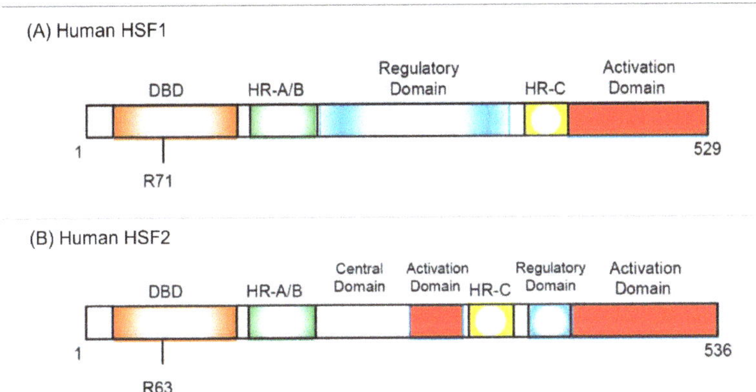

Figure 1. Structure of human HSF1 (hHSF1, (**A**)) and human HSF2 (hHSF2, (**B**)). hHSF1 and hHSF2 have similar domain structures, but several differences exist. R71 and R63 are required for DNA binding in hHSF1 and hHSF2, respectively. The regulatory domain suppresses trimerization of both HSFs and inhibits transcriptional activation. Trimerization is required for the activation of HSFs. DBD, DNA-binding domain; HR-A/B, hydrophobic heptad repeats A and B; HR-C, C-terminal hydrophobic heptad repeat.

2. Discovery of HSF Involvement in Cell Cycle Regulation

The first evidence that HSF1 is involved in cell cycle regulation came from experiments using mutant yeast cells. Smith and Yaffe found that yeast cells containing a mutation in the *mas3* gene display retarded progression through the G_2 stage, and that this *mas3* gene encodes HSF1 [51]. In addition, *mas3* cells showed that the induction of a major heat shock gene, *SSA1*, is defective under heat stress at 37 degrees. These are the first data indicating that HSF1 mediates cell cycle progression.

After this report, other laboratories published papers similarly indicating HSF1's mediation of the cell cycle. Thiele and his colleagues discovered that yeast cells bearing a truncated form of HSF1 in which the C-terminal transcriptional activation domain is deleted undergo reversible cell cycle arrest under heat stress at 37 degrees in the G_2/M phase and exhibit a marked reduction in HSP90 expression [52]. Using wild-type (WT) mouse embryonic fibroblasts (MEFs) and HSF1-deficient MEFs, Dix and his colleagues discovered that a three-fold increase in G_2/M phase-arrested cells occurred in WT–MEFs but not in HSF1-deficient MEFs after conditioning heat stress (43 degrees, non-lethal temperature) followed by lethal heat stress (45 degrees) [53]. However, Li and Martinez reported that human colon cancer HCT116 cells lacking HSF1 did not show G_2/M arrest, and checkpoint activation was lost in these cells [54]. They also showed that p53 is involved in this phenomenon and that the relationship between HSF1 and p53 is critical for cell cycle regulation.

Calderwood and his colleagues reported that overexpression of HSF1 increases the proportion of G_1 cells in HeLa cells under non-heat-stressed conditions [55]. Surprisingly, they observed that HSF1 bound to HSE sequence during the early G_1 phase in the absence of heat stress. In addition, He and Fox reported the behavior of HSF1 during the cell cycle under non-heat-stressed conditions [56]. They measured the binding of HSF1 to the HSE with a gel mobility shift assay using cell extracts from Hoechst 33342-labeled heated Chinese hamster ovary (CHO) cells from the G_1, S, and G_2/M phases. Their study indicated that the binding activity of HSF1 is two-fold higher in the S phase than in the G_1 or G_2/M phases, but the HSF1 expression levels do not vary throughout the cell cycle [56].

Gross' group also examined HSF1 behavior throughout the cell cycle using their original dinucleosomal heat shock promoter model and found that activated HSF1 cannot bind to DNA in G_1-arrested cells but can bind following release from G_1 arrest or after the imposition of either an early S- or late G_2-phase arrest [57]. Their discoveries showed that HSF1 binding activity to the DNA of the target genes changes throughout the cell cycle and is related to G_1, S, and G_2/M phase arrest.

The existence of a relationship between HSF1 and the cell cycle was widely recognized among HSF researchers. Thus, some considered the possibility that HSF1 plays a role in cancer. The first evidence demonstrating this possibility was shown by a study using prostate carcinoma cell lines and carcinomatous prostate tissue sections from patients. In a comparison between the non-metastatic human prostate carcinoma cell line PC-3 and the metastatic variant PC-3M, higher expression of HSF1 was found in both mRNA and protein [58]. Moreover, HSF1 protein was expressed more highly in tumor tissues than in normal sections from the same patient. Calderwood and his colleagues also used PC-3 cells and stably expressed dominant-negative HSF1 (DN-HSF1), which lost transcriptional activities in the cells. In DN-HSF1-expressed PC-3 cells, aneuploidy was inhibited and cyclin B1 degradation was retarded [59], the latter being a key step in the control of mitosis. In addition, p21 expression was increased in DN-HSF1 cells. Although PC-3 is a p53-null cell, DN-HSF1 can induce p21 expression directly or indirectly [59]. The studies using prostate carcinoma cells and sections suggested that HSF1 can positively and negatively regulate cell cycle progression.

The mechanism by which HSF1 is involved in cell cycle regulation has been examined by many researchers. HSF1 was discovered to be a protein binding to HSE sequence found in the promoter of HSPs, the major and heat-inducible targets of HSF1 [43–46], but whether this HSF1–HSP pathway has a pivotal role in cell cycle regulation or another HSF1 pathway is more important was uncertain. At that time, new findings were provided instead by HSF2 research.

Significance in Section 2:

- Mutant HSF1 affects cell cycle progression and arrest.
- Higher expression of HSF1 is related to cancer and aneuploidy.

3. Bookmarking Is the Important Role of HSFs

It was already shown that HSF2 can bind to the *HSP70* promoter, but the reason remained unclear [60–62]. Sarge and his colleagues performed a yeast two-hybrid screen to identify HSF2-interacting proteins and found that a subunit of condensin enzyme called CAP-G protein binds to HSF2 [63]. HSF2 interacts with the C-terminal domain of CAP-G and binds to the *HSP70* promoter in mitotic cells. They previously found that SUMOylation upregulates the DNA binding activity of HSF2 [64], and the SUMOylation also accelerated HSF2 and CAP-G interaction in mitotic G_2/M cells. The level of this interaction between SUMOylated HSF2 and CAP-G was higher in G_2/M cells than G_0/G_1 or S cells [63]. These results suggested that HSF2 binds to the *HSP70* promoter in a mitosis-dependent manner and prevents the compaction of this promoter; thus, they hypothesized that HSF2 has an important role in *HSP70* bookmarking.

The bookmarking mechanism was first proposed by Levens' group [65]. During mitosis, chromatin condenses, transcription is shut off, and most transcription factors are excluded from chromosomes [66,67]. However, several researchers discovered that chromatin is not completely condensed and that chromatin's structure is disturbed in some genes before Levens' group's Nature paper was published [68–70]. Levens and his colleagues showed that chromatin conformational distortion of the TATA box region of the *HSP70* promoter occurred specifically during mitosis using the footprinting technique [65], and they showed the same phenomenon on the *c-myc* and beta-globin promoters. To date, bookmarking has been observed in various gene promoters and transcription factors [71–74]. HSF2 is the first transcription factor to be shown to bind to the gene promoter during mitosis and essential for bookmarking of the *HSP70* gene.

In the bookmarking mechanism on the *HSP70* promoter, interaction between HSF2 and the condensin enzyme CAP-G subunit is required [63]. Sarge and his colleagues discovered that serine–threonine protein phosphatase (PP2A) is recruited to the *HSP70* promoter, and that this recruiting is also necessary for bookmarking [63,75,76]. Condensin activity requires phosphorylation of the CAP-G, CAP-D2, and CAP-H subunits by the mitotic kinase Cdc2–cyclin B complex [77,78]. When condensin interacts with HSF2 via CAP-G, CAP-G is dephosphorylated by PP2A complexed with HSF2. Subsequently, condensin is inactivated and the compaction of DNA on this promoter is prevented [63].

In addition, they discovered that TATA-binding protein (TBP) remains bound to DNA during mitosis and contributes to recruiting PP2A and also that protein regulating cytokinesis 1 (PRC1) also interacts with HSF2 and binds to *HSP70* promoter during mitosis [79,80]. As PRC1 was known to be a CDK substrate that interacts with mitotic spindle and functions in microtubule binding [81,82], they found that HSF2 and PRC1 associate and colocalize during the mitosis phase, whereas PRC1 does not interact with HSF1 [80].

Hayashida reported that HSF2 directly binds to the WD40 repeat protein WDR5, a core component of the Set1 and mixed lineage leukemia (MLL) H3K4 histone methyltransferase complex (Set1/MLL complex), and that HSF2 and major components of the Set1/MLL complex, WDR5, RbBP5, and Ash2L are recruited to the alphaB-crystallin (*CRYAB*) promoter in the same manner as in MEFs [83]. Before this discovery, Vakoc and his colleagues found that MLL1, RbBP5, and Ash2L associate with some gene promoters packaged within condensed mitotic chromosome and thus suggested that the MLL complex also has a mitotic bookmarking function as an additional epigenetic mechanism [84]. The discoveries by Hayashida and Vakoc's groups may support and confirm the existence of the gene bookmarking function of HSF2.

Significance in Section 3:

- HSF2 is the first transcription factor shown to have a gene bookmarking function and to bind to the gene promoter during mitosis.
- Chromatin is not completely condensed. The structure is disturbed in some genes.

4. Change in HSF1 Expression Level Induces Cell Cycle Arrest

4.1. Decreased HSF1 Expression Suppresses Cancer Cell Proliferation

Since a higher expression level of HSF1 in prostate cancer cells and tissue sections from prostate cancer patients was reported, the roles of HSF1 in cancer cells have been examined by many research groups. As already described, overexpression of HSF1 increases the proportion of G_1 cells, probably by induction of G_1 arrest, in cervical carcinoma HeLa cells [55]. The effects of reduced HSF1 expression in cancer cells were examined after the report that dominant-negative HSF1 (DN-HSF1) expression inhibits aneuploidy as well as p53 expression in PC-3 p53-null prostate cancer cells [59].

Lindquist and her colleagues showed that cancer cell survival was strongly inhibited by HSF1 knockdown in five human breast cancer BT-20 cells, BT-474 cells, MCF-7 cells, MDA-MB-231 cells, T47D cells, HeLa cells, PC-3 cells, S462 peripheral nerve sheath tumor cells, 90-8 peripheral nerve sheath tumor cells, and 293T in vitro transformed cells, but WI-38 normal fibroblast cells were not affected at all [85]. Muto and his colleagues also reported that HSF1 knockdown prominently reduced the proliferation of human melanoma MeWo cells but not human normal keratinocyte HaCaT cells [86]. These results indicate that reduced HSF1 expression impairs cancer cell proliferation and survival but not normal cell survival and proliferation. Muto and his colleagues also showed that the protein expression of HSP110, HSP90, HSP70, HSP60, and HSP40 was reduced by HSF1 knockdown, but the HSP expression level was not changed in HaCaT cells except HSP90 [86].

A high expression level of various HSPs in human and mouse cancer cells previously reported in the 1980s [87–90]. The *HSP70* promoter is known to be bookmarked in condensed chromatin during the cell cycle in HeLa cells [63,65], and the HSF–HSP pathway may have more critical roles in cancer cells than in normal cells. We cannot describe here whether both the HSF–HSP pathway and gene bookmarking by HSF are critically important in normal cells; however, recent studies reported that other transcription factors have the gene bookmarking function and that this function has been discovered in normal mammalian cells and in *Caenorhabditis elegans* [91–93]. Therefore, the molecular mechanism of the relationship between the HSF–HSP pathway, cell cycle progression, cell canceration, and gene bookmarking is notable and must be further elucidated.

4.2. Overexpression of HSF1 Also Causes Suppression of Cancer Cell Proliferation

As already described, overexpression of HSF1 increases the proportion of G_1 cells in HeLa cells under non-heat-stressed conditions [55]. Hayashida and his colleagues established that the HeLa cells in which constitutive active HSF1 (caHSF1) [94,95] or DNA binding activity-lacking mutant HSF1 (RgHSF1) are inductively expressed using the tet-off system. The caHSF1 lacks a regulatory domain and thus exhibits constitutive transcriptional activity (Figure 1A). RgHSF1 has an amino acid mutation at R71, and this arginine (R) amino acid is replaced by glycine (G). This R71G mutation diminishes DNA binding activity (Figure 1A, 94,95). They discovered that caHSF1-expressing cells proliferate at a very low speed (Figure 2A, left, [96]). In contrast, both RgHSF1-expressing cells and caHSF1 expression-inhibited cells proliferate normally (Figure 2A, right, 96). They examined the cell population and found that G_1 phase cells were increased 1.5-fold and G_2/M phase cells decreased by 25% in caHSF1-expressing cells (Figure 2B, 96).

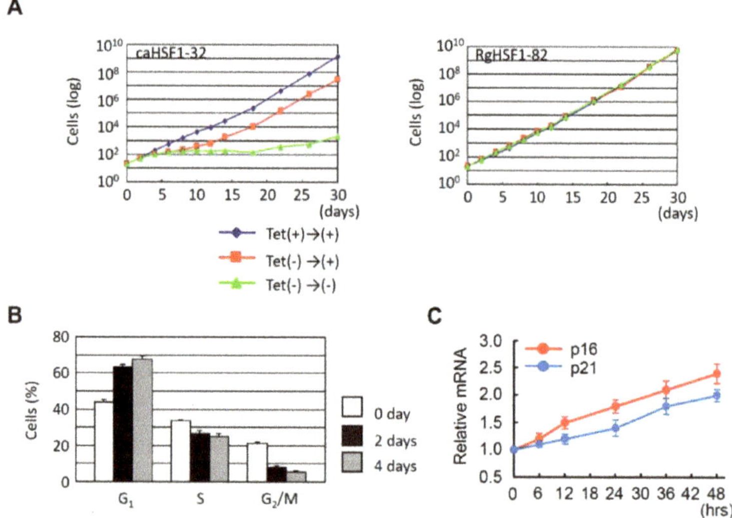

Figure 2. HeLa cell proliferation is inhibited by constitutive active HSF1 (caHSF1). (**A**) caHSF1 expression dramatically inhibits cell proliferation (**left**). In contrast, DNA binding activity lacking mutant HSF1 (RgHSF1) expression does not affect cell proliferation (**right**). (**B**) Change of cell proportion in G_1, S, and G_2/M phases by the induction of caHSF1 expression. G_1 proportion increases and G_2/M proportion decreases. (**C**) mRNA expression of p16 and p21 increases by the induction of caHSF1 expression.

To reveal the mechanism behind this phenomenon, Hayashida and his colleagues examined the expression of p16 and p21 in caHSF1-expressing cells. At 48 h after the start of caHSF1 induction by tetracycline depletion, the expression levels of p16 and p21 were increased 2.3-fold and 1.9-fold, respectively (Figure 2C, 96). It is certain that caHSF1 significantly suppresses cell proliferation though the induction of target genes, but whether p16 and p21 substantially contribute to this phenomenon cannot be identified by only these results. Ilangovan and his colleagues reported that HSP27 binds to p53 and increases p53 transcriptional activity, which increases p21 expression and results in G_2/M cell cycle arrest [97]. Hayashida and his colleagues previously reported that the expression of most HSPs, including HSP27, increases in the same caHSF1-expressing HeLa cells [95]. One hypothesis is that high expression of HSP27 protein may cause cell cycle arrest and suppress cell proliferation through a similar mechanism. Hayashida's group showed that caHSF1 expression induces the increased G_1 cell population (probably through G_1 cell arrest) [96]. According to these data and previous reports, the effects of HSF1 expression may be more complexed than we imagine. The effects of HSF1 expression have been investigated by many HSF researchers, however, this might be a difficult issue and required to investigated in more detail in HSF research field.

Finally, it is surprising that both overexpression and knockdown of HSF1 can induce cell cycle arrest and suppress cell proliferation. The molecular mechanisms that cause these contrasting phenomena are notable, and the analysis of these mechanisms may contribute to the discovery of novel functions of HSF1 in cell cycle regulation.

Significance in Section 4:

-   Reduced HSF1 expression (induced by knockdown) inhibits the proliferation of cancer cells but not normal cells.
-   Constitutive active HSF1 expression also inhibits the proliferation of cancer cells probably through the induction of G_1 cell cycle arrest.

5. Degradation of HSFs during Cell Cycle

5.1. Degradation of Cyclins and CDKs, Subunits of Cyclin-CDK Complex

During the cell cycle, the levels of cyclins and cyclin-dependent kinases (CDKs) are tightly regulated. Several types of cyclin–CDK complexes function during the cell cycle, and the function is indispensable for normal cell cycle progression, appropriate induction of cell cycle arrest, and cell survival. The level of CDKs remains relatively constant, but the level of cyclins oscillates. Cyclins synthesize, bind, and activate CDKs that are then destroyed [98]. Importantly, CDKs cannot be activated without the interaction with cyclins.

As is widely known, proteolytic control in the cell cycle is crucial for cell cycle regulation and is carried out by various cyclin–CDK complexes [99]. In mitosis, the activation of a large ubiquitin-protein ligase, the anaphase-promoting complex (APC, also known as cyclosome and thus frequently described as APC/C) is required for anaphase initiation and exit from mitosis [100]. During anaphase, replicated chromosomes are split and daughter chromatids move to opposite ends of the cell, and the cells proceed to telophase, the last phase. CDC20 and CDH1 (Clb cyclins by dephosphorylating the APC-specificity factor, homolog of CDC20) are activator proteins of APC [100,101] and its cyclin ubiquitination activity.

APC/C is activated not only in anaphase but also other phases. CDC20 activates APC/C in metaphase and CDH1 does so in telophase, and in the G_1 phase of proliferating cells [102–108], and during G_0 in differentiated cells [109]. The APC/C complex targets and degrades proteins related to cell cycle regulation. In the past 10–15 years, HSFs have been found to be degraded by this APC/C complex. Next, we describe the degradation mechanism of HSFs during the cell cycle.

5.2. HSFs Are Degraded by APC/C Complex

As described above, there is growing evidence that HSF1 and HSF2 have important roles in cell cycle regulation. Additionally, whether HSFs and cyclins are degraded by the same mechanism or not has been investigated by many researchers studying HSFs. Among them, Lee and his colleagues reported that HSF1 interacts with CDC20 [110]. This interaction inhibits the association between CDC20 and CDC27, which is one of the canonical subunits of the APC/C complex and is important for mitotic exit and transition into G_1 phase [111–113] and suppresses the phosphorylation and ubiquitination activity of the APC/C complex [110]. Subsequently, they showed that HSF1 is localized to the centromere in mitosis and especially to the spindle poles in metaphase, and that phosphorylated HSF1 undergoes ubiquitin degradation during spindle pole localization [114]. Moreover, they discovered that HSF1 degradation only occurs when HSF1 is phosphorylated, and this phosphorylated HSF1 is released from CDC20. CDC20 binds to the APC/C complex again later. HSF1 is phosphorylated at Ser216 by polo-like kinase (Plk1) in early mitosis, and this phosphorylation is stabilized through interaction with CDC20 [114]. Their observation suggested that the phosphorylated HSF1–CDC20 interaction and subsequent HSF1 degradation may be required for mitotic regulation.

HSF2 is known as the gene bookmarking transcription factor [63]. Sistonen and her colleagues discovered that the APC/C complex ubiquitylates and degrades HSF2 through the interaction as HSF2–CDC20 or HSF2–CDH1 as well as HSF1 [115]. The interaction between HSF2 and the APC/C subunit CDC27 or coactivator CDC20 is enhanced by moderate heat stress, and HSF2 degradation is induced during the acute phase of the heat shock response. Finally, HSF2 is removed from the *HSP70* promoter [115].

Sistonen and her colleagues did not refer to HSF2 phosphorylation, and whether HSF2 is phosphorylated during the cell cycle is uncertain. Sarge and his colleagues reported that HSF2 binds to the PR65 subunit of protein phosphatase 2A (PP2A) and activates PP2A [116]. However, they did not discuss whether HSF2 is dephosphorylated. Phosphorylation is very important for the activation of HSF2 (and HSF1); thus, whether the HSF2 phosphorylation state is stable or not during cell cycle progression needs to be identified.

There are not many papers investigating the molecular mechanism of the degradation of HSF1 and HSF2; thus, we need to further this study and accumulate data. However, the findings of Sistonen's group and Kevin's group are notable because they indicate that HSF1 and HSF2 are probably degraded with the same mechanism as cyclins.

Significance in Section 5:

- Both HSF1 and HSF2 are degraded by the APC/C complex with the same mechanism as cyclins.

6. HSFs Are Important for Meiosis

Meiosis is an essential sexual reproduction process, and meiotic failure seriously affects the production of offspring. Meiosis produces haploid cells from diploid parental cells and reduces the chromosome number by half. The regulation of the cell cycle is different from mitosis; the S phase is followed by two rounds of cell division [117].

HSFs are involved in meiosis as well as mitosis. At the early spermatocyte stage, autosomal chromosomes undergo complete synapsis, but X and Y chromosomes can only undergo incomplete synapsis because of their different sizes and shapes. At the pachytene spermatocyte stage in prophase I, X and Y chromosomes are segregated into a subnuclear compartment called the XY body (also called the sex body) [118]. The unsynapsed regions of X and Y chromosomes are condensed and repressed, and HSF1 localizes to the XY body and contributes to gene silencing [119]. HSF1 is transiently expressed in meiotic spermatocytes and haploid round spermatids in mouse testes, and HSF1-deficient male mice show increased morphological abnormalities in sperm heads [119].

HSF1 also has important roles in oogenesis. Christians and her colleagues found that metaphase II arrest is normally induced in HSF1-deficient mature oocytes but discovered ultrastructural abnormalities and dysfunctional mitochondria in intraovarian HSF1-deficient oocytes [120]. She previously discovered that the embryos produced by HSF1-deficient female mice initiate early development and do not survive even in the oviduct of wild-type mice in spite of the fact that ovulated eggs of HSF1-deficient mice show proper metaphase II arrest [121]. The cell division of these embryos was blocked mainly at the one-cell stage when they were produced by mating with wild-type or HSF1 heterozygous male mice [121]. These results show that HSF1 has crucial roles in meiosis for both spermatogenesis and oogenesis.

Two groups reported that meiosis is affected by HSF2 deficiency in male mice [122,123]. In HSF2-deficient male mice, Mezger and her colleagues discovered that seminiferous tubules are vacuolated and that these tubules are devoid of all meiotic spermatocytes and post-meiotic haploid spermatids [122]. In addition, they found that dying cells increase in HSF2-deficient testes and that 90% of these dying cells are spermatocytes at meiotic divisions, especially at pachytene meiotic prophase. A similar observation was reported by Mivechi and her colleagues [123]. These findings indicate that spermatogenesis is disrupted by HSF2 deficiency.

As well as HSF1 deficiency, oogenesis was suggested to be affected by HSF2 deficiency. Mezger and her colleagues found that 40% of the eggs from HSF2-deficient female mice are fragmented or devoid of polar bodies [122]. They also observed ovarian defects, including the presence of large hemorrhagic follicles with a trapped oocyte; however, they did not show clear data indicating the meiotic defects in oogenesis. No other groups have reported that HSF2 is involved in oogenesis; thus, whether oogenesis is affected or regulated by HSF2 remains to be elucidated.

Significance in Section 6:

- HSF1 and HSF2 are involved in meiosis as well as mitosis.
- Spermatogenesis is affected by both HSF1 and HSF2. It is not clear whether oogenesis is also affected by both HSFs.

7. Conclusions and Perspectives

In this review, we briefly described the early essential findings in cell cycle studies and the discovery of the heat shock response and essential functions of HSFs and referred to the important discoveries to date. In mammalian cells, four major HSFs (HSF1–HSF4) have been discovered, but human HSF3 is a pseudogene and mouse HSF3 has not been well studied yet [124]. There are no reports referring to the relationship between HSF3 and cell cycle regulation.

Concerning HSF4, Mivechi and her colleagues reported that HSF4b (major isoform of HSF4 [125]) binds to the promoters of HSPs with the Brahma-related gene (Brg1) ATPase subunit of SWI/SNF chromatin remodeling complex, and that the HSF4b-Brg1 complex is formed only during the G_1 phase [126]. During the G_1 phase, the chromatin structure is more accessible to transcriptional regulatory proteins; thus, HSF4b may stimulate the expression of HSPs during the cell cycle. However, whether HSF4b has important roles in cell cycle regulation is yet to be elucidated.

The success of molecular cloning of HSF1 and HSF2 gave rise to new experimental techniques and brought about many new discoveries. For example, C-terminal activation domain-lacking mutant HSF1-bearing yeast cells showed to undergo G_2/M arrest, and overexpression of HSF1 increased the population of G_1 cells in HeLa under non-heat-stressed conditions [52,65]. The importance of these results should be noted because these phenomena were observed under non-heat stress. After the success of cloning, several experiments were performed under non-heat-stressed conditions (37 degrees when using mammalian cells); thus, we learned additional information about HSF1 and HSF2. In particular, it was discovered that HSF2 is not activated by heat stress meaning that the success of cloning contributed greatly to the progression of HSF2 research.

p53 is widely recognized as a tumor suppressor gene, and its protein product is also a well-known transcription factor regulating the expression of genes involved in cell cycle arrest, apoptosis, and DNA repair [127,128]. In p53-dependent cell cycle arrest, the cyclin-dependent kinase inhibitor p21 (WAF1) is an important mediator [127,129]. Notably, dominant-negative HSF1 (DN-HSF1) increased p21 expression in p53-null PC-3 prostate carcinoma cells [59]. Robson and his colleagues showed that HSF1 interacts with p53 and that this interaction becomes more stable when the cells are treated with the genotoxic compounds actinomycin D, doxorubicin, or etoposide [130]. In addition, heat or doxorubicin stress induced HSF1 translocation to the nucleus together with p53 and TATA-box binding protein (TBP). They also established p53 knockdown cells and HSF1 knockdown cells and examined whether HSF1 is involved in the expression of p53 target genes. They found that expression of the p53 target genes p21, PUMA, gadd45, and PCNA is reduced, and p21 and PUMA expression in particular is dramatically reduced in both knockdown cells [130]. Further, Hayashida and his colleagues also examined whether p21 and p16 genes are regulated by constitutive active HSF1 in HeLa cells and found increased expression of p21 and p16 and dramatic suppression of cell proliferation [96].

These findings strongly indicate that HSF1 can affect p53 functions directly or indirectly. As it is widely known that p53 has an extremely wide range of effects, HSF1 may affect many biological phenomena more than is currently known. HSF1 knockdown, caHSF1 expression, and normal HSF1 overexpression dramatically reduced cell proliferation [55,86,96]. To understand this paradoxical phenomenon and establish a consistent theory of the molecular mechanism, a more detailed analysis of the formation of HSF1-interacting proteins and various HSF1-containing transcriptional complexes will be required. The analysis should also consider their binding to HSF2.

In this review, we also referred to the important roles of WD40 repeat proteins. Many of these proteins have seven WD domains; the most important function of the WD domain is forming protein–protein interaction [131]. CDC20, CDH1, and other CDC proteins are WD40 repeat proteins [132–135], and these CDC proteins bind to the central regulatory domain of HSF1 [110] and contribute to HSF1 degradation as subunits or core components of the APC/C complex [110,115]. The central domain of HSF2 is not a regulatory function domain [83], but CDC20 and CDH1 bind to this domain of HSF2 and degrade HSF2 [115]. The WD40 repeat protein WDR5 also binds to the HSF2 central domain and contributes to the formation of the HSF2-Set1/MLL histone H3K4 methyltransferase complex [83]. Investigation of the functional complexes formed through WD domains will be valuable and bring about new discoveries in both the cell cycle and HSF research fields. For example, the expression of caHSF1 shows dramatically reduced proliferation, and caHSF1 lacks the central domain to which CDC20 and CDC27 bind. Paradoxically, reduced cell proliferation is also observed in various HSF1 knockdown cancer cells. Investigation of WD domain-mediated complex formation will contribute to determining why these paradoxical phenomena are observed.

We summarized and discussed the cell cycle regulation by HSFs, especially HSF1 and HSF2. For HSF researchers, the most known functions of HSFs are the heat shock response and the maintenance of protein homeostasis (proteostasis [136]); thus, the information on the role of HSF1 and HSF2 during cell cycle and their degradation mechanism is almost similar to that of cyclins during the cell cycle will be very useful. For cell cycle researchers, HSF1 and HSF2 will be unfamiliar terms. Furthermore, although HSF1 is the most well-known HSF, we propose that the biological significance of HSF2 is greater than is currently known and that the importance of HSF2 research will increase.

To date, several groups have shown that HSF1 and HSF2 are involved in both mitosis and meiosis (especially spermatogenesis). These studies might contribute to the development of medicine and therapy. Male infertility plays a role in about half of infertility cases, almost 30% of which are caused by a male factor alone [137]. Konrad and his colleagues analyzed azoospermic patients with defined spermatogenic defects and examined germ cell numbers in the early phase of spermatogenesis with an emphasis on mitosis–meiosis transition [138]. They classify the deficiencies individually by distinguishing between high and low efficiency of spermatogenesis. The patients with maturation arrest in primary spermatocytes show significantly reduced numbers of spermatogonia. In contrast, the patients without histological abnormalities in spermatogenesis can compensate for the number of spermatogonia with a high efficiency of meiotic entry and showed normal numbers of spermatogonia if they did not show maturation arrest [138]. Konrad's group were the first to suggest that a compensatory meiosis mechanism exists in human spermatogenesis and that whether maturation arrest occurs in primary spermatocytes or not is critical.

There is a hypothesis that other HSFs in addition to HSF1 and HSF2 may have important roles in spermatogenesis and oogenesis. In 2004, Nakahori and his colleagues reported that a novel HSF called HSFY exists on the human Y chromosome as multicopies [139]. The human *HSFY* gene located on Yq has a long open reading frame containing an HSF-type DNA-binding domain and is similar to the *LW-1* gene on the human X chromosome, with 53% homology for the amino acid sequences of their presumed DNA-binding domain

(DBD). Unlike HSF1 and HSF2, the 53% homology of DBD is not enough to bind the same DNA promoter sequence; in addition, the DBD homology between HSFY and HSF1 is lower at 31% [139]. The *HSFY* gene generates three different transcripts by alternative splicing, and the corresponding proteins are 401, 203, and 213 amino acids long [140]. Transcript 1 has been detected in testis, brain, pancreas, and sperm tissue but not in other tissues or in Sertoli cells. Transcripts 2 and 3 have been detected only in testes [140]. The DBD exists in only variant 1, which is 401 amino acids long, and no HSFY target genes have been discovered to date. Nakahari's group found the deletion of *HSFY* from men with azoospermia or oligospermia, and Foresta and his colleagues found the same deletion in infertile men [139,140]. These results suggest that HSFY may have important roles in human spermatogenesis. HSFs, including HSF1, HSF2, and HSFY, are most likely involved in the meiotic phase of spermatogenesis and oogenesis, but further research of these phenomena is still needed.

Author Contributions: N.H. prepared the original draft and wrote most of the manuscript. Y.T. also prepared the original draft and arranged the contents of this manuscript. K.-I.O. contributed to the preparation of each section of the original draft and edited the manuscript. All authors have read and agreed to the published version of the manuscript.

Funding: This study was supported by a JSPS KAKENHI Grant-in-Aid for Scientific Research (C) (25430090 to N.H.) and the Okayama University Translational Research Network Program (A33 to N.H.) funded by the Japan Agency of Medical Research (AMED).

Acknowledgments: N.H. and Y.T. would like to thank Yoichi Mizukami and the staff at the Yamaguchi University Center for Gene Research for their support of our research. N.H. also thanks the students of the Yamaguchi Kojo School of Nursing, Ube School of Nursing, and Yamaguchi University School of Medicine. N.H. thanks Shuji Noguchi, Yasuhiro Kuramitsu, Takeshi Honda, and Hiroyuki Kida for their encouragement.

References

1. Hegner, R.W.; Russell, C.P. Differential Mitoses in the Germ-Cell Cycle of Dineutes Nigrior. *Proc. Natl. Acad. Sci. USA* **1916**, *2*, 356–360. [CrossRef]
2. Brooks, F.G. THE Germ Cell Cycle of the Digenetic Trematodes. *Science* **1928**, *68*, 277–278. [CrossRef] [PubMed]
3. Wilson, J.W.; Leduc, E. Stimulation of mitosis in mouse liver. *Anat. Rec.* **1946**, *96*, 534.
4. Koller, P.C. Abnormal mitosis in tumours. *Br. J. Cancer* **1947**, *1*, 38–47. [CrossRef] [PubMed]
5. Elftman, H. The Sertoli cell cycle in the mouse. *Anat. Rec.* **1950**, *106*, 381–392. [CrossRef] [PubMed]
6. O'Connor, R.J. The metabolism of cell division. *Br. J. Exp. Pathol.* **1950**, *31*, 390–396. [PubMed]
7. O'Connor, R.J. The effect on cell division of inhibiting aerobic glycolysis. *Br. J. Exp. Pathol.* **1950**, *31*, 449–453.
8. Roberts, K.B.; Florey, H.W.; Joklik, W.K. The influence of Cortisone on cell division. *Q. J. Exp. Physiol. Cogn. Med. Sci.* **1952**, *37*, 239–257. [CrossRef] [PubMed]
9. Dalton, A.J. Cytoplasmic changes during cell division with reference to mitochondria and the golgi substance. *Ann. N. Y. Acad. Sci.* **1951**, *51*, 1295–1302. [CrossRef] [PubMed]
10. Elliot, E. Formation of new cell walls in cell division. *Nature* **1951**, *168*, 1089. [CrossRef]
11. Bisset, K.A. Spurious mitotic spindles and fusion tubes in bacteria. *Nature* **1952**, *169*, 247. [CrossRef]
12. DeLamater, E.D. Spurious mitotic spindles and fusion tubes in bacteria. *Nature* **1952**, *169*, 248. [CrossRef] [PubMed]
13. Wardrop, A.B. Formation of new cell walls in cell division. *Nature* **1952**, *170*, 329. [CrossRef] [PubMed]
14. King, C.G. Diffuse centromere, and other cytological observations on two desmids. *Nature* **1953**, *171*, 181. [CrossRef] [PubMed]
15. DeLamater, E.D. The mitotic mechanism in bacteria. *Cold Spring Harb. Symp. Quant. Biol.* **1953**, *18*, 99–100. [CrossRef]
16. Boss, J. The spindle and the mechanism of chromosome separation as seen in the living cell. *J. Physiol.* **1953**, *119*, 34–36.
17. Stevens, C.E.; Daoust, R.; Leblond, C.P. The desoxyribonucleic acid of interphase and dividing nuclei. *Can. J. Med. Sci.* **1953**, *31*, 263–278. [CrossRef] [PubMed]
18. Parmentier, R. Production of three-group metaphases in the bone-marrow of the golden hamster. *Nature* **1953**, *171*, 1029–1030. [CrossRef]
19. Fautrez, J.; Fautrez-Firlefyn, N. Deoxyribonucleic acid content of the cell nucleus and mitosis. *Nature* **1953**, *172*, 119–120. [CrossRef]
20. Ogur, M.; Minckler, S.; McClary, D.O. Desoxyribonucleic acid and the budding cycle in the yeasts. *J. Bacteriol.* **1953**, *66*, 642–645. [CrossRef]

21. Dawson, I.M.; Stern, H. Structure in the bacterial cell-wall during cell division. *Biochim. Biophys. Acta* **1954**, *13*, 31–40. [CrossRef]
22. Hamperl, H. Three group metaphases and carcinoma in situ of the cervix uteri. *Acta Unio Int. Contra Cancrum* **1954**, *10*, 128–131.
23. Hsu, T.C. Cytological studies on HeLa, a strain of human cervical carcinoma, I. Observations on mitosis and chromosomes. *Tex. Rep. Biol. Med.* **1954**, *12*, 833–846. [PubMed]
24. Roles, H. Mitosis and deoxyribonucleic acid content of the nucleus. *Nature* **1954**, *173*, 1039–1040. [CrossRef] [PubMed]
25. Reid, B.J.; Culotti, J.G.; Nash, R.S.; Pringle, J.R. Forty-five years of cell-cycle genetics. *Mol. Biol. Cell* **2015**, *26*, 4307–4312. [CrossRef]
26. Hartwell, L.H.; Culotti, J.; Reid, B. Genetic control of the cell-division cycle in yeast. I. Detection of mutants. *Proc. Natl. Acad. Sci. USA* **1970**, *66*, 352–359. [CrossRef] [PubMed]
27. Hartwell, L.H.; Mortimer, R.K.; Culotti, J.; Culotti, M. Genetic Control of the Cell Division Cycle in Yeast: V. Genetic Analysis of *cdc* Mutants. *Genetics* **1973**, *74*, 267–286. [CrossRef]
28. Hereford, L.M.; Hartwell, L.H. Sequential gene function in the initiation of Saccharomyces cerevisiae DNA synthesis. *J. Mol. Biol.* **1974**, *84*, 445–461. [CrossRef]
29. Nasmyth, K.A.; Reed, S.I. Isolation of genes by complementation in yeast: Molecular cloning of a cell-cycle gene. *Proc. Natl. Acad. Sci. USA* **1980**, *77*, 2119–2123. [CrossRef]
30. Nurse, P. Genetic control of cell size at cell division in yeast. *Nature* **1975**, *256*, 547–551. [CrossRef]
31. Nurse, P.; Thuriaux, P. Regulatory genes controlling mitosis in the fission yeast Schizosaccharomyces pombe. *Genetics* **1980**, *96*, 627–637. [CrossRef]
32. Beach, D.; Durkacz, B.; Nurse, P. Functionally homologous cell cycle control genes in budding and fission yeast. *Nature* **1982**, *300*, 706–709. [CrossRef]
33. Evans, T.; Rosenthal, E.T.; Youngblom, J.; Distel, D.; Hunt, T. Cyclin: A protein specified by maternal mRNA in sea urchin eggs that is destroyed at each cleavage division. *Cell* **1983**, *33*, 389–396. [CrossRef]
34. Swenson, K.I.; Farrell, K.M.; Ruderman, J.V. The clam embryo protein cyclin A induces entry into M phase and the resumption of meiosis in Xenopus oocytes. *Cell* **1986**, *47*, 861–870. [CrossRef]
35. Pines, J.; Hunt, T. Molecular cloning and characterization of the mRNA for cyclin from sea urchin eggs. *EMBO J.* **1987**, *6*, 2987–2995. [CrossRef] [PubMed]
36. Ritossa, F. New puffing pattern induced by temperature shock and DNP in Drosophila. *Experientia* **1962**, *18*, 571–573. [CrossRef]
37. Ritossa, F.M. Behavior of RNA and DNA synthesis at the puff level in salivary gland chromosomes of Drosopila. *Exp. Cell Res.* **1964**, *36*, 515–523. [CrossRef]
38. Ritossa, F. Discovery of the heat shock response. *Cell Stress Chaperones* **1996**, *1*, 97–98. [CrossRef]
39. Tissières, A.; Mitchell, H.K.; Tracy, U.M. Protein synthesis in salivary glands of *Drosophila melanogaster*: Relation to chromosome puffs. *J. Mol. Biol.* **1974**, *84*, 389–398. [CrossRef]
40. Schedl, P.; Artavanis-Tsakonas, S.; Steward, R.; Gehring, W.J.; Mirault, M.E.; Goldschmidt-Clermont, M.; Moran, L.; Tissières, A. Two hybrid plasmids with D. melanogaster DNA sequences complementary to mRNA coding for the major heat shock protein. *Cell* **1978**, *14*, 921–929. [CrossRef]
41. Wu, C. Activating protein factor binds in vitro to upstream control sequences in heat shock gene chromatin. *Nature* **1984**, *311*, 81–84. [CrossRef]
42. Wu, C. An exonuclease protection assay reveals heat-shock element and TATA box DNA-binding proteins in crude nuclear extracts. *Nature* **1985**, *317*, 84–87. [CrossRef]
43. Kingston, R.E.; Schuetz, T.J.; Larin, Z. Heat-inducible human factor that binds to a human hsp70 promoter. *Mol. Cell. Biol.* **1987**, *7*, 1530–1534.
44. Sorger, P.K.; Pelham, H.R. Purification and characterization of a heat-shock element binding protein from yeast. *EMBO J.* **1987**, *6*, 3035–3041. [CrossRef]
45. Sorger, P.K.; Lewis, M.J.; Pelham, H.R. Heat shock factor is regulated differently in yeast and HeLa cells. *Nature* **1987**, *329*, 81–84. [CrossRef] [PubMed]
46. Wiederrecht, G.; Shuey, D.J.; Kibbe, W.A.; Parker, C.S. The Saccharomyces and Drosophila heat shock transcription factors are identical in size and DNA binding properties. *Cell* **1987**, *48*, 507–515. [CrossRef]
47. Rabindran, S.K.; Giorgi, G.; Clos, J.; Wu, C. Molecular cloning and expression of a human heat shock factor, HSF1. *Proc. Natl. Acad. Sci. USA* **1991**, *88*, 6906–6910. [CrossRef]
48. Schuetz, T.J.; Gallo, G.J.; Sheldon, L.; Tempst, P.; Kingston, R.E. Isolation of a cDNA for HSF2: Evidence for two heat shock factor genes in humans. *Proc. Natl. Acad. Sci. USA* **1991**, *88*, 6911–6915. [CrossRef]
49. Sarge, K.D.; Zimarino, V.; Holm, K.; Wu, C.; Morimoto, R.I. Cloning and characterization of two mouse heat shock factors with distinct inducible and constitutive DNA-binding ability. *Genes. Dev.* **1991**, *5*, 1902–1911. [CrossRef] [PubMed]
50. Sistonen, L.; Sarge, K.D.; Phillips, B.; Abravaya, K.; Morimoto, R.I. Activation of heat shock factor 2 during hemin-induced differentiation of human erythroleukemia cells. *Mol. Cell. Biol.* **1992**, *12*, 4104–4111. [PubMed]
51. Smith, B.J.; Yaffe, M.P. A mutation in the yeast heat-shock factor gene causes temperature-sensitive defects in both mitochondrial protein import and the cell cycle. *Mol. Cell. Biol.* **1991**, *11*, 2647–2655. [PubMed]
52. Morano, K.A.; Santoro, N.; Koch, K.A.; Thiele, D.J. A trans-activation domain in yeast heat shock transcription factor is essential for cell cycle progression during stress. *Mol. Cell. Biol.* **1999**, *19*, 402–411. [CrossRef]

53. Luft, J.C.; Benjamin, I.J.; Mestril, R.; Dix, D.J. Heat shock factor 1-mediated thermotolerance prevents cell death and results in G_2/M cell cycle arrest. *Cell Stress Chaperones* **2001**, *6*, 326–336. [CrossRef]

54. Li, Q.; Martinez, J.D. Loss of HSF1 results in defective radiation-induced G(2) arrest and DNA repair. *Radiat. Res.* **2011**, *176*, 17–24. [CrossRef] [PubMed]

55. Bruce, J.L.; Chen, C.; Xie, Y.; Zhong, R.; Wang, Y.Q.; Stevenson, M.A.; Calderwood, S.K. Activation of heat shock transcription factor 1 to a DNA binding form during the G(1)phase of the cell cycle. *Cell Stress Chaperones* **1999**, *4*, 36–45. [CrossRef]

56. He, L.; Fox, M.H. Activation of heat-shock transcription factor 1 in heated Chinese hamster ovary cells is dependent on the cell cycle and is inhibited by sodium vanadate. *Radiat. Res.* **1999**, *151*, 283–292. [CrossRef] [PubMed]

57. Venturi, C.B.; Erkine, A.M.; Gross, D.S. Cell cycle-dependent binding of yeast heat shock factor to nucleosomes. *Mol. Cell. Biol.* **2000**, *20*, 6435–6448. [CrossRef]

58. Hoang, A.T.; Huang, J.; Rudra-Ganguly, N.; Zheng, J.; Powell, W.C.; Rabindran, S.K.; Wu, C.; Roy-Burman, P. A novel association between the human heat shock transcription factor 1 (HSF1) and prostate adenocarcinoma. *Am. J. Pathol.* **2000**, *156*, 857–864. [CrossRef]

59. Wang, Y.; Theriault, J.R.; He, H.; Gong, J.; Calderwood, S.K. Expression of a dominant negative heat shock factor-1 construct inhibits aneuploidy in prostate carcinoma cells. *J. Biol. Chem.* **2004**, *279*, 32651–32659. [CrossRef]

60. Morano, K.A.; Thiele, D.J. Heat shock factor function and regulation in response to cellular stress, growth, and differentiation signals. *Gene. Expr.* **1999**, *7*, 271–282.

61. Jolly, C.; Morimoto, R.I. Role of the heat shock response and molecular chaperones in oncogenesis and cell death. *J. Natl. Cancer Inst.* **2000**, *92*, 1564–1572. [CrossRef]

62. Pirkkala, L.; Nykänen, P.; Sistonen, L. Roles of the heat shock transcription factors in regulation of the heat shock response and beyond. *FASEB J.* **2001**, *15*, 1118–1131. [CrossRef]

63. Xing, H.; Wilkerson, D.C.; Mayhew, C.N.; Lubert, E.J.; Skaggs, H.S.; Goodson, M.L.; Hong, Y.; Park-Sarge, O.K.; Sarge, K.D. Mechanism of hsp70i gene bookmarking. *Science* **2005**, *307*, 421–423. [CrossRef]

64. Goodson, M.L.; Hong, Y.; Rogers, R.; Matunis, M.J.; Park-Sarge, O.K.; Sarge, K.D. Sumo-1 modification regulates the DNA binding activity of heat shock transcription factor 2, a promyelocytic leukemia nuclear body associated transcription factor. *J. Biol. Chem.* **2001**, *276*, 18513–18518. [CrossRef] [PubMed]

65. Michelotti, E.F.; Sanford, S.; Levens, D. Marking of active genes on mitotic chromosomes. *Nature* **1997**, *388*, 895–899. [CrossRef] [PubMed]

66. John, S.; Workman, J.L. Bookmarking genes for activation in condensed mitotic chromosomes. *Bioessays* **1998**, *20*, 275–279. [CrossRef]

67. Follmer, N.E.; Francis, N.J. Speed reading for genes: Bookmarks set the pace. *Dev. Cell* **2011**, *21*, 807–808. [CrossRef] [PubMed]

68. Weintraub, H. Assembly of an active chromatin structure during replication. *Nucleic Acids Res.* **1979**, *7*, 781–792. [CrossRef]

69. Struhl, G. A gene product required for correct initiation of segmental determination in Drosophila. *Nature* **1981**, *293*, 36–41. [CrossRef] [PubMed]

70. Groudine, M.; Weintraub, H. Propagation of globin DNAase I-hypersensitive sites in absence of factors required for induction: A possible mechanism for determination. *Cell* **1982**, *30*, 131–139. [CrossRef]

71. Kadauke, S.; Udugama, M.I.; Pawlicki, J.M.; Achtman, J.C.; Jain, D.P.; Cheng, Y.; Hardison, R.C.; Blobel, G.A. Tissue-specific mitotic bookmarking by hematopoietic transcription factor GATA1. *Cell* **2012**, *150*, 725–737. [CrossRef]

72. Kadauke, S.; Blobel, G.A. Mitotic bookmarking by transcription factors. *Epigenetics Chromatin* **2013**, *6*, 6–10. [CrossRef]

73. Caravaca, J.M.; Donahue, G.; Becker, J.S.; He, X.; Vinson, C.; Zaret, K.S. Bookmarking by specific and nonspecific binding of FoxA1 pioneer factor to mitotic chromosomes. *Genes Dev.* **2013**, *27*, 251–260. [CrossRef]

74. Lodhi, N.; Ji, Y.; Tulin, A. Mitotic bookmarking: Maintaining post-mitotic reprogramming of transcription reactivation. *Curr. Mol. Biol. Rep.* **2016**, *2*, 10–16. [CrossRef]

75. Hong, Y.; Lubert, E.J.; Rodgers, D.W.; Sarge, K.D. Molecular basis of competition between HSF2 and catalytic subunit for binding to the PR65/A subunit of PP2A. *Biochem. Biophys. Res. Commun.* **2000**, *272*, 84–89. [CrossRef]

76. Lubert, E.J.; Hong, Y.; Sarge, K.D. Interaction between protein phosphatase 5 and the A subunit of protein phosphatase 2A: Evidence for a heterotrimeric form of protein phosphatase 5. *J. Biol. Chem.* **2001**, *276*, 38582–38587. [CrossRef]

77. Kimura, K.; Hirano, M.; Kobayashi, R.; Hirano, T. Phosphorylation and activation of 13S condensin by Cdc2 in vitro. *Science* **1998**, *282*, 487–490. [CrossRef]

78. Kimura, K.; Cuvier, O.; Hirano, T. Chromosome condensation by a human condensin complex in Xenopus egg extracts. *J. Biol. Chem.* **2001**, *276*, 5417–5420. [CrossRef] [PubMed]

79. Xing, H.; Vanderford, N.L.; Sarge, K.D. The TBP-PP2A mitotic complex bookmarks genes by preventing condensin action. *Nat. Cell Biol.* **2008**, *10*, 1318–1323. [CrossRef] [PubMed]

80. Murphy, L.A.; Wilkerson, D.C.; Hong, Y.; Sarge, K.D. PRC1 associates with the hsp70i promoter and interacts with HSF2 during mitosis. *Exp. Cell Res.* **2008**, *314*, 2224–2230. [CrossRef] [PubMed]

81. Jiang, W.; Jimenez, G.; Wells, N.J.; Hope, T.J.; Wahl, G.M.; Hunter, T.; Fukunaga, R. PRC1: A human mitotic spindle-associated CDK substrate protein required for cytokinesis. *Mol. Cell* **1998**, *2*, 877–885. [CrossRef]

82. Mollinari, C.; Kleman, J.P.; Jiang, W.; Schoehn, G.; Hunter, T.; Margolis, R.L. PRC1 is a microtubule binding and bundling protein essential to maintain the mitotic spindle midzone. *J. Cell Biol.* **2002**, *157*, 1175–1186. [CrossRef] [PubMed]

83. Hayashida, N. Set1/MLL complex is indispensable for the transcriptional ability of heat shock transcription factor 2. *Biochem. Biophys. Res. Commun.* **2015**, *467*, 805–812. [CrossRef]

84. Blobel, G.A.; Kadauke, S.; Wang, E.; Lau, A.W.; Zuber, J.; Chou, M.M.; Vakoc, C.R. A reconfigured pattern of MLL occupancy within mitotic chromatin promotes rapid transcriptional reactivation following mitotic exit. *Mol. Cell* **2009**, *36*, 970–983. [CrossRef]

85. Dai, C.; Whitesell, L.; Rogers, A.B.; Lindquist, S. Heat shock factor 1 is a powerful multifaceted modifier of carcinogenesis. *Cell* **2007**, *130*, 1005–1018. [CrossRef]

86. Nakamura, Y.; Fujimoto, M.; Hayashida, N.; Takii, R.; Nakai, A.; Muto, M. Silencing HSF1 by short hairpin RNA decreases cell proliferation and enhances sensitivity to hyperthermia in human melanoma cell line. *J. Dermatol. Sci.* **2010**, *60*, 187–192. [CrossRef]

87. van Bergen en Henegouwen, P.M.; Linnemans, A.M. Heat shock gene expression and cytoskeletal alterations in mouse neuroblastoma cells. *Exp. Cell Res.* **1987**, *171*, 367–375. [CrossRef]

88. Comolli, R.; Frigerio, M.; Alberti, P. Heat shock, protein synthesis and ribosomal protein S6 phosphorylation in vitro in Yoshida AH 130 ascites hepatoma cells. *Cell Biol. Int. Rep.* **1988**, *12*, 907–917. [CrossRef]

89. Ullrich, S.J.; Moore, S.K.; Appella, E. Transcriptional and translational analysis of the murine 84- and 86-kDa heat shock proteins. *J. Biol. Chem.* **1989**, *264*, 6810–6816. [CrossRef]

90. Cairo, G.; Schiaffonati, L.; Rappocciolo, E.; Tacchini, L.; Bernelli-Zazzera, A. Expression of different members of heat shock protein 70 gene family in liver and hepatomas. *Hepatology* **1989**, *9*, 740–746. [CrossRef]

91. Teves, S.S.; An, L.; Hansen, A.S.; Xie, L.; Darzacq, X.; Tjian, R. A dynamic mode of mitotic bookmarking by transcription factors. *Elife* **2016**, *5*, e22280. [CrossRef]

92. Karagianni, P.; Moulos, P.; Schmidt, D.; Odom, D.T.; Talianidis, I. Bookmarking by Non-pioneer Transcription Factors during Liver Development Establishes Competence for Future Gene Activation. *Cell Rep.* **2020**, *30*, 1319–1328. [CrossRef]

93. Das, S.; Min, S.; Prahlad, V. Gene bookmarking by the heat shock transcription factor programs the insulin-like signaling pathway. *Mol. Cell* **2021**, *81*, 1–18. [CrossRef]

94. Fujimoto, M.; Takaki, E.; Hayashi, T.; Kitaura, Y.; Tanaka, Y.; Inouye, S.; Nakai, A. Active HSF1 significantly suppresses polyglutamine aggregate formation in cellular and mouse models. *J. Biol. Chem.* **2005**, *280*, 34908–34916. [CrossRef] [PubMed]

95. Hayashida, N.; Fujimoto, M.; Tan, K.; Prakasam, R.; Shinkawa, T.; Li, L.; Ichikawa, H.; Takii, R.; Nakai, A. Heat shock factor 1 ameliorates proteotoxicity in cooperation with the transcription factor NFAT. *EMBO J.* **2010**, *29*, 3459–3469. [CrossRef]

96. Momonaka, M.; Hayashida, K.; Hayashida, N. Unexpected Inhibition of cervical carcinoma cell proliferation by expression of heat shock transcription factor 1. *Biomed. Res. Clin. Pract.* **2016**, *1*, 2–6. [CrossRef]

97. Venkatakrishnan, C.D.; Dunsmore, K.; Wong, H.; Roy, S.; Sen, C.K.; Wani, A.; Zweier, J.L.; Ilangovan, G. HSP27 regulates p53 transcriptional activity in doxorubicin-treated fibroblasts and cardiac H9c2 cells: p21 upregulation and G2/M phase cell cycle arrest. *Am. J. Physiol. Heart Circ. Physiol.* **2008**, *294*, H1736–H1744. [CrossRef] [PubMed]

98. Arellano, M.; Moreno, S. Regulation of CDK/cyclin complexes during the cell cycle. *Int. J. Biochem. Cell Biol.* **1997**, *29*, 559–573. [CrossRef]

99. King, R.W.; Deshaies, R.J.; Peters, J.M.; Kirschner, M.W. How proteolysis drives the cell cycle. *Science* **1996**, *274*, 1652–1659. [CrossRef] [PubMed]

100. Fang, G.; Yu, H.; Kirschner, M.W. Control of mitotic transitions by the anaphase-promoting complex. *Philos. Trans. R. Soc. Lond. B Biol. Sci.* **1999**, *354*, 1583–1590. [CrossRef]

101. Kramer, E.R.; Scheuringer, N.; Podtelejnikov, A.V.; Mann, M.; Peters, J.M. Mitotic regulation of the APC activator proteins CDC20 and CDH1. *Mol. Biol. Cell* **2000**, *11*, 1555–1569. [CrossRef]

102. Schwab, M.; Lutum, A.S.; Seufert, W. Yeast Hct1 is a regulator of Clb2 cyclin proteolysis. *Cell* **1997**, *90*, 683–693. [CrossRef]

103. Sigrist, S.J.; Lehner, C.F. *Drosophila* fizzy-related down-regulates mitotic cyclins and is required for cell proliferation arrest and entry into endocycles. *Cell* **1997**, *90*, 671–681. [CrossRef]

104. Visintin, R.; Prinz, S.; Amon, A. CDC20 and CDH1: A family of substrate-specific activators of APC-dependent proteolysis. *Science* **1997**, *278*, 460–463. [CrossRef]

105. Fang, G.; Yu, H.; Kirschner, M.W. Direct binding of CDC20 protein family members activates the anaphase-promoting complex in mitosis and G1. *Mol. Cell* **1998**, *2*, 163–171. [CrossRef]

106. Kramer, E.R.; Gieffers, C.; Hölzl, G.; Hengstschläger, M.; Peters, J.M. Activation of the human anaphase-promoting complex by proteins of the CDC20/Fizzy family. *Curr. Biol.* **1998**, *8*, 1207–1210. [CrossRef]

107. Zachariae, W.; Schwab, M.; Nasmyth, K.; Seufert, W. Control of cyclin ubiquitination by CDK-regulated binding of Hct1 to the anaphase promoting complex. *Science* **1998**, *282*, 1721–1724. [CrossRef]

108. Jaspersen, S.L.; Charles, J.F.; Morgan, D.O. Inhibitory phosphorylation of the APC regulator Hct1 is controlled by the kinase Cdc28 and the phosphatase Cdc14. *Curr. Biol.* **1999**, *9*, 227–236. [CrossRef]

109. Gieffers, C.; Peters, B.H.; Kramer, E.R.; Dotti, C.G.; Peters, J.M. Expression of the CDH1-associated form of the anaphase-promoting complex in postmitotic neurons. *Proc. Natl. Acad. Sci. USA* **1999**, *96*, 11317–11322. [CrossRef]

110. Lee, Y.J.; Lee, H.J.; Lee, J.S.; Jeoung, D.; Kang, C.M.; Bae, S.; Lee, S.J.; Kwon, S.H.; Kang, D.; Lee, Y.S. A novel function for HSF1-induced mitotic exit failure and genomic instability through direct interaction between HSF1 and Cdc20. *Oncogene* **2008**, *27*, 2999–3009. [CrossRef]

111. King, R.W.; Peters, J.M.; Tugendreich, S.; Rolfe, M.; Hieter, P.; Kirschner, M.W. A 20S complex containing CDC27 and CDC16 catalyzes the mitosis-specific conjugation of ubiquitin to cyclin B. *Cell* **1995**, *81*, 279–288. [CrossRef]
112. Yu, H.; Peters, J.M.; King, R.W.; Page, A.M.; Hieter, P.; Kirschner, M.W. Identification of a cullin homology region in a subunit of the anaphase-promoting complex. *Science* **1998**, *279*, 1219–1222. [CrossRef]
113. Kazemi-Sefat, G.E.; Keramatipour, M.; Talebi, S.; Kavousi, K.; Sajed, R.; Kazemi-Sefat, N.A. Mousavizadeh, K.; The importance of CDC27 in cancer: Molecular pathology and clinical aspects. *Cancer Cell Int.* **2021**, *21*, 1–11. [CrossRef]
114. Lee, Y.J.; Kim, E.H.; Lee, J.S.; Jeoung, D.; Bae, S.; Kwon, S.H.; Lee, Y.S. HSF1 as a mitotic regulator: Phosphorylation of HSF1 by Plk1 is essential for mitotic progression. *Cancer Res.* **2008**, *68*, 7550–7560. [CrossRef] [PubMed]
115. Ahlskog, J.K.; Björk, J.K.; Elsing, A.N.; Aspelin, C.; Kallio, M.; Roos-Mattjus, P.; Sistonen, L. Anaphase-promoting complex/cyclosome participates in the acute response to protein-damaging stress. *Mol. Cell. Biol.* **2010**, *30*, 5608–5620. [CrossRef] [PubMed]
116. Hong, Y.; Sarge, K.D. Regulation of protein phosphatase 2A activity by heat shock transcription factor 2. *J. Biol. Chem.* **1999**, *274*, 12967–12970. [CrossRef] [PubMed]
117. Lee, J. Roles of cohesin and condensin in chromosome dynamics during mammalian meiosis. *J. Reprod. Dev.* **2013**, *59*, 431–436. [CrossRef]
118. Cloutier, J.M.; Turner, J.M. Meiotic sex chromosome inactivation. *Curr. Biol.* **2010**, *20*, R962–R963. [CrossRef]
119. Akerfelt, M.; Vihervaara, A.; Laiho, A.; Conter, A.; Christians, E.S.; Sistonen, L.; Henriksson, E. Heat shock transcription factor 1 localizes to sex chromatin during meiotic repression. *J. Biol. Chem.* **2010**, *285*, 34469–34476. [CrossRef]
120. Bierkamp, C.; Luxey, M.; Metchat, A.; Audouard, C.; Dumollard, R.; Christians, E. Lack of maternal Heat Shock Factor 1 results in multiple cellular and developmental defects, including mitochondrial damage and altered redox homeostasis, and leads to reduced survival of mammalian oocytes and embryos. *Dev. Biol.* **2010**, *339*, 338–353. [CrossRef]
121. Christians, E.; Davis, A.A.; Thomas, S.D.; Benjamin, I.J. Maternal effect of Hsf1 on reproductive success. *Nature* **2000**, *407*, 693–694. [CrossRef] [PubMed]
122. Kallio, M.; Chang, Y.; Manuel, M.; Alastalo, T.P.; Rallu, M.; Gitton, Y.; Pirkkala, L.; Loones, M.T.; Paslaru, L.; Larney, S.; et al. Brain abnormalities, defective meiotic chromosome synapsis and female subfertility in HSF2 null mice. *EMBO J.* **2002**, *21*, 2591–2601. [CrossRef] [PubMed]
123. Wang, G.; Zhang, J.; Moskophidis, D.; Mivechi, N.F. Targeted disruption of the heat shock transcription factor (hsf)-2 gene results in increased embryonic lethality, neuronal defects, and reduced spermatogenesis. *Genesis* **2003**, *36*, 48–61. [CrossRef] [PubMed]
124. Fujimoto, M.; Hayashida, N.; Katoh, T.; Oshima, K.; Shinkawa, T.; Prakasam, R.; Tan, K.; Inouye, S.; Takii, R.; Nakai, A. A novel mouse HSF3 has the potential to activate nonclassical heat-shock genes during heat shock. *Mol. Biol. Cell* **2010**, *21*, 106–116. [CrossRef] [PubMed]
125. Tanabe, M.; Sasai, N.; Nagata, K.; Liu, X.D.; Liu, P.C.; Thiele, D.J.; Nakai, A. The mammalian HSF4 gene generates both an activator and a repressor of heat shock genes by alternative splicing. *J. Biol. Chem.* **1999**, *274*, 27845–27856. [CrossRef] [PubMed]
126. Tu, N.; Hu, Y.; Mivechi, N.F. Heat shock transcription factor (Hsf)-4b recruits Brg1 during the G1 phase of the cell cycle and regulates the expression of heat shock proteins. *J. Cell Biochem.* **2006**, *98*, 1528–1542. [CrossRef]
127. Bunz, F.; Dutriaux, A.; Lengauer, C.; Waldman, T.; Zhou, S.; Brown, J.P.; Sedivy, J.M.; Kinzler, K.W.; Vogelstein, B. Requirement for p53 and p21 to sustain G2 arrest after DNA damage. *Science* **1998**, *282*, 1497–1501. [CrossRef]
128. Zhou, B.B.; Elledge, S.J. The DNA damage response: Putting checkpoints in perspective. *Nature* **2000**, *408*, 433–439. [CrossRef]
129. El-Deiry, W.S.; Tokino, T.; Velculescu, V.E.; Levy, D.B.; Parsons, R.; Trent, J.M.; Lin, D.; Mercer, W.E.; Kinzler, K.W.; Vogelstein, B. WAF1, a potential mediator of p53 tumor suppression. *Cell* **1993**, *75*, 817–825. [CrossRef]
130. Logan, I.R.; McNeill, H.V.; Cook, S.; Lu, X.; Meek, D.W.; Fuller-Pace, F.V.; Lunec, J.; Robson, C.N. Heat shock factor-1 modulates p53 activity in the transcriptional response to DNA damage. *Nucleic Acids Res.* **2009**, *37*, 2962–2973. [CrossRef]
131. Xu, C.; Min, J. Structure and function of WD40 domain proteins. *Protein Cell* **2011**, *2*, 202–214. [CrossRef]
132. Page, A.M.; Hieter, P. The anaphase-promoting complex: New subunits and regulators. *Annu. Rev. Biochem.* **1999**, *68*, 583–609. [CrossRef]
133. Higa, L.A.; Zhang, H. Stealing the spotlight: CUL4-DDB1 ubiquitin ligase docks WD40-repeat proteins to destroy. *Cell Div.* **2007**, *2*, 5. [CrossRef]
134. Yu, H. Cdc20: A WD40 activator for a cell cycle degradation machine. *Mol. Cell* **2007**, *27*, 3–16. [CrossRef] [PubMed]
135. Schapira, M.; Tyers, M.; Torrent, M.; Arrowsmith, C.H. WD40 repeat domain proteins: A novel target class? *Nat. Rev. Drug Discov.* **2017**, *16*, 773–786. [CrossRef] [PubMed]
136. Balch, W.E.; Morimoto, R.I.; Dillin, A.; Kelly, J.W. Adapting proteostasis for disease intervention. *Science* **2008**, *319*, 916–919. [CrossRef] [PubMed]
137. Brugh, V.M.; Lipshultz, L.I. Male factor infertility: Evaluation and management. *Med. Clin. N. Am.* **2004**, *88*, 367–385. [CrossRef]
138. Borgers, M.; Wolter, M.; Hentrich, A.; Bergmann, M.; Stammler, A.; Konrad, L. Role of compensatory meiosis mechanisms in human spermatogenesis. *Reproduction* **2014**, *148*, 315–320. [CrossRef] [PubMed]

139. Shinka, T.; Sato, Y.; Chen, G.; Naroda, T.; Kinoshita, K.; Unemi, Y.; Tsuji, K.; Toida, K.; Iwamoto, T.; Nakahori, Y. Molecular characterization of heat shock-like factor encoded on the human Y chromosome, and implications for male infertility. *Biol. Reprod.* **2004**, *71*, 297–306. [CrossRef]
140. Tessari, A.; Salata, E.; Ferlin, A.; Bartoloni, L.; Slongo, M.L.; Foresta, C. Characterization of *HSFY*, a novel *AZFb* gene on the Y chromosome with a possible role in human spermatogenesis. *Mol. Hum. Reprod.* **2004**, *10*, 253–258. [CrossRef]

Communication

A Novel Hyperactive Nud1 Mitotic Exit Network Scaffold Causes Spindle Position Checkpoint Bypass in Budding Yeast

Michael Vannini [1], Victoria R. Mingione [2], Ashleigh Meyer [3], Courtney Sniffen [4], Jenna Whalen [5] and Anupama Seshan [6,*]

[1] Boston University School of Medicine, Boston, MA 02118, USA; vanninim@bu.edu
[2] Department of Pharmacological Sciences, Stony Brook University School of Medicine, Stony Brook, NY 11794, USA; victoria.mingione@stonybrook.edu
[3] Dana Farber Cancer Institute, Boston, MA 02215, USA; ashleigh_meyer@dfci.harvard.edu
[4] Renaissance School of Medicine, Stony Brook University Hospital, Stony Brook, NY 11794, USA; courtney.sniffen@stonybrookmedicine.edu
[5] Department of Molecular, Cell, and Cancer Biology, University of Massachusetts Medical School, Worcester, MA 01605, USA; jenna.whalen@umassmed.edu
[6] Department of Biology, Emmanuel College, Boston, MA 02115, USA
[*] Correspondence: seshana@emmanuel.edu

Abstract: Mitotic exit is a critical cell cycle transition that requires the careful coordination of nuclear positioning and cyclin B destruction in budding yeast for the maintenance of genome integrity. The mitotic exit network (MEN) is a Ras-like signal transduction pathway that promotes this process during anaphase. A crucial step in MEN activation occurs when the Dbf2-Mob1 protein kinase complex associates with the Nud1 scaffold protein at the yeast spindle pole bodies (SPBs; centrosome equivalents) and thereby becomes activated. This requires prior priming phosphorylation of Nud1 by Cdc15 at SPBs. Cdc15 activation, in turn, requires both the Tem1 GTPase and the Polo kinase Cdc5, but how Cdc15 associates with SPBs is not well understood. We have identified a hyperactive allele of *NUD1*, *nud1-A308T*, that recruits Cdc15 to SPBs in all stages of the cell cycle in a *CDC5*-independent manner. This allele leads to early recruitment of Dbf2-Mob1 during metaphase and requires known Cdc15 phospho-sites on Nud1. The presence of *nud1-A308T* leads to loss of coupling between nuclear position and mitotic exit in cells with mispositioned spindles. Our findings highlight the importance of scaffold regulation in signaling pathways to prevent improper activation.

Keywords: Nud1; Cdc15; MEN; mitotic exit; Dbf2; Mob1; spindle position checkpoint

Citation: Vannini, M.; Mingione, V.R.; Meyer, A.; Sniffen, C.; Whalen, J.; Seshan, A. A Novel Hyperactive Nud1 Mitotic Exit Network Scaffold Causes Spindle Position Checkpoint Bypass in Budding Yeast. *Cells* 2022, *11*, 46. https://doi.org/10.3390/cells11010046

Academic Editor: Zhixiang Wang

Received: 1 December 2021
Accepted: 23 December 2021
Published: 24 December 2021

Publisher's Note: MDPI stays neutral with regard to jurisdictional claims in published maps and institutional affiliations.

1. Introduction

Mitotic cyclins, such as cyclin A and cyclin B, complexed with cyclin-dependent kinases (CDKs), are necessary and sufficient for cells to enter the mitotic phase of the cell division cycle [1]. Therefore, it follows that exit from mitosis and preparation for the subsequent cell cycle requires the inactivation of mitotic cyclin–CDK complexes [2]. This is achieved in large part by the destruction of mitotic cyclins during anaphase through ubiquitin-mediated proteolysis initiated by the anaphase promoting complex or cyclosome (APC/C) ubiquitin ligase [3–6]. In budding yeast, the essential phosphatase Cdc14 triggers the ubiquitination of Clb2, the cyclin B homolog, by the APC/C and its late mitotic activator Cdh1. Cdc14 dephosphorylates Cdh1, which leads to Cdh1-APC/C activation. Cdc14 also promotes the accumulation of the mitotic CDK inhibitor Sic1 to trigger the exit from mitosis [7–9].

The regulated activation of Cdc14 is critical for both cell cycle progression and the maintenance of proper ploidy. In the absence of Cdc14 activation, cells remain arrested in anaphase with elongated spindles and high Clb2 levels [7–9]. On the other hand, in cells where Cdc14 is activated prematurely during the cell cycle by *GAL*-induced overexpression, exit from mitosis also occurs prematurely [8]. The timing of Cdc14 activation is particularly crucial for mitotic checkpoints, such as the spindle assembly checkpoint (SAC) and the spindle position checkpoint (SPoC), to function properly. The SAC is required to restrain cell cycle progression during the metaphase-to-anaphase transition until all chromosomes have successfully attached to mitotic spindle microtubules in a bipolar manner, which prevents errors in chromosome segregation and aneuploidy [10]. The SPoC functions in anaphase to ensure the coupling of nuclear position and mitotic exit. In cells with mispositioned spindles, where anaphase occurs entirely within the mother cell, the SPoC is essential to preserve ploidy in the daughter cells [11,12]. Cdc14 activation is controlled by a Ras-like signaling pathway known as the mitotic exit network (MEN). The core MEN components that function upstream of Cdc14 include the small GTPase Tem1, the Hippo-like kinase Cdc15, the polo kinase Cdc5, the LATS/NDR kinase Dbf2-Mob1, and the spindle pole body (SPB) scaffold protein Nud1 [13]. Regulation of MEN activators and inhibitors to ensure the coordinate and timely execution of mitotic exit is required for error-free cell division to occur.

The SPBs, which are yeast centrosome equivalents, act as a hub for the assembly of activated MEN components (see Figure 1A). The small Ras-like GTPase Tem1 functions at the top of the MEN pathway and provides an essential link between nuclear position and Cdc14 activation. Tem1 first accumulates on SPBs during anaphase upon mitotic spindle elongation, and preferentially associates with the daughter cell-bound SPB (dSPB) [14–17]. Prior to spindle elongation, the localization of Tem1 to the SPBs is countered by the MEN inhibitor Kin4, which is localized to the mother cell cortex and mother SPB (mSPB) [18,19]. Kin4 activates the two-component GTPase Bub2-Bfa1, which inhibits Tem1 if the SAC or the SPoC are triggered [20–23]. Movement of the dSPB into the bud compartment during anaphase relieves the inhibition on Tem1 localization to this organelle due to multiple functions of the MEN activator Lte1. Lte1 is localized to the bud cortex and antagonizes Kin4 in this cellular compartment [14,15,24–26]. This in turn allows Tem1 to associate with the dSPB, likely by binding to Bfa1 and Nud1 on the outer plaque of the dSPB [15,27,28]. Lte1 likely has additional unknown Tem1-activating functions [26]. Importantly, the spatial segregation of the Tem1-inhibitor Kin4 in the mother cell compartment, and the Tem1-activator Lte1 in the daughter cell compartment creates an MEN-inhibitory and MEN-activating zone, respectively, and promotes the coupling of nuclear migration with mitotic exit [29,30]. Prior work has shown that tethering Tem1 to both SPBs by creating fusions with SPB outer plaque components such as *SPC72* or *CNM67* or to the dSPB by tethering with *BFA1* leads to SPoC bypass and can accelerate anaphase [28,31]. On the other hand, removal of Tem1 from SPBs by the attachment of a CAAX plasma membrane-targeting signal or by overexpression of its SPB receptor *BFA1* can inhibit MEN activation [28,32]. Therefore, restriction of Tem1's SPB association to anaphase cells with correct spindle position is essential for cell cycle control and ploidy maintenance (Figure 1A).

Immediately downstream of Tem1 in the MEN pathway is the Hippo/PAK-like kinase Cdc15 [33–35]. Tem1 and the polo kinase Cdc5 act coordinately to recruit Cdc15 to SPBs where it becomes concentrated first at the dSPB and subsequently at the mSPB [17,36]. Cdc15 then phosphorylates Nud1 on the outer plaque of the SPBs at S63 and T78, creating SPB phospho-docking sites for Dbf2-Mob1 where it can be phosphorylated and activated by Cdc15 [37,38]. The Dbf2-Mob1 complex associates first with the mSPB, and subsequently with the dSBP [17]. The active Dbf2-Mob1 complex then translocates into the nucleolus to participate directly in the activation of Cdc14 by liberating the protein from its nucleolar inhibitor, Cfi1/Net1, and allowing the protein phosphatase to reach its targets in the nucleus and cytoplasm [39–41]. In sum, the inactivation of mitotic CDK activity and resulting mitotic exit is achieved by (1) the release of Cdc14 from the nucleolus that leads to reversal of the phosphorylation imposed by CDK, (2) mitotic cyclin degradation, and (3) the accumulation of the mitotic CDK-inhibitor Sic1.

Just as Tem1 regulation provides spatial coordination for MEN activation, the regulation of the protein kinases Cdc5, Cdc15, and Dbf2-Mob1 is important for temporal coordination. Cdc5 kinase is activated, and the protein localizes to SPBs from S phase through mitosis, which poses some temporal restrictions on MEN activation [23,42–44]. However, the MEN remains inactive with Cdc14 sequestered in the nucleolus even when the SPBs makes brief forays into the bud during cells undergoing a prolonged mitotic metaphase arrest [45]. This is because the activity and SPB localization of both Cdc15 and Dbf2-Mob1 is inhibited by high mitotic CDK activity [45–47]. Therefore, MEN activity is restricted to anaphase (Figure 1A). It is only during this stage of mitosis where both a pool of mitotic CDKs is degraded following the metaphase–anaphase transition and a Tem1-bearing SPB enters the bud compartment that MEN can be activated [14,15,45,48]. When Cdc15 is forced to localize to SPBs outside of anaphase, premature MEN activity occurs, bypassing spatial and cell cycle progression checkpoints. For example, the tethering of Cdc15 to SPBs by fusing Cdc15 to the SPB component *CNM67* can lead to premature Dbf2-Mob1 kinase activity [36]. A *CDC15* allele, Cdc15(1-750), in which the resulting mutant protein is lacking its inhibitory C-terminus, similarly leads to SPB localization regardless of cell cycle stage and premature Dbf2-Mob1 kinase activity [36,49]. Both hyperactivated alleles of *CDC15* circumvent the requirement for the spatial sensor Tem1 and the cell cycle clock sensor Cdc5 to generate an active pool of Cdc15 [36]. Thus, restricting Cdc15 SPB localization to anaphase is necessary to coordinate MEN activation with spatial and temporal signals. Despite its importance, the question of how Cdc15 associates with SPBs is not yet well understood.

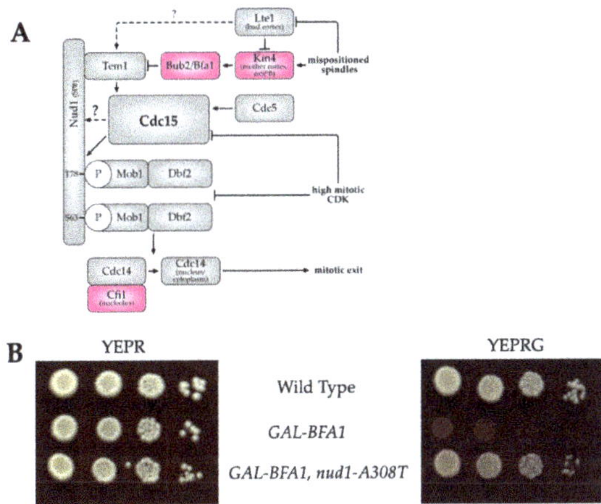

Figure 1. The *nud1-A308T* allele suppresses the growth defect of *GAL-BFA1* cells. (**A**) The mitotic exit network components and regulators are shown. Proteins that inhibit mitotic exit are depicted in red boxes, while activators are colored as grey boxes. The majority of the components assemble on the SPBs. The cellular locations of components that are not on this structure are indicated in parentheses underneath the protein names. See text for details. (**B**) Wild-type (AS3), *GAL-BFA1* (AS5), and *GAL-BFA1 nud1-A308T* (AS53) strains were grown in YEP + Raffinose (YEPR) overnight and 10-fold serial dilutions were spotted on YEPR (left) or YEPR + Galactose (YEPRG; right). The plates were incubated at 25 °C for 3 days.

In this study, we have identified and characterized an allele of the MEN scaffold *NUD1*, *nud1-A308T*, as a spontaneous suppressor of *GAL-BFA1* lethality. This allele exhibits a partial bypass of the SAC and a robust bypass of the SPoC. In addition, we show that cells containing *nud1-A308T* exhibit a shortened anaphase length. Our data suggest that the A308T mutation in *NUD1* results in a hyperactive scaffold, causing the recruitment of the MEN kinase Cdc15 in all phases of the cell cycle, and the early recruitment of Dbf2-Mob1 in metaphase cells. Furthermore, *nud1-A308T* cells display a reduced requirement for the polo kinase Cdc5 in MEN activation. Our data highlight the critical importance of scaffold regulation in preserving the integrity of scaffold-assisted signaling pathways.

2. Materials and Methods

2.1. Yeast Strains and Growth Conditions

All yeast strains used in this study were derivatives of W303 (AS3). The relevant genotypes of the strains used in this study were indicated in Table 1 below. The *GAL-nud1-T78A*, *GAL-nud1-S53A, S63A, T78A*, and *GAL-mob1-77* constructs were described in [38]. Culturing conditions were described in the figure legends.

Table 1. Strains used in this study. All strains are derivatives of W303 (AS3).

Number	Genotype
3	*MAT a ade2-1, leu2-3, ura3, trp1-1, his3-1115, can1-100, GAL, [phi+] (W303)*
5	*MAT a GAL-GFP-BFA1::HIS3MX6*
15	*MAT a TEM1-GFP:HIS3MX6*
53	*MAT a nud1-A308T*
79	*MAT a TEM1-GFP:HIS3MX6, HIS3MX6::GAL-BFA1*
112	*MAT a bub2::HIS3MX6*
138	*MAT a CDC14-3HA*
228	*MAT a cdc14-3, MOB1-6HA:HIS3MX6, NUD1-3V5::kanMX6*
296	*MAT a TEM1-GFP:HIS3MX6, HIS3MX6::GAL-BFA1, nud1(A308T)-3V5::kanMX6*
303	*MAT a nud1(A308T)-3V5::kanMX6, CDC14-3HA*
326	*MAT a NUD1-3V5::kanMX6*
383	*MAT a NUD1-3V5::kanMX6, trp1::YIP204-HIS3MX6::GAL-nud1(S53A,T78A)-3V5::TRP1*
384	*MAT a NUD1-3V5::kanMX6, trp1::YIP204-HIS3MX6::GAL-nud1(T78A)-3V5::TRP1*
385	*MAT a NUD1-3V5::kanMX6, trp1:: YIP204-HIS3MX6::GAL-nud1(S63A,T78A)-3V5::TRP1*
387	*MAT a nud1(A308T)-3V5::kanMX6*
388	*MAT a NUD1-3V5::kanMX6, trp1::YIP204-HIS3MX6::GAL-nud1(S53A,S63A,T78A)-3V5::TRP1*
390	*MAT a nud1(A308T)-3V5::kanMX6, trp1::YIP204-HIS3MX6::GAL-nud1(S53A,S63A,T78A)-3V5::TRP1*
392	*MAT a nud1(A308T)-3V5::kanMX6, trp1::YIP204-HIS3MX6::GAL-nud1(S63A,T78A)-3V5::TRP1*
395	*MAT a nud1(A308T)-3V5::kanMX6, trp1::YIP204-HIS3MX6::GAL-nud1(T78A)-3V5::TRP1*
397	*MAT a nud1(A308T)-3V5::kanMX6, trp1::YIP204-HIS3MX6::GAL-nud1(S53A,T78A)-3V5::TRP1*
402	*MAT a kanMX6::GAL-mob1-77*
407	*MAT a cdc14-3, MOB1-6HA:HIS3MX6*
417	*MAT a dyn1::HIS3MX6,ura3::pAFS125-TUB1p-GFP-TUB1::URA3*
439	*MAT a cdc14-3, MOB1-6HA:HIS3MX6, nud1(A308T)-3V5::kanMX6*
447	*MAT a cdc15-2, nud1(A308T)-3V5::kanMX6*
474	*MAT a dyn1::HIS3MX6,ura3::pAFS125-TUB1p-GFP-TUB1::URA3, nud1(A308T)-3V5::kanMX6*
476	*MAT a DBF2-eGFP::HIS3MX6, NUD1-3V5::kanMX6,ura3::pRS306-TUB1-mCherry-tub1::URA3*
477	*MAT a DBF2-eGFP::HIS3MX6, nud1(A308T)-3V5::kanMX6,ura3::pRS306-TUB1-mCherry-tub1::URA3*
478	*MAT a nud1(A308T)-3V5::kanMX6, kanMX6::GAL-mob1-77*
513	*MAT a CDC15-eGFP::kanMX6, NUD1-3V5::kanMX6,ura3::pRS306-TUB1-mCherry-tub1::URA3*
517	*MAT a CDC15-eGFP::kanMX6, nud1(A308T)-3V5::kanMX6,ura3::pRS306-TUB1-mCherry-tub1::URA3*
564	*MAT a cdc5-1*
570	*MAT a cdc5-1, nud1(A308T)-3V5::kanMX6*

2.2. Fixed-Cell Imaging and Analysis

Indirect immunofluorescence was performed as described previously for anti-tubulin (YOL1/34, Abcam) [50]. Images for Supplementary Figure S1 were acquired using a Zeiss Axioplan 2 (Zeiss, Thornwood, NY, USA) with a Hamamatsu Orca-R2 camera (Hamamatsu, Middlesex, NJ, USA) and a 63× objective. Cells for Figures 3 and 5 were fixed in a 4% paraformaldehyde, 3.4% sucrose solution for 15 min. Cells were washed once in potassium phosphate/sorbitol buffer (1.2 M sorbitol, 0.1 M potassium phosphate, pH 7.5), and treated with 1% Triton X-100 for 5 min. Cells were washed again in potassium phosphate/sorbitol buffer and resuspended in potassium phosphate/sorbitol buffer containing 4′, 6-diamidino-2-phenylindole dihydrochloride before imaging. These cells were imaged using a DeltaVision Elite microscope (GE Healthcare Bio-Sciences, Pittsburgh, PA, USA) with an InsightSSI solid-state light source, a CoolSNAP HQ2 camera, and a 60× plan-ApoN objective. Cells for budding and re-budding analyses were fixed in 3.7% formaldehyde, 0.1 M potassium phosphate, pH 6.4 prior to brief sonication and analysis.

2.3. Live-Cell Imaging and Analysis

Live cell movies for Figures 4C,D and 7C were done on agarose pads (2% agarose, synthetic complete (SC) medium containing 2% glucose. Cells for Figures 6 and 7A,B were imaged directly from log phase cultures. These cells were imaged using a DeltaVision Elite microscope (GE Healthcare Bio-Sciences, Pittsburgh, PA, USA) with an InsightSSI solid-state light source, an UltimateFocus hardware autofocus system, and a model IX-71 Olympus microscope (Center Valley, PA, USA) controlled by SoftWoRx software. A 60× Plan APO 1.42NA objective and CoolSNAP HQ2 camera (Teledyne Photometrics, Tuscon, AZ, USA) were used for image acquisition. Fiji software was used for image processing and analysis. To quantify anaphase length, image processing was performed by Volocity (PerkinElmer, Waltham, MA, USA) software.

2.4. Protein Extraction, Immunoprecipitation, and Western Blot Analysis

Approximately 25ODs of log phase cells grown in indicated conditions (see Figure 9) were pelleted and cell pellets were resuspended in cold NP40 buffer, containing 2X HALT Protease Inhibitor Cocktail (Promega, Madison, WI, USA), 60 mM β-glycerophosphate, 15 mM PNPP, 1 mM DTT, and 100 μM sodium orthovanadate. Cells were lysed by bead-beating using a chilled MiniBeadbeater (BioSpec Products, Bartlesville, OK, USA; 4 rounds of 60 s each) and glass beads. Protein extracts were cleared using a microcentrifuge and the protein concentration in each lysate was determined by Bradford assay (Bio-Rad, Hercules, CA, USA). Equilibrated anti-HA agarose beads (Sigma-Aldrich, St. Louis, MO, USA) were used to immunoprecipitate Mob1-6HA for 2 h at 4 °C. The reactions were washed with NP40 buffer six times before boiling in SDS PAGE protein buffer (Bio-Rad) for 5 min. Proteins were resolved on a 4–20% gradient gel (Bio-Rad) prior to transfer onto a nitrocellulose membrane. Mob1-6HA was detected using mouse anti-HA (Invitrogen, Waltham, MA, USA) at a 1:1000 dilution and Nud1-3V5 was detected using mouse anti-V5 (Invitrogen) at a 1:2000 dilution. Secondary antibodies were at 1:2000 and 1:5000, respectively. Blots were imaged using the ECL Plus system (GE Healthcare Bio-sciences).

3. Results

3.1. Nud1-A308T Partially Suppresses the Cell Cycle Defects of GAL-BFA1

Tem1 localization to the spindle pole body is highly dynamic and is regulated by the two component GTPase activating protein Bfa1-Bub2 [15–17,28,31]. It was previously shown that the overexpression of *BFA1* using a galactose-inducible promoter caused inhibition of mitotic exit in a *BUB2*-independent manner [51]. We demonstrated that this defect is due to the role of Bfa1 in Tem1-localization to the SPB and that the defect could be suppressed by concomitant overexpression of *TEM1* [32]. We sought to uncover new modes of MEN regulation by selecting for and characterizing spontaneous suppressors of *GAL-BFA1* lethality. Several of the spontaneous suppressors that we obtained by plating

GAL-BFA1 cells on galactose-containing plates caused decreased expression of *GAL-BFA1*. However, one suppressor strain did not (see Supplementary Figure S1). We sequenced the genome of this strain and identified a point mutation in the *NUD1* gene locus that changed the alanine 308 residue into a threonine (*nud1-A308T*). As seen in Figure 1B, when cells containing *GAL-BFA1* were plated on galactose containing media, this caused a severe growth defect that was largely suppressed by the presence of *nud1-A308T*.

3.2. Nud1-A308T Exhibits Weak Bypass of the Spindle Assembly Checkpoint

Upon treatment of cells with a microtubule poison, such as nocodazole, the SAC becomes activated and results in indefinite arrest in metaphase. This arrest is maintained by inhibition of mitotic exit. It has been previously shown that hyper-activation of the MEN, such as by deletion of the inhibitor *BUB2*, results in bypass of the SAC-induced metaphase arrest, leading to inappropriate exit from mitosis and re-budding [22]. The SAC can also be bypassed using a hyperactive version of an MEN activator. For example, overexpression of *TEM1* through a galactose promoter has been shown to bypass the SAC resulting in re-budding [52]. We suspected that this novel allele of *NUD1* caused a hyper-activated MEN phenotype, and we were interested in determining the level of hyperactivity exhibited by this allele. We decided to test the ability of *nud1-A308T* to bypass the SAC in the metaphase.

Alpha factor arrested G1 cells were synchronously released into nocodazole-containing media. Samples were taken approximately every 30 min and analyzed for re-budding. The percentage of budded and re-budded cells was graphed. As expected, almost no re-budding was observed in the wild-type and *NUD1-3V5* cells (Figure 2A,B). In *bub2Δ* cells, where SAC is bypassed, re-budded cells appeared beginning at 120 min and plateaued at around 80% (Figure 2C). In *nud1-A308T* cells, an increase in re-budded cells occurred at 140 min, but re-budding plateaued at around 30%. Therefore, we concluded that *nud1-A308T* caused a weak bypass of the SAC, unlike cells that completely lack the MEN inhibitor *BUB2*.

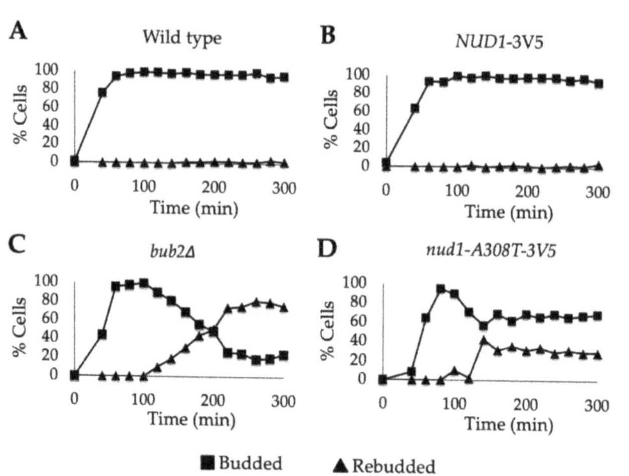

Figure 2. *nud1-A308T* partially bypasses the spindle assembly checkpoint (SAC). Log phase wild-type (AS3), *NUD1-3V5* (AS326), *bub2Δ* (AS112), and *nud1-A308T* (AS387) cells were arrested in the presence of alpha-factor pheromone (5 ug/mL) for a total of 3 h at 25 °C. Cells were then released into medium containing nocodazole (15 ug/mL) and samples were taken at the indicated times for budding analysis. After 120 min and 240 min, 7.5 ug/mL nocodazole was re-added to maintain the metaphase block. The percentage of budded (squares) and re-budded (triangles) cells observed at each time point in wild-type (**A**), *NUD1-3V5* (**B**), *bub2Δ* (**C**), and *nud1-A308T-3V5* (**D**) cells was plotted (*n* = 100).

3.3. Nud1-A308T Suppresses the Spindle Position Checkpoint

In addition to looking at the ability of Nud1-A308T to bypass the SAC, we also looked at the ability of Nud1-A308T to bypass SPoC. It was previously shown that hyperactive MEN components, such as *TEM1-eGFP*, harbor the ability to bypass the SPoC and complete mitotic exit despite misaligned spindles [52]. Spindle mispositioning can be induced by growing cells lacking the *DYN1* spindle positioning motor protein at low temperatures [11]. Wild-type and *nud1-A308T* cells containing a deletion of the *DYN1* locus were cultured at 14 °C to induce spindle mispositioning. Samples were taken after 24 h and imaged. As expected, we observed cells with properly aligned spindles going through anaphase and cells with misaligned spindles arresting in anaphase. We also saw a third category of cells with multiple nuclei. This is indictive of cells with misaligned spindles inappropriately exiting from mitosis and therefore bypassing the SPoC (Figure 3A). The percentage of cells from wild-type and *nud1-A308T* strains with arrested or bypassed morphology was plotted. In wild-type cells, 20% of cells were arrested in anaphase, while approximately 2.5% of cells bypassed the SPoC, resulting in multiple nuclei (Figure 3B). In *nud1-A308T* cells, 2.5% of cells were arrested in anaphase, while 20% of cells bypassed the SPoC (Figure 3B). We concluded that *nud1-A308T* exhibits a robust bypass of the SPoC.

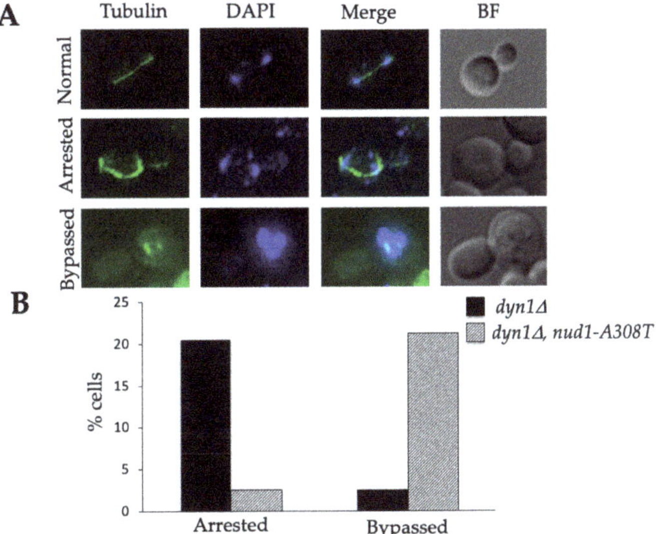

Figure 3. The *nud1-A308T* allele causes bypass of the Spindle Position Checkpoint (SPoC). Log phase wild-type (AS417) and *nud1-A308T* (AS474) cells containing both *dyn1Δ* and *GFP-TUB1* were grown at 14 °C for 25 h to induce spindle mispositioning. Samples were collected and the cells were fixed with paraformaldehyde and stained with DAPI prior to imaging. (**A**) Representative images of a cell undergoing a proper anaphase (normal), arrested in anaphase with a mispositioned spindle (arrested), and with multiple nuclei and spindles indicating SPoC bypass (bypassed) are shown. (**B**) The percentage of cells from each strain with arrested or bypassed morphology was plotted ($n > 500$).

3.4. Nud1-A308T Cells Have a Shortened Anaphase

Previous studies demonstrated that hyperactive alleles of MEN mutants exhibited decreased anaphase length [17]. Therefore, we wanted to analyze the length of anaphase in individual cells harboring *nud1-A308T*. Both wild-type and *nud1-A308T* cells were synchronously released from a G1 alpha factor arrest into media lacking pheromone.

Samples were taken at indicated time points and processed for tubulin immunofluorescence in order to visualize spindle morphology. Anaphase lasted for approximately 70 min in wild-type cells (Figure 4A) but for 60 min in *nud1-A308T* cells (Figure 4B). In order to more precisely quantify this difference in anaphase length, we performed live cell imaging of wild-type and *nud1-A308T* cells proceeding through the cell cycle and analyzed the length of anaphase in individual cells. Anaphase onset was determined by a spindle length of 3 μm or longer and anaphase completion was determined by spindle breakdown (Figure 4D). We quantified the length of anaphase in 19 or more cells from each genotype and observed that wild-type cells have an average anaphase length of 23.3 min, while *nud1-A308T* cells have an average anaphase length of 19.36 min (Figure 4C). These results indicated that the presence of the Nud1-A308T mutant protein has an effect on anaphase progression. This could be due to cells having the ability to pass through the metaphase to anaphase transition without satisfying the checkpoint, as is suggested by the data in Figure 2.

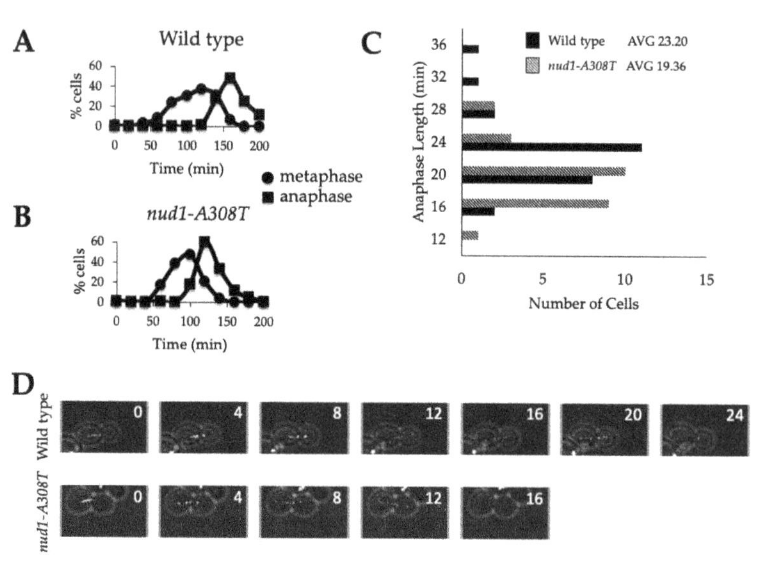

Figure 4. Mutants containing *nud1-A308T* have a shortened anaphase duration. (**A**) and (**B**): wild-type (AS138) and *nud1-A308T* (AS303) cells were grown to log phase and arrested in 5 ug/mL alpha-factor pheromone for 2.5 h. The cells were then released into medium lacking pheromone and samples were taken at the indicated times for tubulin immunofluorescence. The percentage of cells with metaphase (circles) or anaphase (squares) morphology in wild-type (**A**) and *nud1-A308T* (**B**) strains was quantified and plotted (*n* = 100). (**C**) and (**D**): Wild-type (AS476) and *nud1-A308T* (AS477) cells harboring Tub1-mCherry and Dbf2-eGFP fusions were grown to log phase in SC + glucose medium and living cells growing on an agar pad were imaged every 4 min for a total of 4 h. (**C**) The start and completion of anaphase was marked and the number of cells with anaphase duration of 16, 20, 24, or 28 min in each strain was plotted (*n* > 19). The average duration of anaphase in each strain is indicated above the graph. (**D**) A representative wild-type and *nud1-A308T* cell is shown.

3.5. Nud1-A308T Does Not Promote Association of Tem1 to SPBs

If *nud1-A308T* cells progress from metaphase to anaphase without satisfying the SAC, as our data indicate, this could be caused by early activation and recruitment of an MEN component by the mutant Nud1-A308T scaffold. One possible mechanism of action would be early or increased recruitment of Tem1 to the SPBs, as this has been shown to hyperactivate the MEN [17,29,31]. The *nud1-A308T* allele was discovered as a suppressor of lethality in *GAL-BFA1* cells, in which Tem1 localization is disrupted (Figure 1B; [32]). Therefore, we looked at Tem1 localization in *GAL-BFA1 nud1-A308T* cells to determine whether Tem1 recruitment was responsible for the suppression phenotype.

We examined Tem1 localization by fixing samples of wild-type, *GAL-BFA1*, and *GAL-BFA1 nud1-A308T* cells containing Tem1-GFP before and after the induction of *BFA1* overexpression using galactose. The percentage of anaphase cells with Tem1 localized to one or both SPB before and after galactose addition was plotted for each strain (Figure 5A). In wild-type anaphase cells, Tem1 was localized to the SPBs as expected regardless of culture medium. In control *GAL-BFA1* cells, Tem1 only localized to SPBs when galactose was not present. Interestingly, cells containing *GAL-BFA1 nud1-A308T* appeared similar to *GAL-BFA1* alone, and Tem1 was not localized to SPBs during anaphase when *BFA1* was overexpressed (Figure 5). These data showed that the mutant Nud1-A308T does not function by restoring the Tem1 localization defect in cells overexpressing *BFA1*. This was surprising since Tem1 is known to bind to the Nud1 MEN scaffold [28,53].

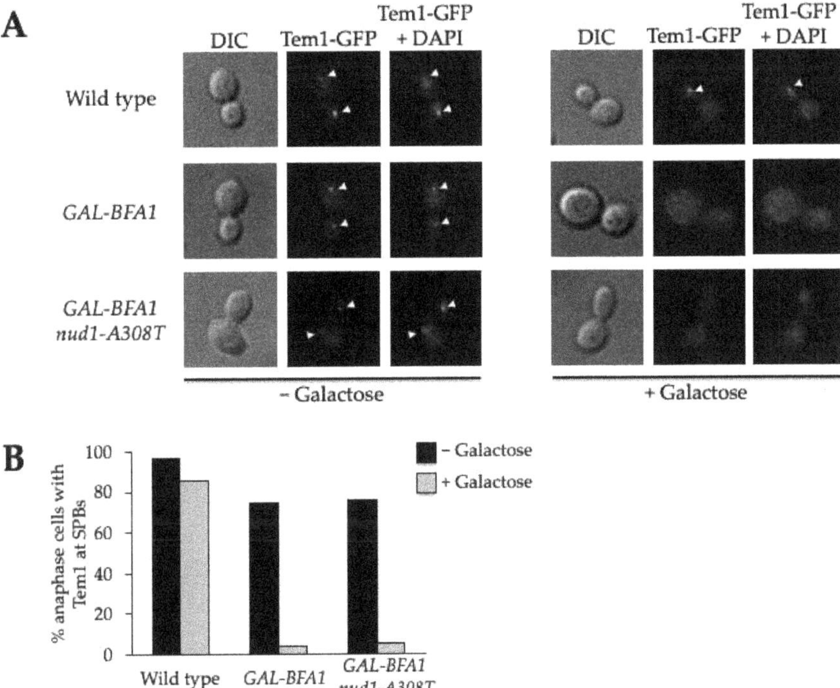

Figure 5. The *nud1-A308T* allele does not promote association of Tem1 to SPBs. Wild-type (AS15), *GAL-BFA1* (AS79), and *GAL-BFA1 nud1-A308T* (AS296) cells all harboring a Tem1-GFP fusion were grown to log phase at 30 °C in YEPR; 2% galactose was then added to induce the overexpression of *BFA1* for a total of 3 h. Samples were taken just prior to galactose induction and after 3 h induction and the cells were fixed in paraformaldehyde prior to imaging. (**A**) Representative composite images of anaphase cells for each strain prior to galactose addition, and following a 3 h galactose induction are shown. The DNA stained with DAPI is shown in blue and Tem1-GFP is in green. Arrowhead indicates Tem1-GFP localization to the SPB. (**B**) The percentage of anaphase cells with Tem1-GFP localized to one or both SPBs in the absence of galactose (black) and 3 h (gray) after galactose addition for each strain was plotted. (*n* > 100 cells).

3.6. Cdc15 Is Recruited to SPBs in All Cell Cycle Phases in the Majority of Nud1-A308T Cells

After concluding that Nud1-A308T was not acting by recruiting Tem1, we were interested in looking at Cdc15 SPB localization in *nud1-A308T* cells. Cdc15 is known to localize to SPBs upon its activation [33–35]. Cdc15 is also responsible for creating Nud1 phospho-docking sites to recruit and activate Dbf2-Mob1, but the mechanism of Cdc15 interaction with SPBs is not currently understood [38]. In order to analyze Cdc15 localization, we imaged wild-type and *nud1-A308T* cells harboring Cdc15-eGFP and Tub1-mCherry and quantified the percentage of G1/S, metaphase, and anaphase cells with Cdc15 localized to one or more SPB (Figure 6). Spindle morphology was used to determine which stage of the cell cycle cells were in. In wild-type cells, Cdc15 was localized in almost 100% of anaphase cells, while only 17% of G1/S cells and 27% of metaphase cells exhibited SPB localization (Figure 6B). In contrast, in the presence of *nud1-A308T* 84% of G1/S cells and 87% of metaphase cells exhibited Cdc15 SPB localization (Figure 6A,B). These data indicated that the *nud1-A308T* allele was hyperactive in MEN activation due to unscheduled recruitment of Cdc15 SPBs.

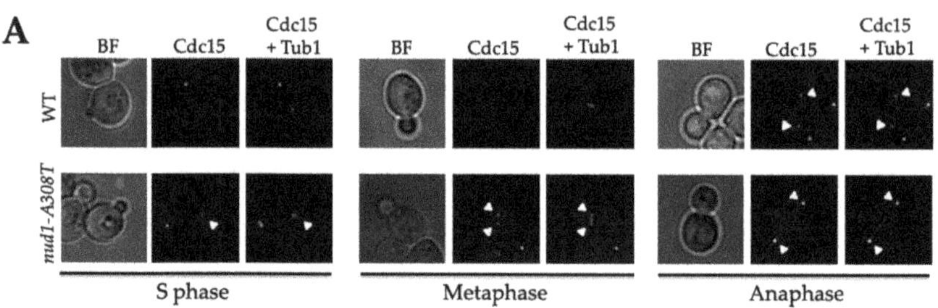

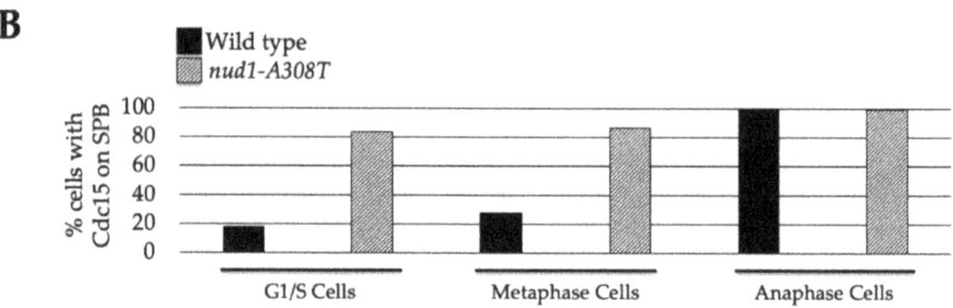

Figure 6. Cdc15 is recruited to SPBs in all cell cycle phases in *nud1-A308T* cells. Wild-type (AS513) and *nud1-A308T* (AS517) cells harboring Cdc15-eGFP and Tub1-mCherry fusions were grown to log phase in SC + glucose medium and living cells were imaged. (**A**) Brighfield (BF), Cdc15, and Cdc15 and Tub1 composite representative wild-type and *nud1-A308T* cells from S phase, metaphase, and anaphase are shown. Arrows indicate Cdc15 localization to SPBs. (**B**) The percentage of G1/S, metaphase, and anaphase cells containing Cdc15 on one or more SPBs in wild-type (black) or *nud1-A308T* (striped) cells is plotted ($n \geq 80$ cells for each stage).

3.7. Nud1-A308T Prematurely Recruits Dbf2 to the Mother SPB in Metaphase

It had previously been shown that cells with premature Cdc15 SPB localization also resulted in premature association of Dbf2-Mob1 to the SPBs [36,38]. After seeing the increase in Cdc15 localization, we wanted to determine whether Dbf2 localized to the SPB earlier in *nud1-A308T* cells. To carry this out, we imaged wild-type and *nud1-A308T* cells harboring Dbf2-eGFP and Tub1-mCherry. We then quantified the percentage of cells in metaphase and anaphase with Dbf2 localized to one or two SPBs (Figure 7). In wild-type cells, 1% of metaphase cells had Dbf2 localized to the SPB and 93% of anaphase cells had Dbf2 localized to the SPB (Figure 7A). In *nud1-A308T* cells, 82% of metaphase cells had

Dbf2 localized to one SPB and 100% of anaphase cells had Dbf2 localized to both SPBs (Figure 7B). We wanted to identify whether Dbf2 localized to the mSPB or the dSPB during metaphase in *nud1-A308T* cells. To do this we performed live cell imaging of *nud1-A308T* cells growing on agar pads. Images were taken every 4 min and spindle morphology and Dbf2 localization were analyzed. In 100% of metaphase cells analyzed ($n = 20$), *nud1-A308T* cells had Dbf2 localized to the mSPB only (Figure 7C). These results indicated that the mutant Nud1-A308T caused premature recruitment of Dbf2 to the mSPB during metaphase.

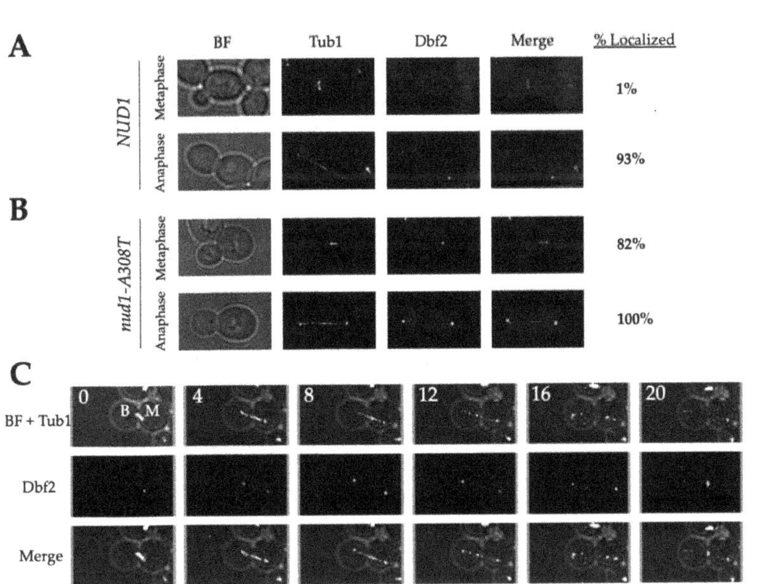

Figure 7. *nud1-A308T* recruits Dbf2 asymmetrically to the mSPB in metaphase. (**A**) and (**B**): wild-type (AS476) and *nud1-A308T* (AS477) cells harboring Dbf2-eGFP and Tub1-mCherry fusions were grown to log phase in SC + glucose media. Live cells were imaged. Representative metaphase and anaphase cells for each genotype are shown. The percentage of cells in each phase with Dbf2-eGFP localized to one or two SPBs is indicated in the column to the right of the images ($n = 100$). (**C**) *nud1-A308T* (AS477) cells harboring Dbf2-eGFP and Tub1-mCherry fusions were grown on an SC + glucose agar pad and imaged every 4 min for a total of 4 h. The start of metaphase was determined by identifying spindles that were 2 microns in length and this time was marked as 0 ($n = 20$). A representative cell is shown. The bud compartment is labeled "B" and the mother compartment is labeled "M".

3.8. Nud1-A308T Function Requires CDC15 and MOB1, but Not CDC5

Now that we had established that Cdc15 was recruited in all cell cycle phases, and that this resulted in early recruitment of Dbf2-Mob1 in metaphase, we wanted to determine whether *CDC15*, *MOB1*, and *CDC5* were required for the MEN activating function of *nud1-A308T*. As described previously, the localization of Cdc15 to SPBs in wild-type cells requires the activity of both Tem1 and Cdc5 [36]. We had already shown that the Nud1-A308T mutant was not acting by recruiting Tem1 (Figure 5), but we wondered whether this mutant scaffold could also bypass the requirement for Cdc5 activity in Cdc15 recruitment. On the other hand, we expected that if the mechanism of action of *nud1-A308T* was through Cdc15, then both *CDC15* and *MOB1* would be required for MEN activation in this mutant.

To determine if *CDC5*, *CDC15*, and *MOB1* were required for *nud1-A308T*, we crossed the *nud1-A308T* allele into cells containing temperature-sensitive mutations in each of these MEN components. When cells that were double mutant for the kinase-defective *cdc5-1* allele and *nud1-A308T* were grown at the restrictive temperature of 37 °C, we were surprised to find that they were able to grow almost as well as wild-type (Figure 8A). These data demonstrated that *nud1-A308T* recruits Cdc15 to SPBs independently of both *TEM1* and *CDC5*. As expected, when *cdc15-2 nud1-A308T* double mutants were grown at the restrictive temperature of 37 °C, the cells resembled the phenotype of *cdc15-2* alone (Figure 8B). This result confirmed that the mechanism by which the Nud1-A308T mutant scaffold is acting involves Cdc15 recruitment.

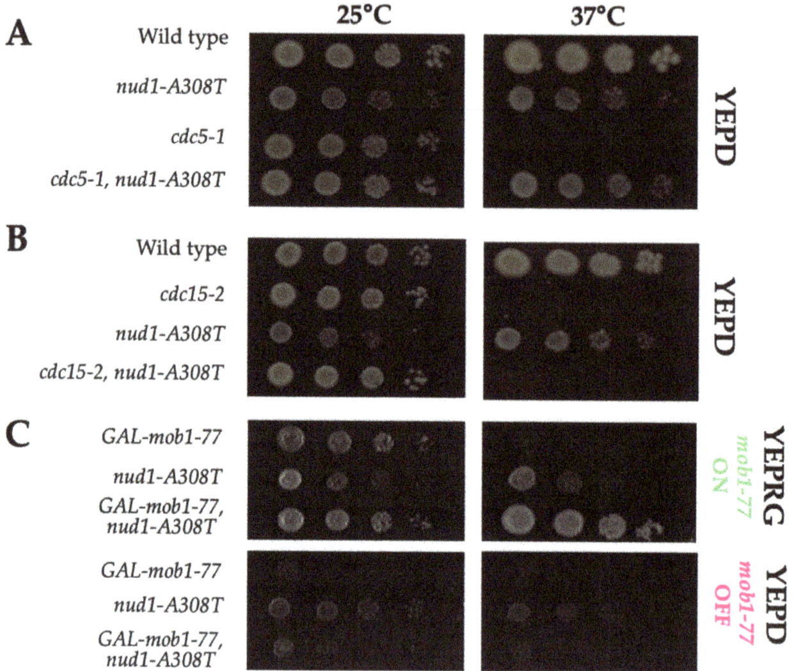

Figure 8. Nud1-A308T function requires *CDC15* and *MOB1* but not *CDC5*. (**A**) Ten-fold dilutions of cells containing *NUD1* (AS3), *nud1-A308T* (AS387), *cdc5-1* (AS564), and *nud1-A308T*, *cdc5-1* (AS570) were spotted onto YEPD plates that were incubated at 25 °C or 37 °C for 2 days before imaging. (**B**) Ten-fold dilutions of cells containing *NUD1* (AS3), *cdc15-2* (AS179), *nud1-A308T* (AS387), or *nud1-A308T, cdc15-2* (AS447) were spotted onto YEPD plates and incubated at 25 °C or 37 °C for 2–3 days before imaging. (**C**) Ten-fold dilutions of cells containing *GAL-mob1-77* (AS402), *nud1-A308T* (AS387), or both *GAL-mob1-77* and *nud1A308T* (AS478) were spotted onto YEPRG (top row) or YEPD (bottom row) plates to activate or to repress the expression of *mob1-77*. The plates were incubated at 25 °C or 37 °C for 2 days before imaging.

To confirm that Mob1, which acts downstream of Cdc15, was also required for MEN activation in the *nud1-A308T* background, we combined this mutation with the *GAL-mob1-77* allele. In the W303 strain background, the *mob1-77* allele causes a less severe growth defect at 37 °C than in the S288C strain background, where this allele was originally characterized [38,54]. Therefore, we examined the double mutant in cells where the *mob1-77* allele was placed under the *GAL*-inducible promoter to allow for complete depletion of *MOB1* by growth in medium containing glucose [38]. In the presence of galactose at 25 °C, all strains were able to grow. However, at 37 °C, when galactose was present, cells with *GAL-mob1-77* were unable to grow while the presence of *nud1-A308T* suppressed this growth defect (Figure 8C, top YEPRG panels). In the presence of glucose, where the expression of *GAL-mob1-77* was repressed, the *nud1-A308T* allele was unable to suppress the growth defect at either temperature (Figure 8C, bottom YEPD panels). These results demonstrated that the presence of Mob1 was also required for Nud1-A308T to activate the MEN. However, interestingly, the mutant scaffold was able to suppress the MEN defect of W303 cells containing *mob1-77* when the protein was overexpressed.

3.9. Nud1-A308T Recruits Mob1 in Metaphase and Suppresses the Dominant GAL-nud1-T78A

Taken together, our data suggested a likely model where the mutant Nud1-A308T scaffold led to hyperactivation of the MEN by recruitment of Cdc15 and a resulting early phosphorylation of Nud1 phospho-docking sites; whereby, Dbf2-Mob1 could then be recruited and activated. We decided to further test this by examining whether the Nud1-A308T mutant protein could associate with Mob1 in metaphase by co-immunoprecipitation analysis. It had been previously demonstrated that higher migrating forms of Nud1 corresponding to phosphorylation accumulated in anaphase and that Nud1 could associate with Mob1 specifically in anaphase [38]. We arrested *MOB1-6HA* cells containing Nud1 tagged with a 3V5 epitope or the mutant Nud1-A308T-3V5 protein in metaphase using nocodazole or in anaphase using the *cdc14-3* temperature-sensitive allele. Upon immunoprecipitating Mob1-6HA, we observed that both Nud1-3V5 and Nud1-A308T-3V5 were pulled down in cells arrested in anaphase (Figure 9, left panel). In contrast, only Nud1-A308T-3V5 could associate with immunoprecipitated Mob1-6HA in metaphase-arrested cells, while the wild-type Nud1 protein did not associate (Figure 9, right panel). Interestingly, higher migrating forms of Nud1-A308T were associated with Mob1 in both metaphase and anaphase-arrested cells. Therefore, we next decided to explore the role of Nud1 phosphorylation on Nud1-A308T function more directly.

Normally, when Cdc15 interacts with SPBs, three important Nud1 residues are phosphorylated. These sites are serine 53, serine 63, and threonine 78 (hereafter S53, S63, and T78). Only S63 and T78 are Cdc15-dependent, and T78 is the phosphorylation site most vital for MEN activation [38]. We were interested in determining whether *nud1-A308T* required the presence of these known phosphorylation sites. To do this we introduced *nud1-A308T* into a strain with the *GAL-nud1-S53A,S63A,T78A* allele integrated at the *TRP1* locus. This is a version of Nud1 with all three key phosphorylation sites muted to alanine, which prevents the formation of the Dbf2-Mob1 phospho-docking sites [38]. This allele is under the *GAL*-promoter allowing overexpression of Nud1 with nonfunctional phosphorylation sites in the presence of galactose. We also utilized the *GAL-nud1-T78A* allele, in which only site T78 is mutate to alanine [38]. As expected, when grown for several hours in the presence of galactose 80% of *GAL-nud1-S53A, S63A* and *T78A* cells, also containing wild-type *NUD1*, were arrested in anaphase. The *GAL-nud1-T78A* allele containing wild-type *NUD1* had a less severe anaphase arrest phenotype as previously demonstrated ([38]; Figure 10, closed circles and closed triangles). In cells containing both *GAL-nud1-S53A,S63A,T78A*, and *nud1-A308T*, there was no suppression of the dominant anaphase arrest phenotype (Figure 10, open circles). Intriguingly, while 55% of *GAL-nud1-T78 NUD1* cells arrested in anaphase after 6 h growth in galactose-containing medium, only 30% of *GAL-nud1-T78A nud1-A308T* double mutant cells arrested (Figure 10, triangles). Taken together, these data indicated that at least one of the phosphorylation sites was needed after Cdc15 recruitment to allow for MEN activation when the mutant scaffold was present. However, the increased Cdc15 recruitment to SPBs as a result of the mutant Nud1-A308T protein could suppress the defects associated with overexpression of *NUD1* containing a non-functional T78 phosphorylation site.

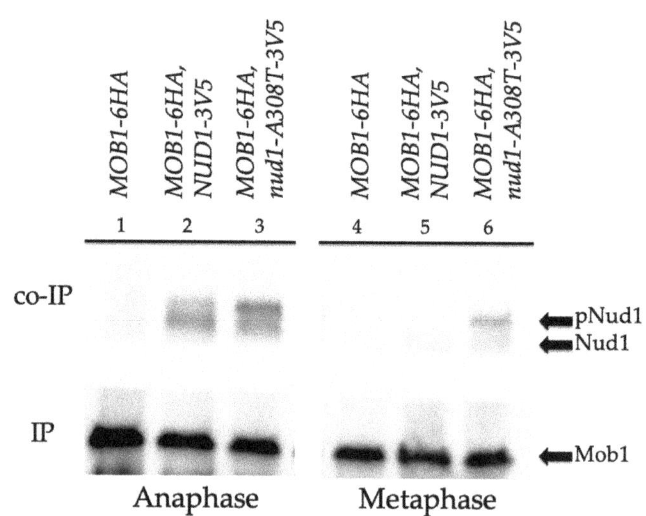

Figure 9. The mutant Nud1-A308T protein recruits Mob1 in both metaphase and anaphase cells. *MOB1-6HA* (AS407), *MOB1-6HA NUD1-3V5* (AS228), and *MOB1-6HA nud1-A308T-V5* (AS439) cells harboring the *cdc14-3* mutation were grown to log phase. Cell cultures were split, and half of each culture was incubated at 37 °C for two hours to inactivate Cdc14 and produce an anaphase arrest (samples 1–3). The other half of each culture was arrested with 15 µg/mL nocodazole at 25 °C to produce a metaphase arrest (samples 4–6). Protein extracts were produced from each sample and Mob1-6HA was immunoprecipitated using anti-HA agarose beads. The samples were analyzed by SDS-PAGE and Western blotting using anti-HA antibody to detect the amount of Mob1-6HA immunoprecipitated in each sample and anti-V5 antibodies to detect the amount of Nud1-3V5 or Nud1-A308T-3V5 co-immunoprecipitated. Mob1, Nud1 (samples 2 and 5), and Nud1-A308T (samples 3 and 6) are labeled on the blot. Higher migrating phosphorylated species of Nud1 and Nud1-A308T are labeled (pNud1).

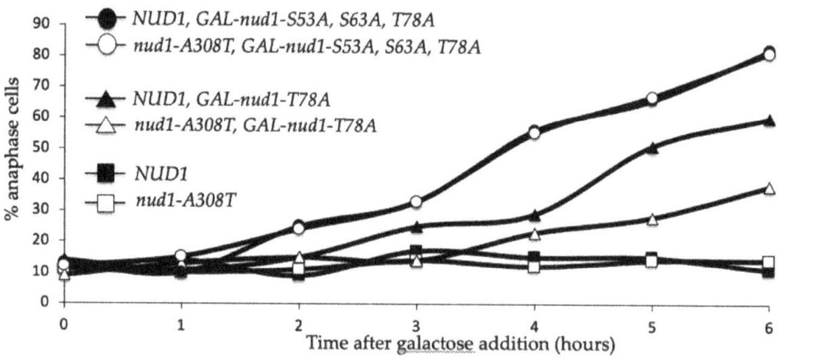

Figure 10. *nud1-A308T* suppresses the dominant GAL-nud1-T78A. *NUD1* (AS326), *nud1-A308T* (AS387), *NUD1 GAL-nud1-T78A* (AS384), *nud1-A308T GAL-nud1-T78A* (AS395), *NUD1 GAL-nud1-S53A, S63A, T78A* (AS388), and *nud1-A308T GAL-nud1-S53A, S63A, T78A* (AS390) cells were grown to log phase in YEPR and subsequently induced with 2% galactose addition. Samples were taken at the indicated times and processed for tubulin immunofluorescence. The percentage of cells with anaphase spindles at each time point in each strain was determined and plotted (*n* = 100–200).

4. Discussion

The MEN is essential in budding yeast to execute the exit from mitosis transition during the cell division cycle. Proper regulation of MEN components is necessary to ensure that exit from mitosis is coordinated with both cell cycle clock progression and migration of the daughter cell nucleus into the bud compartment. In this study, using a spontaneous genetic suppressor screen of *GAL-BFA1* lethality, we have identified a hyperactive allele of the MEN scaffold *NUD1*. The *nud1-A308T* allele exhibited bypass of the SPoC in anaphase cells with mispositioned spindles, resulting in the accumulation of anucleate and multinucleate cells (Figure 3). In addition, *nud1-A308T* exhibited a mild bypass of the SAC in metaphase-arrested cells (Figure 2), demonstrating that proper regulation of Nud1 is important to maintain genome integrity when mitotic checkpoints are activated.

The *nud1-A308T* allele could suppress the mitotic exit defects of cells impaired for Cdc5 and Dbf2-Mob1 kinase activity. However, cells that were impaired for Cdc15 activity and cells completely lacking *MOB1* expression were not rescued by the mutant Nud1 scaffold (Figure 8). These data are consistent with the model that the *nud1-A308T* allele is hyperactive due to unscheduled recruitment of Cdc15 and resulting phosphorylation of the scaffold at S63 and T78, which prematurely creates Dbf2-Mob1 phospho-docking sites. Several lines of evidence support this model. The hyperactive Nud1 scaffold resulted in localization of Cdc15 to the SPB in all phases of the cell cycle, and therefore led to premature SPB recruitment and activation of the downstream Dbf2-Mob1 kinase in metaphase (Figures 6 and 7). In addition, in cells containing a single-copy of *nud1-A308T*, the dominant negative effects of overexpressing non-phosphorylatable Nud1 mutants, such as *GAL-nud1-T78A*, *GAL-nud1-S53A, T78A*, and *GAL-nud1-S63A, T78A* are partially suppressed ([38]; Figure 10, Supplementary Figure S1). Lastly, Mob1 is associated with higher migrating forms of Nud1-A308T in both metaphase- and anaphase-arrested cells by co-immunoprecipitation analysis (Figure 9).

It is notable that the presence of *nud1-A308T* causes cells to grow more slowly at both 25 °C and 37 °C. However, with the presence of temperature-sensitive mutations in MEN components, such as *cdc5-1* and *mob1-77*, the growth defects at 37 °C were ameliorated (see Figure 8A,C). These defects were likely due to the premature localization of Cdc15 to SPBs in the mutant, which led to MEN hyperactivation. The expression of the CDC15-SPB allele, in which *CDC15* was fused to the SPB outer plaque component *CNM67*, was found to be lethal unless the allele was placed under the control of the low-strength *MET3* promoter, indicating that cells cannot tolerate the constitutive localization of Cdc15 to SPBs well [36]. An additional non-mutually exclusive possibility for the poor growth of the *nud1-A308T* mutant could be that there is increased spindle mispositioning in this mutant. Nud1 is important for astral microtubule organization and spindle position, as shown by defects in both these processes that become apparent in the presence of the *nud1-44* allele and the *nud1-42A* allele. This is due to the role of Nud1 in localization of the astral microtubule anchor Spc72 [23,38,53]. We noticed the slightly increased presence of mispositioned spindles in all strains with the *nud1-A308T* allele, indicating that this could be another reason for decreased viability (AS unpublished observations).

The *nud1-A308T* allele causes an interesting Dbf2-Mob1 phenotype—the asymmetric recruitment of the complex to the mSPB in metaphase cells (Figure 7C). Rock et al. demonstrated that expression of the *CDC15-SPB* allele also led to early recruitment of Mob1, in that case, in all phases of the cell cycle [38]. Therefore, we conclude that the *nud1-A308T* allele causes a weaker Cdc15 and downstream kinase activation phenotype than *CDC15-SPB*, since, in the former, Dbf2 is detected at the mSPB only during the metaphase. We ask the following question: if Cdc15 can localize to the SPBs in all cell cycle phases in *nud1-A308T* cells, what prevents Dbf2-Mob1 from becoming activated until metaphase? Most likely, this is due to inhibition by mitotic CDKs. Both Cdc15 and Mob1 are subject to inhibitory Clb/CDK phosphorylation [46,47]. Removal of Clb/CDK inhibition on Cdc15, using the *CDC15-7A* allele in metaphase-arrested cells, led to mitotic exit in a small percentage of cells where the spindle moved into the bud. Preventing mitotic CDK inhibition of Mob1 using the *MOB1-2A* allele had a comparatively much stronger effect [45]. It is possible that the alanine to threonine mutation in Nud1 causes a phenotype similar to *CDC15-7A* alone, where the inhibition on Cdc15 activity by Clb/CDKs is somehow bypassed by *nud1-A308T*, but the inhibitory phosphorylation of Mob1 remains.

The localization of Dbf2-Mob1 to only the mSPB in metaphase, and then later to both dSPB and mSPB, is in agreement with prior work, which demonstrated that Dbf2-Mob1 localizes first to the mSPB in anaphase and then to the dSPB. In this way, the ultimate MEN kinase is distinct from Tem1 and Cdc15, which both accumulate at the dSPB before the mSPB [17]. The reason that Dbf2-Mob1 associates first with the mSPB in anaphase upon activation is not yet known, but the authors suggested that this could be due to lowered CDK inhibitory activity at the mSPB than at the dSPB. Therefore, we ask the following question: what could be occurring in the metaphase, which is, prior to the reduction in Clb/CDKs at the metaphase-to-anaphase transition, to allow Dbf2-Mob1 to localize to the mSPB early in mitosis, but not in prior cell cycle phases in *nud1-A308T* cells? It was recently shown that Cdc5 phosphorylation of Cfi1/Net1 in metaphase was important for Cdc14 activation in early and late anaphase [41]. However, we observed that the *NUD1* allele acts independently of *CDC5*. Therefore, this remains an interesting question that may shed light on further MEN-activating conditions during metaphase.

A further question is as follows: does the *nud1-A308T* allele impact other SPB-mediated cellular processes? Nud1 is a critical component of the SPB inheritance network (SPIN), which causes the pre-existing "old" SPB to be inherited by the daughter cell, while the "new" SPB is inherited by the mSPB [55–58]. Specifically, Nud1 is modified by both the Swe1 kinase and NuA4 acetyltransferase complex and this allows the asymmetric segregation of the old SPB into the daughter cell, while the new SPB remains in the mother cell. This is due to the restriction of the spindle-positioning motor protein Kar9 to the old SPB in an MEN-dependent manner [56,57]. Interestingly, in mutants such as the *BFA1-SPC72* fusion and the GTPase-defective *TEM1-Q79L* allele, symmetric and premature loading of both the Tem1-GTPase and of Cdc15 led to increased symmetric Kar9 loading on both SPBs and resulting defects in mitotic spindle orientation and proper SPB inheritance [31,57]. Recent findings have suggested that constitutive disruption of proper SPB inheritance has negative consequences on replicative life span in budding yeast [59]. It will be interesting to see whether the early recruitment of Cdc15 and Dbf2 observed in the presence of *nud1-A308T* causes any alterations in SPB inheritance through the SPIN, and on aging.

Our data leaves the possibility open, that the SPB component that Cdc15 binds to is the Nud1 MEN scaffold itself. An important question is the following: how does the mutation of alanine 308 to a threonine in Nud1 lead to Cdc5-independent recruitment of Cdc15 to SPBs? This is an important question because Cdc15 is homologous to PAK-kinases that are known to be effectors in Ras-GTPase cascades and contain an auto-inhibitory C-terminal domain [60]. Cdc15 is no different, since prior work demonstrated that deletion of its C-terminal domain led to constitutive localization of the protein to SPBs and hyperactivation of Dbf2-Mob1 activity [36,49]. One possibility is that the mutation creates a new phosphorylation site in Nud1 that leads to the recruitment of Cdc15 to the Nud1 scaffold. Nud1 is phosphorylated by the mitotic CDK Clb/Cdc28, so it is possible that a new docking site for Cdc15 is created by the mitotic kinase [61]. We do not favor this possibility; however, since the mutation does not create a site that fits consensus phosphorylation site for CDKs S/T-P-x-K/R (see Supplementary Figure S1).

Another intriguing possibility is that the A308T alteration affects the stability of Nud1. It was demonstrated that Nud1 is subject to ubiquitin-mediated degradation in late mitosis by the E3 ubiquitin ligase Dma1/2. This overexpression of *DMA2* led to an inability to recruit Bfa1 and Mob1 to SPBs in anaphase, both of which are known to interact with Nud1. It also led to an inability to recruit Cdc15 to SPBs. The deletion of both *DMA1* and *DMA2* led to elevated levels of Bfa1, Cdc15, and Mob1 to SPBs in anaphase cells, while the tethering of Dma2 to SPBs led to cell death due to mitotic exit failure [62]. It is therefore possible that the mutation in *nud1-A308T* inhibits Dma1/2-mediated degradation, perhaps by preventing

Dma1/2 association with Nud1 itself or with another SPB component, and this leads to recruitment of Cdc15 and subsequently Dbf2-Mob1 to Nud1. Interestingly, the sequence surrounding alanine 308 in Nud1 is highly conserved within the *Saccharomyces* clade, and weakly conserved in the mammalian homolog of Nud1, Centriolin (see Supplementary Figure S1). Centriolin is also required for late mitotic events, such as cytokinesis, and its loss can result in cytokinesis failure [63]. It is also important for proper mitotic spindle orientation during mammalian development, and loss of this protein from centrosomes is associated with severe developmental defects [64]. These findings highlight the possibility that further studies of the *nud1-A308T* allele may shed light on the regulation of Centriolin and the mechanisms by which normal embryonic development occurs in humans.

Supplementary Materials: The following are available online at https://www.mdpi.com/article/10.3390/cells11010046/s1, Figure S1: The mutant Nud1-A308T protein recruits Mob1 in both metaphase and anaphase cells—original blots.

Author Contributions: Conceptualization, A.S.; methodology, A.S., M.V., V.R.M., A.M., C.S. and J.W.; investigation, M.V., V.R.M., A.M., C.S., J.W. and A.S.; data curation, M.V., V.R.M., A.M., C.S. and A.S.; writing—original draft preparation, M.V. and A.S.; writing—review and editing, A.S. All authors have read and agreed to the published version of the manuscript.

Funding: This research received no external funding.

Institutional Review Board Statement: Not applicable.

Informed Consent Statement: Not applicable.

Acknowledgments: This study was completed with financial support from the Paul Shannahan Wenger CURE Laboratory Fund to Emmanuel College. We are grateful to Angelika Amon and members of her laboratory for many helpful discussions and for generous sharing of strains and equipment. We are grateful to Rajiv Raman for technical assistance with figure preparation.

Conflicts of Interest: The authors declare no conflict of interest.

References

1. Murray, A.W.; Kirschner, M.W. Cyclin synthesis drives the early embryonic cell cycle. *Nat. Cell Biol.* **1989**, *339*, 275–280. [CrossRef] [PubMed]
2. Sullivan, M.; Morgan, D. Finishing mitosis, one step at a time. *Nat. Rev. Mol. Cell Biol.* **2007**, *8*, 894–903. [CrossRef]
3. Irniger, S.; Piatti, S.; Michaelis, C.; Nasmyth, K. Genes involved in sister chromatid separation are needed for b-type cyclin proteolysis in budding yeast. *Cell* **1995**, *81*, 269–277. [CrossRef]
4. King, R.W.; Peters, J.-M.; Tugendreich, S.; Rolfe, M.; Hieter, P.; Kirschner, M.W. A 20s complex containing CDC27 and CDC16 catalyzes the mitosis-specific conjugation of ubiquitin to cyclin B. *Cell* **1995**, *81*, 279–288. [CrossRef]
5. Sudakin, V.; Ganoth, D.; Dahan, A.; Heller, H.; Hershko, J.; Luca, F.C.; Ruderman, J.V.; Hershko, A. The cyclosome, a large complex containing cyclin-selective ubiquitin ligase activity, targets cyclins for destruction at the end of mitosis. *Mol. Biol. Cell* **1995**, *6*, 185–197. [CrossRef]
6. Tugendreich, S.; Tomkiel, J.; Earnshaw, W.; Hieter, P. CDC27Hs colocalizes with CDC16Hs to the centrosome and mitotic spindle and is essential for the metaphase to anaphase transition. *Cell* **1995**, *81*, 261–268. [CrossRef]
7. Jaspersen, S.L.; Charles, J.; Tinker-Kulberg, R.L.; Morgan, D.O. A Late Mitotic Regulatory Network Controlling Cyclin Destruction in Saccharomyces cerevisiae. *Mol. Biol. Cell* **1998**, *9*, 2803–2817. [CrossRef] [PubMed]
8. Visintin, R.; Craig, K.; Hwang, E.S.; Prinz, S.; Tyers, M.; Amon, A. The Phosphatase Cdc14 Triggers Mitotic Exit by Reversal of Cdk-Dependent Phosphorylation. *Mol. Cell* **1998**, *2*, 709–718. [CrossRef]
9. Zachariae, W.; Schwab, M.; Nasmyth, K.; Seufert, W. Control of Cyclin Ubiquitination by CDK-Regulated Binding of Hct1 to the Anaphase Promoting Complex. *Science* **1998**, *282*, 1721–1724. [CrossRef]
10. Musacchio, A.; Salmon, E.D. The spindle-assembly checkpoint in space and time. *Nat. Rev. Mol. Cell Biol.* **2007**, *8*, 379–393. [CrossRef]
11. Yeh, E.; Skibbens, R.V.; Cheng, J.W.; Salmon, E.D.; Bloom, K. Spindle dynamics and cell cycle regulation of dynein in the budding yeast, Saccharomyces cerevisiae. *J. Cell Biol.* **1995**, *130*, 687–700. [CrossRef]
12. Caydasi, A.K.; Ibrahim, B.; Pereira, G. Monitoring spindle orientation: Spindle position checkpoint in charge. *Cell Div.* **2010**, *5*, 28. [CrossRef] [PubMed]
13. Stegmeier, F.; Amon, A. Closing Mitosis: The Functions of the Cdc14 Phosphatase and Its Regulation. *Annu. Rev. Genet.* **2004**, *38*, 203–232. [CrossRef] [PubMed]

14. Bardin, A.J.; Visintin, R.; Amon, A. A Mechanism for Coupling Exit from Mitosis to Partitioning of the Nucleus. *Cell* **2000**, *102*, 21–31. [CrossRef]
15. Pereira, G.; Höfken, T.; Grindlay, J.; Manson, C.; Schiebel, E. The Bub2p Spindle Checkpoint Links Nuclear Migration with Mitotic Exit. *Mol. Cell* **2000**, *6*, 1–10. [CrossRef]
16. Molk, J.N.; Schuyler, S.C.; Liu, J.Y.; Evans, J.G.; Salmon, E.D.; Pellman, D.; Bloom, K. The Differential Roles of Budding Yeast Tem1p, Cdc15p, and Bub2p Protein Dynamics in Mitotic Exit. *Mol. Biol. Cell* **2004**, *15*, 1519–1532. [CrossRef]
17. Campbell, I.W.; Zhou, X.; Amon, A. Spindle pole bodies function as signal amplifiers in the Mitotic Exit Network. *Mol. Biol. Cell* **2020**, *31*, 906–916. [CrossRef]
18. D'Aquino, K.E.; Monje-Casas, F.; Paulson, J.; Reiser, V.; Charles, G.M.; Lai, L.; Shokat, K.M.; Amon, A. The Protein Kinase Kin4 Inhibits Exit from Mitosis in Response to Spindle Position Defects. *Mol. Cell* **2005**, *19*, 223–234. [CrossRef]
19. Pereira, G.; Schiebel, E. Kin4 Kinase Delays Mitotic Exit in Response to Spindle Alignment Defects. *Mol. Cell* **2005**, *19*, 209–221. [CrossRef] [PubMed]
20. Alexandru, G.; Zachariae, W.; Schleiffer, A.; Nasmyth, K. Sister chromatid separation and chromosome re-duplication are regulated by different mechanisms in response to spindle damage. *EMBO J.* **1999**, *18*, 2707–2721. [CrossRef]
21. Fesquet, D.; Fitzpatrick, P.J.; Johnson, A.L.; Kramer, K.M.; Toyn, J.H.; Johnston, L.H. A Bub2p-dependent spindle checkpoint pathway regulates the Dbf2p kinase in budding yeast. *EMBO J.* **1999**, *18*, 2424–2434. [CrossRef]
22. Fraschini, R.; Formenti, E.; Lucchini, G.; Piatti, S. Budding Yeast Bub2 Is Localized at Spindle Pole Bodies and Activates the Mitotic Checkpoint via a Different Pathway from Mad2. *J. Cell Biol.* **1999**, *145*, 979–991. [CrossRef]
23. Maekawa, H.; Priest, C.; Lechner, J.; Pereira, G.; Schiebel, E. The yeast centrosome translates the positional information of the anaphase spindle into a cell cycle signal. *J. Cell Biol.* **2007**, *179*, 423–436. [CrossRef] [PubMed]
24. Bertazzi, D.T.; Kurtulmus, B.; Pereira, G. The cortical protein Lte1 promotes mitotic exit by inhibiting the spindle position checkpoint kinase Kin4. *J. Cell Biol.* **2011**, *193*, 1033–1048. [CrossRef]
25. Falk, J.; Chan, L.; Amon, A. Lte1 promotes mitotic exit by controlling the localization of the spindle position checkpoint kinase Kin4. *Proc. Natl. Acad. Sci. USA* **2011**, *108*, 12584–12590. [CrossRef] [PubMed]
26. Falk, J.E.; Campbell, I.W.; Joyce, K.; Whalen, J.; Seshan, A.; Amon, A. LTE1 promotes exit from mitosis by multiple mech-anisms. *Mol. Biol. Cell* **2016**, *27*, 3991–4001. [CrossRef] [PubMed]
27. Kim, J.; Jang, S.S.; Song, K. Different Levels of Bfa1/Bub2 GAP Activity Are Required to Prevent Mitotic Exit of Budding Yeast Depending on the Type of Perturbations. *Mol. Biol. Cell* **2008**, *19*, 4328–4340. [CrossRef] [PubMed]
28. Valerio-Santiago, M.; Monje-Casas, F. Tem1 localization to the spindle pole bodies is essential for mitotic exit and impairs spindle checkpoint function. *J. Cell Biol.* **2011**, *192*, 599–614. [CrossRef]
29. Chan, L.; Amon, A. Spindle Position Is Coordinated with Cell-Cycle Progression through Establishment of Mitotic Exit-Activating and -Inhibitory Zones. *Mol. Cell* **2010**, *39*, 444–454. [CrossRef]
30. Falk, J.E.; Tsuchiya, D.; Verdaasdonk, J.; Lacefield, S.; Bloom, K.; Amon, A. Spatial signals link exit from mitosis to spindle position. *eLife* **2016**, *5*, e14036. [CrossRef] [PubMed]
31. Scarfone, I.; Venturetti, M.; Hotz, M.; Lengefeld, J.; Barral, Y.; Piatti, S. Asymmetry of the Budding Yeast Tem1 GTPase at Spindle Poles Is Required for Spindle Positioning But Not for Mitotic Exit. *PLoS Genet.* **2015**, *11*, e1004938. [CrossRef]
32. Whalen, J.; Sniffen, C.; Gartland, S.; Vannini, M.; Seshan, A. Budding Yeast BFA1 Has Multiple Positive Roles in Directing Late Mitotic Events. *G3 Genes Genomes Genet.* **2018**, *8*, 3397–3410. [CrossRef] [PubMed]
33. Asakawa, K.; Yoshida, S.; Otake, F.; Toh-E, A. A Novel Functional Domain of Cdc15 Kinase Is Required for Its Interaction with Tem1 GTPase in Saccharomyces cerevisiae. *Genetics* **2001**, *157*, 1437–1450. [CrossRef] [PubMed]
34. Lee, S.E.; Frenz, L.M.; Wells, N.J.; Johnson, A.L.; Johnston, L.H. Order of function of the budding-yeast mitotic exit-network proteins Tem1, Cdc15, Mob1, Dbf2, and Cdc5. *Curr. Biol.* **2001**, *11*, 784–788. [CrossRef]
35. Visintin, R.; Amon, A. Regulation of the Mitotic Exit Protein Kinases Cdc15 and Dbf2. *Mol. Biol. Cell* **2001**, *12*, 2961–2974. [CrossRef]
36. Rock, J.M.; Amon, A. Cdc15 integrates Tem1 GTPase-mediated spatial signals with Polo kinase-mediated temporal cues to activate mitotic exit. *Genes Dev.* **2011**, *25*, 1943–1954. [CrossRef] [PubMed]
37. Mah, A.S.; Jang, J.; Deshaies, R.J. Protein kinase Cdc15 activates the Dbf2-Mob1 kinase complex. *Proc. Natl. Acad. Sci. USA* **2001**, *98*, 7325–7330. [CrossRef]
38. Rock, J.M.; Lim, D.; Stach, L.; Ogrodowicz, R.W.; Keck, J.M.; Jones, M.H.; Wong, C.C.L.; Yates, J.R.; Winey, M.; Smerdon, S.J.; et al. Activation of the Yeast Hippo Pathway by Phosphorylation-Dependent Assembly of Signaling Complexes. *Science* **2013**, *340*, 871–875. [CrossRef]
39. Mohl, D.A.; Huddleston, M.J.; Collingwood, T.S.; Annan, R.S.; Deshaies, R.J. Dbf2–Mob1 drives relocalization of protein phosphatase Cdc14 to the cytoplasm during exit from mitosis. *J. Cell Biol.* **2009**, *184*, 527–539. [CrossRef]
40. Manzoni, R.; Montani, F.; Visintin, C.; Caudron, F.; Ciliberto, A.; Visintin, R. Oscillations in Cdc14 release and sequestration reveal a circuit underlying mitotic exit. *J. Cell Biol.* **2010**, *190*, 209–222. [CrossRef]
41. Zhou, X.; Li, W.; Liu, Y.; Amon, A. Cross-compartment signal propagation in the mitotic exit network. *eLife* **2021**, *10*, 63645. [CrossRef] [PubMed]
42. Cheng, L.; Hunke, L.; Hardy, C.F.J. Cell Cycle Regulation of the Saccharomyces cerevisiae Polo-Like Kinase Cdc5p. *Mol. Cell. Biol.* **1998**, *18*, 7360–7370. [CrossRef] [PubMed]

43. Park, J.E.; Park, C.J.; Sakchaisri, K.; Karpova, T.; Asano, S.; McNally, J.; Lee, K.S. Novel functional dissection of the lo-calization-specific roles of budding yeast polo kinase Cdc5p. *Mol. Cell. Biol.* **2004**, *24*, 9873–9886. [CrossRef] [PubMed]

44. Botchkarev, V.V.; Rossio, V.; Yoshida, S. The budding yeast Polo-like kinase Cdc5 is released from the nucleus during anaphase for timely mitotic exit. *Cell Cycle* **2014**, *13*, 3260–3270. [CrossRef]

45. Campbell, I.W.; Zhou, X.; Amon, A. The Mitotic Exit Network integrates temporal and spatial signals by distributing regulation across multiple components. *eLife* **2019**, *8*, 41139. [CrossRef]

46. Jaspersen, S.L.; Morgan, D. Cdc14 activates Cdc15 to promote mitotic exit in budding yeast. *Curr. Biol.* **2000**, *10*, 615–618. [CrossRef]

47. König, C.; Maekawa, H.; Schiebel, E. Mutual regulation of cyclin-dependent kinase and the mitotic exit network. *J. Cell Biol.* **2010**, *188*, 351–368. [CrossRef]

48. Shirayama, M.; Tóth, A.; Gálová, M.; Nasmyth, K. APCCdc20 promotes exit from mitosis by destroying the anaphase inhibitor Pds1 and cyclin Clb5. *Nat. Cell Biol.* **1999**, *402*, 203–207. [CrossRef]

49. Bardin, A.J.; Boselli, M.G.; Amon, A. Mitotic Exit Regulation through Distinct Domains within the Protein Kinase Cdc15. *Mol. Cell. Biol.* **2003**, *23*, 5018–5030. [CrossRef]

50. Visintin, R.; Hwang, E.S.; Amon, A. Cfi1 prevents premature exit from mitosis by anchoring Cdc14 phosphatase in the nucleolus. *Nat. Cell Biol.* **1999**, *398*, 818–823. [CrossRef]

51. Ro, H.-S.; Song, S.; Lee, K.S. Bfa1 can regulate Tem1 function independently of Bub2 in the mitotic exit network of Saccharomyces cerevisiae. *Proc. Natl. Acad. Sci. USA* **2002**, *99*, 5436–5441. [CrossRef] [PubMed]

52. Chan, L.Y.; Amon, A. The protein phosphatase 2A functions in the spindle position checkpoint by regulating the checkpoint kinase Kin4. *Genes Dev.* **2009**, *23*, 1639–1649. [CrossRef] [PubMed]

53. Gruneberg, U.; Campbell, K.; Simpson, C.; Grindlay, J.; Schiebel, E. Nud1p links astral microtubule organization and the control of exit from mitosis. *EMBO J.* **2000**, *19*, 6475–6488. [CrossRef]

54. Luca, F.C.; Winey, M. MOB1, an Essential Yeast Gene Required for Completion of Mitosis and Maintenance of Ploidy. *Mol. Biol. Cell* **1998**, *9*, 29–46. [CrossRef]

55. Pereira, G.; Tanaka, T.; Nasmyth, K.; Schiebel, E. Modes of spindle pole body inheritance and segregation of the Bfa1p-Bub2p checkpoint protein complex. *EMBO J.* **2001**, *20*, 6359–6370. [CrossRef]

56. Hotz, M.; Lengefeld, J.; Barral, Y. The MEN mediates the effects of the spindle assembly checkpoint on Kar9-dependent spindle pole body inheritance in budding yeast. *Cell Cycle* **2012**, *11*, 3109–3116. [CrossRef]

57. Lengefeld, J.; Hotz, M.; Rollins, M.; Baetz, K.; Barral, Y. Budding yeast Wee1 distinguishes spindle pole bodies to guide their pattern of age-dependent segregation. *Nat. Cell Biol.* **2017**, *19*, 941–951. [CrossRef]

58. Lengfeld, J.; Barral, Y. Asymmetric Segregation of Aged Spindle Pole Bodies During Cell Division: Mechanisms and Relevance Beyond Budding Yeast? *BioEssays* **2018**, *40*, e1800038. [CrossRef]

59. Manzano-Lopez, J.; Matellán, L.; Álvarez-Llamas, A.; Blanco-Mira, J.C.; Monje-Casas, F. Asymmetric inheritance of spindle microtubule-organizing centres preserves replicative lifespan. *Nat. Cell Biol.* **2019**, *21*, 952–965. [CrossRef] [PubMed]

60. Kumar, R.; Sanawar, R.; Li, X.; Li, F. Structure, biochemistry, and biology of PAK kinases. *Gene* **2017**, *605*, 20–31. [CrossRef] [PubMed]

61. Park, C.J.; Park, J.-E.; Karpova, T.S.; Soung, N.-K.; Yu, L.-R.; Song, S.; Lee, K.H.; Xia, X.; Kang, E.; Dabanoglu, I.; et al. Requirement for the Budding Yeast Polo Kinase Cdc5 in Proper Microtubule Growth and Dynamics. *Eukaryot. Cell* **2008**, *7*, 444–453. [CrossRef] [PubMed]

62. Tamborrini, D.; Juanes, M.A.; Ibanes, S.; Rancati, G.; Piatti, S. Recruitment of the mitotic exit network to yeast centrosomes couples septin displacement to actomyosin constriction. *Nat. Commun.* **2018**, *9*, 4308. [CrossRef]

63. Gromley, A.; Jurczyk, A.; Sillibourne, J.; Halilovic, E.; Mogensen, M.; Groisman, S.; Blomberg, M.; Doxsey, S. A novel human protein of the maternal centriole is required for the final stages of cytokinesis and entry into S phase. *J. Cell Biol.* **2003**, *161*, 535–545. [CrossRef] [PubMed]

64. Chen, C.-T.; Hehnly, H.; Yu, Q.; Farkas, D.; Zheng, G.; Redick, S.D.; Hung, H.-F.; Samtani, R.; Jurczyk, A.; Akbarian, S.; et al. A Unique Set of Centrosome Proteins Requires Pericentrin for Spindle-Pole Localization and Spindle Orientation. *Curr. Biol.* **2014**, *24*, 2327–2334. [CrossRef] [PubMed]

Article

Effects of Dithiothreitol on Fertilization and Early Development in Sea Urchin

Nunzia Limatola [1], Jong Tai Chun [2,*], Sawsen Cherraben [3], Jean-Louis Schmitt [3], Jean-Marie Lehn [3] and Luigia Santella [1,*]

[1] Department of Research Infrastructures for Marine Biological Resources, Stazione Zoologica Anton Dohrn, 80121 Napoli, Italy; nunzia.limatola@szn.it

[2] Department of Biology and Evolution of Marine Organisms, Stazione Zoologica Anton Dohrn, 80121 Napoli, Italy

[3] Laboratory of Supramolecular Chemistry, Institut de Science et d'Ingénierie Supramoléculaires ISIS—Université de Strasbourg, 8 allée Gaspard Monge, 67000 Strasbourg, France; sawsen.cherraben@sorbonne-universite.fr (S.C.); jlschmitt@unistra.fr (J.-L.S.); lehn@unistra.fr (J.-M.L.)

* Correspondence: chun@szn.it (J.T.C.); santella@szn.it (L.S.); Tel.: +39-0815833407 (J.T.C.); +39-0815833289 (L.S.)

Abstract: The vitelline layer (VL) of a sea urchin egg is an intricate meshwork of glycoproteins that intimately ensheathes the plasma membrane. During fertilization, the VL plays important roles. Firstly, the receptors for sperm reside on the VL. Secondly, following cortical granule exocytosis, the VL is elevated and transformed into the fertilization envelope (FE), owing to the assembly and crosslinking of the extruded materials. As these two crucial stages involve the VL, its alteration was expected to affect the fertilization process. In the present study, we addressed this question by mildly treating the eggs with a reducing agent, dithiothreitol (DTT). A brief pretreatment with DTT resulted in partial disruption of the VL, as judged by electron microscopy and by a novel fluorescent polyamine probe that selectively labelled the VL. The DTT-pretreated eggs did not elevate the FE but were mostly monospermic at fertilization. These eggs also manifested certain anomalies at fertilization: (i) compromised Ca^{2+} signaling, (ii) blocked translocation of cortical actin filaments, and (iii) impaired cleavage. Some of these phenotypic changes were reversed by restoring the DTT-exposed eggs in normal seawater prior to fertilization. Our findings suggest that the FE is not the decisive factor preventing polyspermy and that the integrity of the VL is nonetheless crucial to the egg's fertilization response.

Keywords: vitelline layer; fertilization; sea urchin eggs; plasticity; Ca^{2+} signaling; actin; DTT; TCEP; BPA-C8-Cy3; electron microscopy

Citation: Limatola, N.; Chun, J.T.; Cherraben, S.; Schmitt, J.-L.; Lehn, J.-M.; Santella, L. Effects of Dithiothreitol on Fertilization and Early Development in Sea Urchin. *Cells* **2021**, *10*, 3573. https://doi.org/10.3390/cells10123573

Academic Editor: Zhixiang Wang

Received: 11 November 2021
Accepted: 15 December 2021
Published: 17 December 2021

1. Introduction

For a cell like the sea urchin egg, the presence of a well-developed extracellular matrix is of particular importance. The extracellular matrix not only protects the cell by covering its surface but also plays multiple roles in cell−cell communication. For example, sperm-activating peptides released from the jelly coat of sea urchin eggs serve as a chemoattractant for the sperm [1–3]. On the other hand, the vitelline layer (VL) of a sea urchin egg, which intimately covers the plasma membrane, has long been recognized as the primary subcellular site of sperm attachment during fertilization [4–6]. Indeed, bindin, a protein isolated from the acrosomal process of sea urchin sperm, was shown to mediate species-specific attachment to the egg's VL, and the egg's receptor for sperm or bindin has also been isolated from the VL [7–14]. Nevertheless, it is also true that the VL has often been removed or circumvented in order to facilitate the experimental procedure [15,16]. It has been generally assumed that such a practice does not affect the fertilization process and embryonic development, while its effect on the cytological and biochemical changes characterizing egg activation has not been addressed sufficiently.

The fertilization process comprises a series of sequential events such as sperm chemotaxis and activation, sperm adhesion and fusion with the oolemma, egg penetration, and finally the fusion of the male and female pronuclei [11,17,18]. One of the hallmarks of fertilization in virtually all animal species is the increase of Ca^{2+} inside the eggs [19–23]. In sea urchin, a fertilized egg exhibits two modes of Ca^{2+} increase: immediate Ca^{2+} influx and slow propagation of a Ca^{2+} wave (see below) [20,24]. In parallel to this, some cortical actin filaments translocate to the inner cytoplasm, while others elongate the microvilli in the perivitelline space [25–29]. These progressive changes of the Ca^{2+} signals and the reorganization of the cortical actin cytoskeleton are believed to play diverse roles in egg activation and in early development [30–33].

The mobilization of intracellular Ca^{2+} and actin filaments at fertilization should be under tight control because their deregulation is often accompanied by developmental problems [18,27,28,30,34]. The present study is on the extension of a series of our previous studies on egg quality and the roles of the cortical actin cytoskeleton in fertilization [18,27–33]. This time, we focused on the effect of reducing agents. As aforementioned, DTT has been utilized to "safely" remove the extracellular matrix, such as the egg VL, which has facilitated a number of experiments. However, DTT may affect a host of proteins involved in fertilization and egg activation. Indeed, the sea urchin egg's receptor itself for sperm binding at the VL is a multisubunit complex linked by disulfide bonds [35,36] that are sensitive to DTT. Furthermore, in view of the fact that the extracellular matrix is an important part of the cell surface that constantly receives chemical and mechanical signals from the external space [37,38], it is conceivable that even a modest alteration of the structure may influence the aforementioned biological processes related to fertilization and egg activation.

In this context, our specific question here is whether and how the reducing condition involving DTT, which is mild enough not to remove the VL of the eggs, would affect the cell physiology of sea urchin eggs. It was found that treatment of sea urchin eggs (*Lytechinus pictus*) with DTT results in the removal of VL in a time- and pH-dependent manner, without affecting the fertilization response and embryo development [15]. In that study, 10 mM DTT (7–10 min) was effective at removing the VL only above a certain pH, such as 9.2. In our experiment, the NSW containing 10 mM DTT exhibited pH 7.57, which is within the range that would not effect the removal of the VL within 10 min. After examining how the incubation condition (NSW with 10 mM DTT, 10 min) alters the VL and other ultrastructures of *Paracentrotus lividus* eggs through the use of electron microscopy, we assessed the effect of the same reducing conditions on the cytoskeletal properties of the eggs, as well as on its physiological responses to fertilizing sperm. In addition, we tested whether the physiological effect of DTT could be reversed by restoring DTT-exposed eggs in the normal seawater prior to fertilization.

2. Materials and Methods

2.1. Gametes Collection, Fertilization Procedure, and Embryos Observation

Sea urchins (*Paracentrotus lividus*) were collected from October to May in the Gulf of Naples and were maintained at 16 °C in circulating seawater. Spawning was induced by intracoelomic injection of 0.5 M KCl, and the resulting eggs were collected in natural seawater (NSW) filtered with a Millipore membrane of 0.2 μm pore size (Nalgene vacuum filtration system, Thermo Fisher Scientific, Rochester, NY, USA). For fertilization, dry sperm collected by pipetting on the male animal's body were kept at 4 °C and diluted in NSW only a few minutes before fertilization. The final sperm concentration for egg insemination was 1.84×10^6 cells/mL. The subsequent embryonic development was observed with a Leica DMI6000 B inverted microscope. By default, the number of zygotes examined for each experiment was 100, and three independent experiments were conducted with different batches.

2.2. Visualization of the Egg-Incorporated Sperm

P. lividus sperm were prepared afresh and stained with 5 μM Hoechst-33342 (Sigma-Aldrich, St. Louis, MO, USA) for 30 s prior to fertilization. Diluted sperm (10 μL) were added to the media containing the eggs (1 mL). The number of egg-integrated sperm was counted 5 min after insemination by epifluorescence microscopy with a cooled CCD (charge-coupled device) camera (MicroMax, Princeton Instruments, Inc., Trenton, NJ, USA) mounted on a Zeiss Axiovert 200 inverted microscope (Carl Zeiss AG, Oberkochen, Germany) with a Plan-Neofluar 40×/0.75 objective and a UV laser. The Hoechst-33342 solution used in this condition was able to visualize both male and female pronuclei in the zygote when viewed in a Leica TCS SP8X confocal laser scanning microscope equipped with a Diode 405 laser and hybrid detectors (Leica Microsystem, Wetzlar, Germany). The number of fertilized eggs being examined (n) and the number of the independently repeated experiments (N) for each condition are specified in Tables 1 and 2.

Table 1. Frequency of polyspermy (%) in *P. lividus* pretreated with various conditions.

Frequency (%)	NSW	10mM DTT	DTT/WASH	0.5mM TCEP	TCEP/WASH
Mean	4	11	60 *,#	0	0
SD	4.18	21.9	35.9	0	0
N	5	5	5	1	1

Note: Each trial in a given condition comprises 20 eggs. * $p < 0.01$ in Tukey test in comparison with NSW. # $p < 0.05$, with DTT.

Table 2. Number of egg-incorporated sperm inside the *P. lividus* eggs fertilized in various treatments.

Sperm Per Egg	NSW	10mM DTT	DTT/WASH	0.5mM TCEP	TCEP/WASH
Mean	1.04	1.13	2.38 *	1	1
SD	0.2	0.42	1.62	0	0
n	100	100	100	20	20

Note: * $p < 0.00001$ in U-test in comparison with the control (fertilization in natural seawater, NSW).

2.3. Scanning Electron Microscopy (SEM)

P. lividus eggs from three animals were fixed directly in NSW containing 0.5% glutaraldehyde (pH 8.1) for 1 h at room temperature, and post-fixed with 1% osmium tetroxide for an additional hour. The specimens obtained before and after fertilization (at different time points) were dehydrated in increasing concentrations of ethanol and were subjected to critical point drying (LEICA EM CP300). The samples were then coated with a thin layer of gold using a LEICA ACE200 sputter coater, and at least five different eggs for each condition were observed with a JEOL 6700F scanning electron microscope (Akishima, Tokyo, Japan).

2.4. Transmission Electron Microscopy (TEM)

Eggs from three animals were fixed before and at different time points after fertilization, directly in NSW containing 0.5% glutaraldehyde (pH 8.1) for 1 h at room temperature, and were post-fixed with 1% osmium tetroxide and 0.8% $K_3Fe(CN)_6$ for another hour at 4 °C. After washing in NSW for 10 min twice, the samples were rinsed in distilled water for 10 min twice and subsequently treated with 0.15% tannic acid for 1 min at room temperature. After extensive rinsing in distilled water (three times, 10 min each), the specimens were dehydrated in increasing concentrations of ethanol followed by propylene oxide for embedding in Epon 812. Ultrathin sections (70 nm) were stained with UAR-EMS (Uranyl Acetate Replacement Stain, Electron Microscope Sciences, Hatfield, PA, USA) for 30 min, and with 0.3% lead citrate for 30 s. Then, at least five different eggs for each condition were observed with a transmission electron microscope (Zeiss LEO 912 AB).

2.5. Chemicals and Reagents

DL-Dithiothreitol (DTT) (Sigma-Aldrich, St. Louis, MO, USA) and Tris-(2-carboxyethyl)phosphine hydrochloride (TCEP) (ThermoFisher, Waltham, MA, USA) were dissolved in distilled water (DW) and used for bath incubation at the concentrations specified in the text. Hoechst-33342 and all other unspecified materials were purchased from Sigma Aldrich. BPA-C8-Cy3 was synthesized by following the procedure specified in the Supplementary Material File S1.

2.6. Microinjection, Ca^{2+} Imaging, and Confocal Microscopy

Intact eggs were microinjected using an air pressure transjector (Eppendorf FemtoJet, Hamburg, Germany), as previously described [39]. To monitor changes in the intracellular Ca^{2+} levels at fertilization, 500 µM Calcium Green 488 conjugated with 10 kDa dextran was mixed with 35 µM Rhodamine Red (Molecular Probes, Eugene, OR, USA) in the injection buffer (10 mM Hepes, 0.1 M potassium aspartate, pH 7.0) and microinjected into the eggs before insemination. The fluorescence signals of the cytosolic Ca^{2+} increases were captured with a cooled CCD camera (Micro-Max, Princeton Instruments) mounted on a Zeiss Axiovert 200 microscope with a Plan-Neofluar $40\times/0.75$ objective at about 3 s intervals, and the data were analyzed with MetaMorph (Universal Imaging Corporation, Molecular Devices, LLC, San Jose, CA, USA). Following the formula $F_{rel} = [F - F_0]/F_0$, where F represents the average fluorescence level of the entire egg and F_0 the baseline fluorescence, the overall Ca^{2+} signals were quantified for each moment and F_{rel} was expressed as RFU (relative fluorescence unit) for plotting the Ca^{2+} trajectories. Applying the formula $F_{inst} = [F_t - F_{(t-1)}]/F_{(t-1)}$, the instantaneous increments of the Ca^{2+} level was analyzed to locate the specific area of momentary Ca^{2+} increase. The values of Ca^{2+} signals were obtained from four independent experiments (N), and the number of the eggs (n) being analyzed for each condition is specified in the Results.

To visualize F-actin in living eggs, 10 µM (pipette concentration in methanol) of AlexaFluor568-phalloidin (Molecular Probes) was microinjected into the eggs in three independent experiments utilizing as many female animals. To visualize the plasma membrane and the extracellular layers, eggs from two different females were incubated with 5 µM FM 1-43 (ThermoFisher Scientific) or 50 µM of a branched fluorescent polyamine (BPA-C8-Cy3) for 10 min in two independent experiments. Both probes were dissolved in distilled water. The eggs treated with the fluorescent probes were observed with a Leica TCS SP8X confocal laser scanning microscope equipped with a white light laser and hybrid detectors (Leica Microsystem, Wetzlar, Germany). The number of eggs examined for each condition is specified in the Results.

2.7. Statistical Analysis

The numerical MetaMorph data were compiled and analyzed with Excel (Microsoft Office 2010) and reported as mean $\pm$ standard deviation (SD in all cases in this manuscript. Oneway ANOVA and U-test were performed through Prism 8.0 (GraphPad Software), and $p < 0.05$ was considered to be statistically significant. For ANOVA results showing $p < 0.05$, statistical significance of the difference between the two groups was assessed by Tukey's post hoc tests. The two groups of data showing significant differences from each other were marked with brackets and symbols indicating the p values. The pairwise comparison that produced insignificant p values (>0.05) were not mentioned for the sake of simplicity in the description.

2.8. Ethics Statement

Sea urchins *P. lividus* used for the present study were collected according to the Italian legislation (DPR 1639/68, 19 September 1980 and confirmed on 1 October 2000). All the experimental procedures were carried out in accordance with the guidelines of the European Union (Directive 609/86).

3. Results

3.1. DTT Induces Ultrastructural Changes on the Surface of Unfertilized Sea Urchin Eggs

As shown in Figure 1, the pretreatment of unfertilized *P. lividus* eggs had subtle but evident effects on the egg surface. Although the topography of the egg surface, which is characterized by the presence of a myriad of microvilli, was not drastically changed by DTT treatment, the individual microvilli became more irregular in shape and occasionally thicker (Figure 1B). In addition, the SEM and TEM images revealed the formation of numerous blebs as big as 0.5 µm on the surface of the eggs treated with DTT (Figure 1B,D arrows), which is in line with earlier findings with *S. purpuratus* eggs treated in the same condition [40]. These blebs were previously interpreted as expelled cortical granules [40]. More importantly, the enlarged view of the TEM images revealed that the continuous contour of the VL, which was evident in the control eggs, was intermittently interrupted or fuzzy in the eggs treated with 10 mM DTT for 10 min (Figure 1E,F). Thus, it appears that the given condition of the egg preincubation with DTT did not remove the VL but modified it, and eventually induced additional changes in the ultrastructure of the cell surface, such as the microvilli.

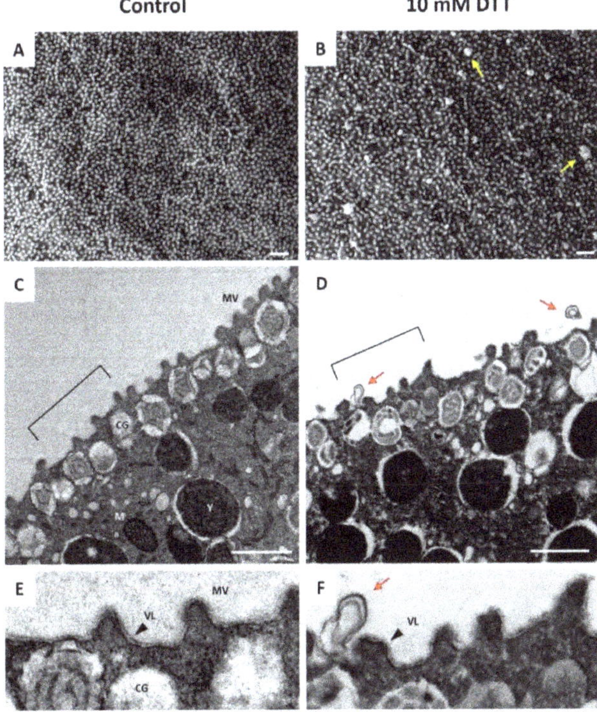

Figure 1. Ultrastructure of unfertilized sea urchin eggs at the surface after brief treatment with dithiothreitol. *P. lividus* eggs were exposed to 10 mM dithiothreitol (DTT) in seawater for 10 min and were subjected to analyses by electron microscopy, following the procedures described in Materials and Methods. (**A,B**) Images obtained by SEM. (**C,D**) Images obtained by TEM. Note that, after the DTT treatment, the shape of the individual microvilli (MV) became irregular and slightly thicker, and that numerous blebs were formed on the egg surface (yellow and red arrows in **B,D,F**). (**E,F**) For closer examination, the areas of the TEM images marked by brackets in panels (**C,D**) are enlarged. Scale bars, 1 µm. CG—cortical granules; MV—microvilli; M—mitochondria; Y—yolk granules; VL—vitelline layer.

3.2. Effect of DTT Pretreatment of the Eggs on the Fertilization Process

Given that the DTT pretreatment partially disrupts the VL and modifies the microvilli, we examined how the same treatment would influence the fertilization process. To this end, *P. lividus* eggs preincubated in NSW for 10 min in the presence or absence (control) of 10 mM DTT were fertilized and fixed at certain intervals and were subjected to SEM and TEM analyses (Figure 2). By 25 s after insemination, it was already evident that the VL of the control egg had started to be elevated around the area where the fertilizing sperm was fused with the oolemma (Figure 2A). This change, induced by partially elevated nascent FE, created a dimple-like surface ([41,42], see Supplemental Video S1).

The detachment of the VL from the plasma membrane and the subsequent elevation of the VL to form the FE is a wave-like process resulting from the coordinated exocytosis of the cortical granules from the underlying egg cortex. By 1 min, the entire surface of the control egg was covered with FE (Figure 2B). On the other hand, fertilization in the eggs pretreated with 10 mM DTT appeared to proceed with some notable anomaly. By 25 s, the sperm was attached to the microvilli of the egg, but the initial elevation of the FE did not take place (Figure 2D). Although a more prominent fertilization cone was formed in the eggs fertilized in the presence of DTT, the eggs were covered with the thin and patchy looking vitelline layer (VL) (Figure 2E, also see the enlarged image in Figure S1). In agreement with this finding, the TEM image of the control egg obtained 1 min after fertilization exhibited a thick layer of FE (Figure 2C, arrow), which was absent in the eggs fertilized 10 min after the preincubation in 10 mM DTT (Figure 2F, see Supplemental Video S2). Judging from the discontinuous appearance of the VL (Figure 1F), the failure of VL to elevate and form FE in the DTT-pretreated eggs is thought to be related to loss of the integrity of the VL due to the breakage of the disulfide bonds in and among the molecules in the VL. In the latter case, the extruded contents of the cortical granules failed to accumulate in the perivitelline space as they were eventually lost to the media through the loose holes on the VL [7].

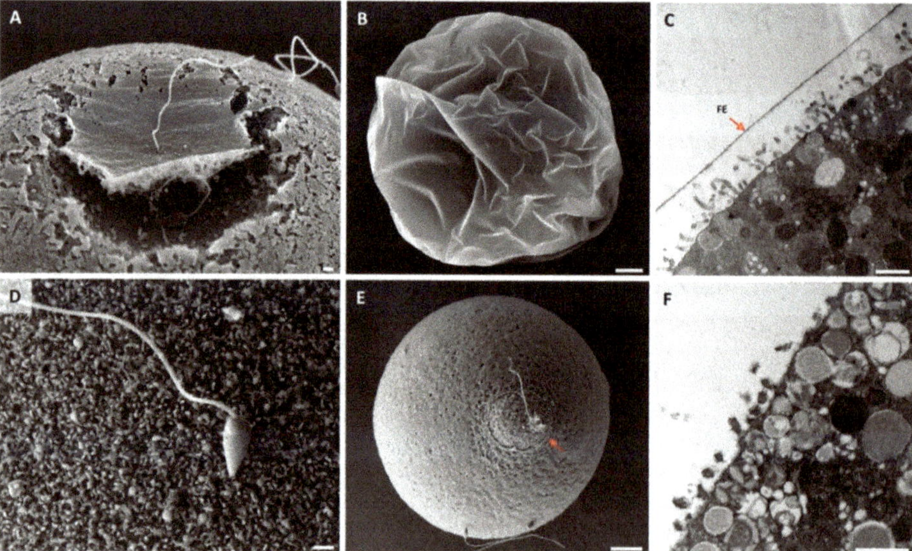

Figure 2. Ultrastructure of the egg surface immediately after fertilization. *P. lividus* eggs were fertilized and fixed for SEM (**A,B,D,E**) and TEM (**C,F**) analyses. (**A–C**) Control eggs fertilized in NSW. (**D–F**) Eggs preincubated (10 min) and fertilized in seawater containing 10 mM DTT. Images in (**A,D**) were captured by fixing the eggs 25 s after insemination. The images in all the other panels show the ultrastructure of the egg surface 1 min after insemination. Note that, in panel (**E**), most parts of a fertilizing sperm were not internalized yet, but were surrounded by the egg protrusion through the ruptured fertilization envelope (red arrow). Scale bars: panels (**B,E**), 10 µm; all other panels, 1 µm. FE, fertilization envelope.

3.3. Effect of DTT Pretreatment of the Eggs on the Later Stage of Fertilization

A fertilized sea urchin egg undergoes drastic reorganization at the cortex as the contents of the cortical granules are being extruded to the perivitelline space by exocytosis. Microvilli and the subplasmalemmal actin network are also rapidly reorganized at this time [26,43,44]. To test whether and how DTT affects the pattern of cortical rearrangement after fertilization, we preincubated *P. lividus* eggs in the presence or absence of 10 mM DTT prior to fertilization, and the resulting zygotes were fixed for SEM and TEM analyses 5 min after insemination (Figure 3). The whole-cell view of the SEM image showed that the entire surface of the egg fertilized in NSW was covered with a thick layer of FE (Figure 3A,C). By contrast, the eggs preincubated and fertilized in 10 mM DTT seawater did not form the normal FE (Figure 3B). Instead, numerous holes were observed on the surface, and the VL was barely elevated, so that the underlying microvilli were clearly visible within the holes (Figure 3D, arrows) and through the thin veiling VL. By 5 min, exocytosis of cortical granules and their extruded contents were heavily deposited to form the hyaline layer underneath the FE (Figure 3E). The exocytosis of the cortical granules was thought to be exhaustive by this time, as judged by their absence in the subplasmalemmal region. Indeed, the depth of the newly formed perivitelline space seen in the light microscopy was so immense that it often collapsed during the procedure of fixation, which was applied 5 min after insemination (Figure 3E). Similarly, most cortical granules also underwent exocytosis in the eggs pretreated and fertilized in the presence of 10 mM DTT, but the VL did not swell. There was only a faint remnant of elevated VL (Figure 3F, arrow). Notably, however, these eggs exhibited extensive elongation of the microvilli that stretched out to the VL (Figure 3F).

3.4. DTT Pretreatment of the Eggs Alters the Ca^{2+} Response at Fertilization

The physiological consequence of the DTT pretreatment was assessed by examining the eggs' Ca^{2+} response at fertilization. In the normal condition, fertilized sea urchin eggs displayed two modes of intracellular Ca^{2+} response (Figure 4A,B): (i) the synchronized Ca^{2+} increase near the plasma membrane of the entire egg surface (cortical flash (CF), and (ii) the Ca^{2+} wave that locally originates at the sperm-interaction site and propagates to the antipode. The rapid CF taking place immediately after the fertilizing sperm makes meaningful contact with the egg due to the Ca^{2+} influx into the egg through the voltage-gated calcium channels that can be modulated by the surrounding subcellular structures such as the microvilli [45–47]. On the other hand, the Ca^{2+} wave depends on the synthesis of Ca^{2+} mobilizing second messengers [20,22,24] and on the structural modification of cortical granules and vesicles [47,48].

The three groups of eggs (control, DTT, and WASH) responded to insemination with Ca^{2+} signals without showing any significant difference in latency: around 40 s after insemination in the given conditions. This observation suggests that the sperm receptivity in this regard was not changed much by the treatment with DTT. However, as expected from the modification of the egg surface, the amplitude of CF in the eggs preincubated and fertilized in 10 mM DTT seawater was significantly lower than that of the control eggs (Figure 4C,D). Similarly, the average peak amplitude of the Ca^{2+} wave in the eggs pretreated with DTT was significantly lower (0.55 ± 0.15 RFU, $n = 24$) in comparison with the control eggs (0.74 ± 0.15 RFU, $n = 23$, $p < 0.01$). In addition, the Ca^{2+} waves propagated more slowly in the eggs pretreated with 10 mM DTT, as judged by the time spans from the first appearance of the localized Ca^{2+} signal at the sperm interaction site to the moment of wave's arrival at the antipode (Figure 4E, traverse time). While the average traverse time in the control eggs was 22.3 ± 3.0 s ($n = 23$), the corresponding value in the eggs pretreated with 10 mM DTT was 26.0 ± 2.8 s ($n = 24$, $p < 0.01$). Curiously, when the DTT-pretreated eggs were fertilized after being rinsed and restored in normal seawater, some of the Ca^{2+} signaling patterns altered by DTT were significantly reversed to assimilate those in the control eggs. Such a trend was found in the CF (Figure 4C), whose average amplitude in the eggs restored in NSW (WASH, 0.06 ± 0.05 RFU, $n = 15$) was comparable to the control

(0.07 ± 0.03 RFU, $n = 23$), but significantly higher than the values in the eggs pretreated and fertilized in 10 mM DTT (0.23 ± 0.02 RFU, $n = 19$, $p < 0.01$). A similar reversal of the DTT-induced alteration in Ca^{2+} signaling pattern was also found in terms of the traverse time, which was shortened again (Figure 4E, WASH, 21.9 ± 3.7 s, $n = 21$, $p < 0.01$), but not in respect to the peak amplitude of the Ca^{2+} wave (Figure 4D, WASH, 0.53 ± 0.12 RFU, $n = 21$).

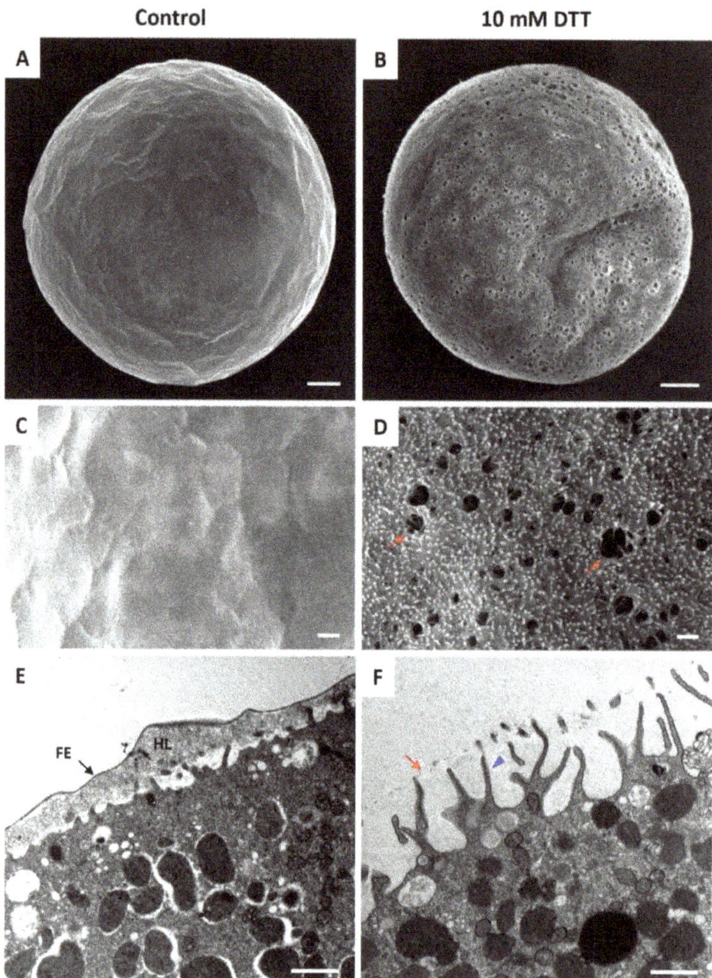

Figure 3. Ultrastructure of the egg surface at a later stage of fertilization. *P. lividus* eggs were fertilized and fixed 5 min later for SEM and TEM analyses. (**A,C,E**) Control eggs fertilized in NSW. (**B,D,F**) Eggs preincubated (10 min) and fertilized in seawater containing 10 mM DTT. Panels (**A–D**) show SEM images, while panels (**E,F**) represent results of TEM. Red arrows in panels (**D,F**) indicate the ruptured fertilization envelope. A blue arrowhead in panel F indicates the over-elongated microvilli. Scale bars: panels (**A,B**), 10 μm; all other panels, 1 μm. FE, fertilization envelope (arrows); HL, hyaline layer.

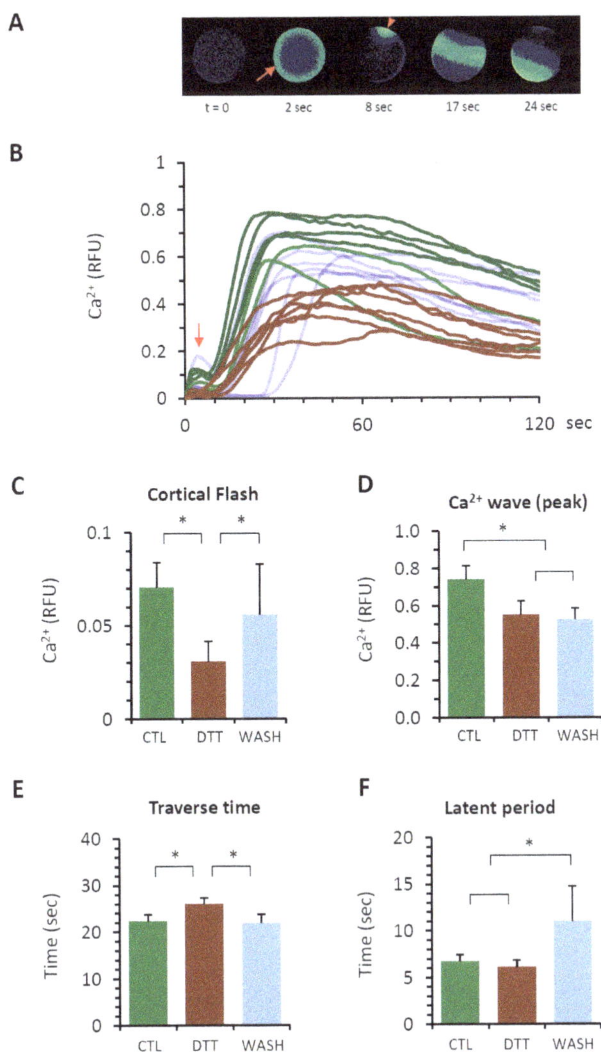

Figure 4. DTT pretreatment of the eggs alters the Ca^{2+} response at fertilization. *P. lividus* eggs were microinjected with calcium dye and preincubated in seawater with 10 mM DTT (10 min) prior to fertilization. As a control (CTL), eggs from the same batch were prepared in parallel and fertilized in NSW. For comparison, some eggs preincubated in the presence of 10 mM DTT were restored in normal seawater and fertilized (WASH). (**A**) Instantaneous increment of Ca^{2+} signals inside a control egg visualized in pseudocolour. The time point immediately before the first detectable Ca^{2+} signal was set as t = 0. The arrow indicates cortical flash (CF), and the arrowhead pinpoints the sperm interaction site on the egg where the Ca^{2+} wave is initiated. (**B**) Changes of the Ca^{2+} signals: green curves (CTL), brown curves (eggs fertilized in 10 mM DTT), and curves in pale blue (eggs fertilized after restoration in NSW, 10 min). The vertical arrow indicates the CF. Out of four independent experiments conducted, the result of one representative experiment was plotted for Ca^{2+} trajectories. Histograms in panels C−F represent comparisons of the egg groups with respect to the CF amplitude (**C**), the peak amplitude of the Ca^{2+} wave (**D**), the duration of the Ca^{2+} wave traversing the egg from the sperm interaction site to the antipode (**E**), and the time interval between the CF and the initiation of the Ca^{2+} wave (**F**). * Tukey's post hoc test, $p < 0.01$. RFU, relative fluorescence unit.

Another noteworthy observation is about the time interval between the CF and the first detectable Ca^{2+} signal at the site of sperm interaction, which is referred to as the "latent period" in this work (Figure 4F). This is the period when the fertilized egg already transmitting the intracellular signals such as membrane depolarization and Ca^{2+} influx prepares itself for the generation of the Ca^{2+} wave [24,31,47,49]. Despite the notable structural changes on the egg surface, the latent period of the Ca^{2+} response in the eggs preincubated and fertilized in seawater containing 10 mM DTT (6.1 ± 1.4 s, $n = 19$) was not much different from that of the control eggs (6.7 ± 1.5 s, $n = 23$). When the DTT-preincubated eggs were rinsed and restored in NSW before fertilization, however, the latent period for the Ca^{2+} response in these eggs was nearly doubled (11.3 ± 7.6 s, $n = 15$, $p < 0.01$) in comparison with both CTL and DTT. These results suggest that the alteration of the Ca^{2+}-release systems induced by DTT pretreatment is reversed to some extent by the restoration of the eggs in NSW, but the recovery path may not be the same as the path of modification induced by DTT. In other words, the eggs in the group WASH are in another physical state and show a distinct response to the fertilizing sperm.

3.5. Effect of DTT Pretreatment of the Eggs on Monospermic Fertilization

As the DTT pretreatment significantly altered the topography of the egg surface, it was conceivable that such structural changes may induce polyspermy at fertilization. Indeed, the porous and patchy appearance of the egg's VL at fertilization and its failure to form a strong FE all raised the possibility of supernumerary sperm entry. However, the results pooled from five different batches of *P. lividus* suggested that this is not the case. When the eggs were preincubated and fertilized in seawater containing 10 mM DTT, most eggs were still monospermic. The frequency of polyspermic fertilization was not significantly increased in comparison with the control eggs fertilized in NSW (Table 1). Furthermore, the number of egg-incorporated sperm had only a marginal increase in comparison with the control eggs, which was not statistically significant (Table 2). However, when the DTT-pretreated eggs were restored and then fertilized, the frequency of polyspermy went up to 60% (Table 1), and the number of egg-incorporated sperm at fertilization was greatly increased (2.38 sperm per egg, Table 2). This result was not the effect of prolonged incubation or agitation of the eggs, because in another control experiment, the eggs continuously exposed to 10 mM DTT, following the same protocol (i.e., 10 min incubation, media change, and another incubation for 10 min) exhibited mostly monospermic fertilization. The polyspermy rate in these eggs was merely 5%, which was virtually the same as that of the eggs fertilized in NSW.

The observation that polyspermy is induced in the DTT-pretreated eggs after the restoration in NSW may also be related to the fact that DTT can traverse the cell membrane and spread inside the cell. Indeed, a similar experiment with another reducing agent, TCEP, which does not penetrate the cell membrane [50–52], did not induce polyspermy in its presence. The eggs restored and fertilized in normal seawater after being preincubated with TCEP were also all monospermic at fertilization (Table 2). Hence, the cause of the polyspermy in these eggs might be attributable to some internal factors. One explanation could be the possibility that an abrupt shift of the pH of the external media (from pH 7.57 of DTT seawater to pH 8.1 of NSW) and the accompanying changes in microvillar morphology (elongation) due to the cytoplasmic alkalinization might have contributed to polyspermy [53], in a sense that elongated microvilli could increase the capacity of the egg to fuse with sperm [54].

3.6. F-Actin Mobilization during Egg Activation Is Inhibited by DTT

In a series of previous studies on starfish and sea urchin eggs, we demonstrated that polyspermy is closely linked to an anomaly in the actin cytoskeleton of the egg cortex. Indeed, when the actin cytoskeleton was artificially altered through the use of actin-targeting toxins and chemicals, an actin-binding protein cofilin, anti-depactin antibody, nicotine, or even by "ageing", the eggs showed a strong tendency of polyspermic fertilization [27,29,39,55–58]. At variance with these experimental conditions that caused cytoskeletal alterations, the pretreatment of *P. lividus* eggs with DTT or TCEP in our experimental condition did not induce discernable changes in the structure of the cortical actin cytoskeleton (data not shown). This may be one of the reasons why the DTT- and TCEP-pretreated eggs were not prominently polyspermic (Tables 1 and 2). Nonetheless, it is possible that the concerted rearrangement of the actin cytoskeleton taking place in the fertilized eggs during egg activation [25,28,29] may not be influenced by the DTT pretreatment. In other words, the reducing agent may not have a profound effect on the actin filaments inside a quiescent egg, but it may still affect the mobilization kinetics in post-fertilization eggs. We tested this idea by microinjecting the eggs with a fluorescent F-actin probe (AlexaFluor 568-Phalloidin) before the DTT and TCEP pretreatment and fertilization (Figure 5).

In the control eggs fertilized in NSW, the subplasmalemmal F-actin started to take an orderly form and left the area to move toward the inner cytoplasm ($n = 15$). This was more evident when the locations of the thick ring-shaped F-actin clusters were compared with the border of the egg surface in the egg images 30 min after fertilization. In the control egg, by then, there was at least 10–15 µm distance between the F-actin clusters and the boundary of the egg surface (marked by dotted curve, Figure 5). Hence, the actin filaments underneath the plasma membrane were evacuated from the subplasmalemmal zone. However, this characteristically concerted centripetal migration of the cortical actin filaments was mostly blocked in the eggs preincubated and fertilized in 10 mM DTT ($n = 12$), as the ring-shaped F-actin clusters had a strong tendency to remain in the vicinity of the plasma membrane from 5 min and on. At 30 min, they were still very close to the plasma membrane (Figure 5, see the dotted curve for the DTT-pretreated eggs). This inhibitory effect of DTT on the inward movement of the cortical actin filaments may be attributed to its reducing power inside the egg's cytoplasm, but not on the external surface of the egg. A similar reducing agent unable to permeate the cell membrane, TCEP, did not have such an effect. In these eggs ($n = 8$), the ring-shaped F-actin clusters were relocated toward the inner cytoplasm just like in the control eggs (Figure 5, dotted curve). The lack of the inhibitory effect on F-actin mobilization displayed by TCEP is not a matter of the low efficacy of the agent. When added to the media, 0.5 mM TCEP blocked the formation of the FE as effectively as the 10 mM DTT did (Figure 5, see the egg images in the light transmission view). These results suggest that the orderly migration of the egg's cortical actin filaments following fertilization is sensitive to the redox status of the cytoplasm.

3.7. Effect of DTT Pretreatment on the Late Events of Fertilization

In the eggs preincubated and fertilized in 10 mM DTT seawater, the anomaly of the actin cytoskeleton was also observed in the fertilization cone, the F-actin-based cytoskeletal apparatus specializing in the capture and engulfing of the fertilizing sperm. On the confocal plane where the fertilization cone was located, the control eggs fertilized in NSW ($n = 15$) after microinjection of AlexaFlour 568-Phalloidin displayed an array of thick actin fibers oriented perpendicular to the plasma membrane (Figure 6A). In the eggs preincubated and fertilized in the presence of 10 mM DTT ($n = 12$), however, the morphological symmetry of the fertilization cone was lost, and the aggregation of the thick actin fibers was visibly exaggerated amid other thick actin fibers in the inner cytoplasm, which were similar to the stress fibers (Figure 6B, arrowhead). On the other hand, in the eggs preincubated with DTT but washed ($n = 12$) and restored in NSW prior to fertilization, there were often multiple fertilization cones (Figure 6C).

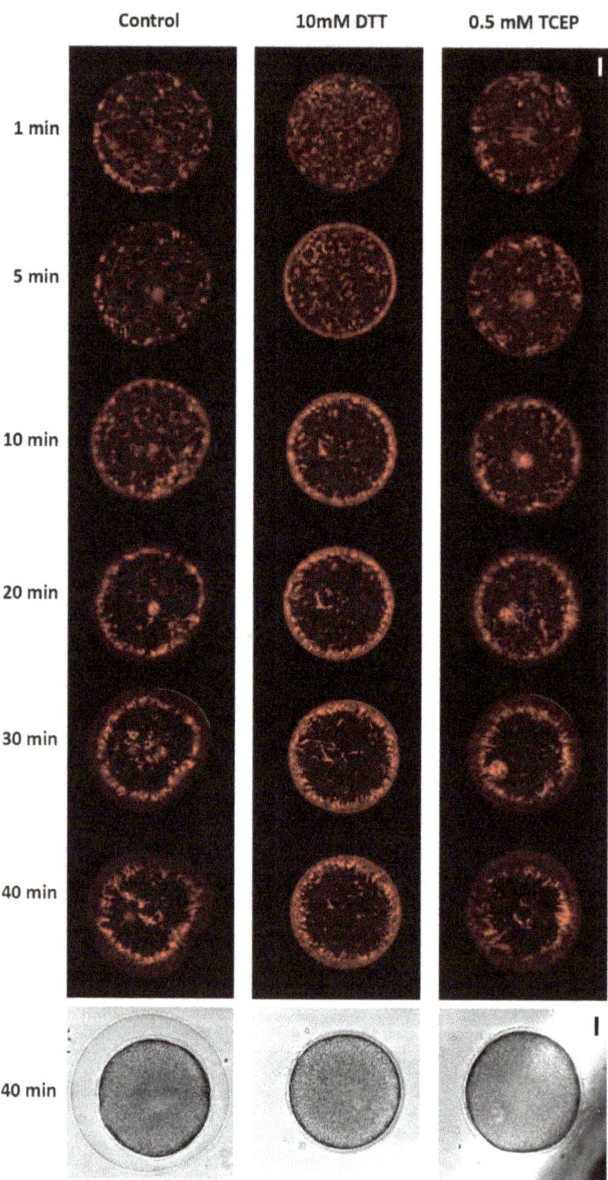

Figure 5. F-actin mobilization during egg activation is inhibited by DTT, the membrane-permeant reducing agent. *P. lividus* eggs were microinjected with F-actin dye, AlexaFluor 568-phalloidin, and preincubated in seawater in the presence or absence (control) of 10 mM DTT for 10 min prior to fertilization. For comparison, some of the control eggs were exposed to another type of reducing agent that does not penetrate cell membrane, tris(2-carboxyethyl)phosphine (TCEP). The changes of the actin cytoskeleton following fertilization were monitored by confocal microscopy at intervals. The moment of insemination was set as t = 0. The white dotted curves drawn on the images at 30 min delineate the borders of the egg surface. To visualize the fertilization envelope, the same AlexaFluor 568-phalloidin eggs were shown in the bright field view 40 min after insemination. These images are representative of at least two independent experiments. Scale bar, 10 μm.

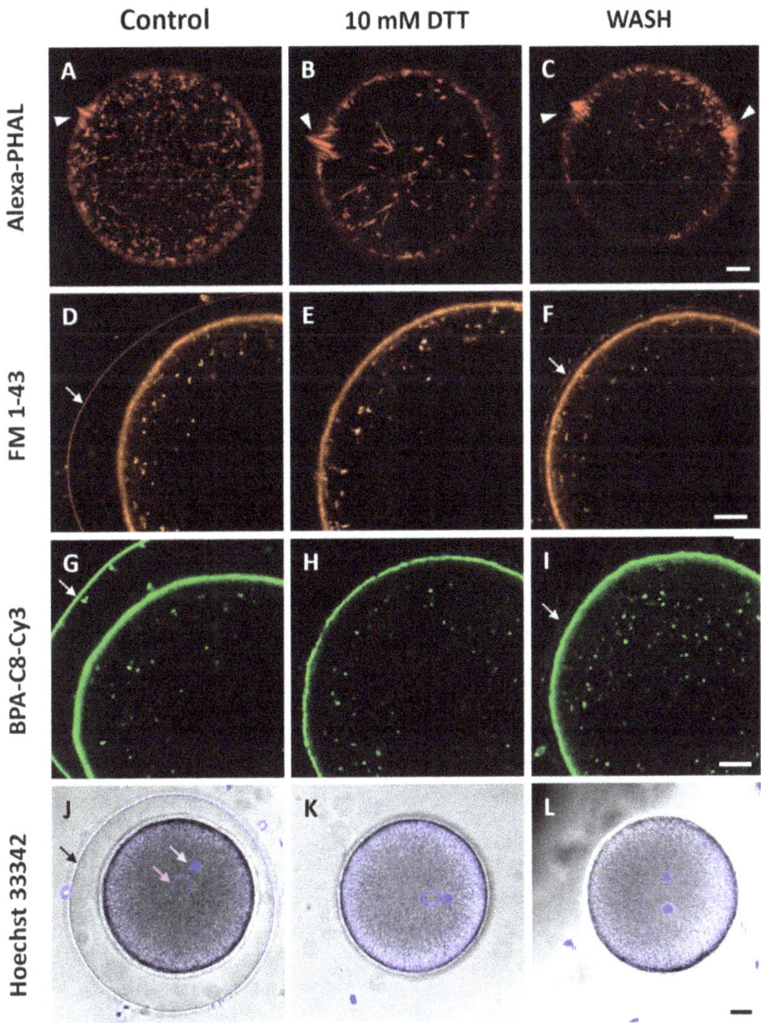

Figure 6. Effect of DTT pretreatment on the late events of fertilization. (**A–C**) *P. lividus* eggs were microinjected with F-actin dye, AlexaFluor 568-phalloidin, and preincubated in seawater in the presence or absence (control) of 10 mM DTT for 10 min prior to fertilization. After 10 min, the actin cytoskeleton of the fertilized eggs was visualized by confocal microscopy. The fertilization cones are marked by arrowheads. (**D–I**) Eggs pretreated in the same conditions as above (but without F-actin dyes) were fertilized and incubated in the presence of FM 1-43 (**D–F**) or BPA-C8-Cy3 (**G–I**). White arrows indicate the fertilization envelope and its remnants. (**J–L**) Hoechst 33,342 introduced with sperm visualized the male and female pronuclei of the zygotes in the given conditions, as indicated by pale blue and pink arrows, respectively. The black arrow indicates the fertilization envelope. These images are representative of at least two independent experiments. Scale bar, 10 μm.

Although the vitelline layer was partially disrupted and certain aspects of the dynamic property of the egg's cortical actin cytoskeleton were altered by DTT (e.g., centripetal migration of actin filaments), the endocytotic activity at the plasma membrane that retrieves the membranes following cortical granule exocytosis [59,60] appeared to be normal. When the eggs were stained in the media by adding the fluorescent membrane probe FM 1–43 that sticks to the outer leaflet of the plasma membrane [61,62], numerous internalized vesicles appeared in the inner cytoplasm of the eggs 10 min after fertilization (Figure 6D, $n = 7$). Evidently, the eggs preincubated and fertilized in the presence of 10 mM DTT ($n = 7$) also displayed virtually the same results (Figure 6E), and DTT washing and restoration in NSW ($n = 7$) did not make much difference to the membrane retrieval after fertilization. Nonetheless, it is noteworthy that, in these DTT-preincubated eggs restored in NSW before fertilization, there was some modest formation of FE (Figure 6F, arrow). This result was also corroborated by another fluorescent probe that is derivative of synthetic polyamine [63], which is thought to have a tendency to bind to negatively charged macromolecular surfaces such as the plasma membrane and actin filament meshwork [64,65]. Applied in the media around echinoderm eggs, this fluorescent probe appears to stain predominantly the jelly coat of starfish oocytes (Santella et al. unpublished data). For *P. lividus* eggs at fertilization, this novel probe BPA-C8-Cy3 exhibited intense fluorescent labelling of the VL, FE, and subplasmalemmal regions, as well as the membraneous structures within the cytoplasm and the perivitelline space (Figure 6G, $n = 10$). In the DTT-preincubated eggs that were restored in NSW before fertilization ($n = 10$), modest but evident formation of the FE was detected, as with FM 1-43 (Figure 6I, arrow). Hence, these results suggest that certain aspects of cellular changes induced by DTT can be reversed to some extent, but not completely.

Importantly, we noted that BPA-C8-Cy3 produced evidently thinner and occasionally interrupted labelling at the junction of the plasma membrane and the VL in the eggs pretreated and fertilized in the seawater containing 10 mM DTT ($n = 10$, compare Figure 6H with Figure 6G,I). Again, after washing and restoration in NSW, the fluorescent labelling at the VL and plasma membrane came back to the appearance of the control eggs (Figure 6I). By contrast, such reversed variations in the level of fluorescence were not observed in the eggs labelled with the membrane-specific probe FM 1-43 (Figure 6D–F). Subtle as it might look, this was the striking difference between FM 1-43 and BPA-C8-Cy3. These observations in comparison with FM 1-43 are compatible with the idea that BPA-C8-Cy3 selectively visualizes the VL, which was occasionally interrupted in the eggs pretreated with DTT.

We then tested whether or not DTT pretreatment of the eggs affected the migration of male and female pronuclei. Following the experimental procedure described in Materials and Methods, the male and female pronuclei were fluorescently labelled in the zygote (Figure 6J). The densely labelled male pronucleus appeared more compact than the female pronucleus, and the distance between the two pronuclei was not much different in the zygotes obtained from DTT-preincubated eggs either with or without restoration in NSW (Figure 6K,L). Thus, taken together with the results in Figure 5, these observations suggest that exposure of the eggs to DTT in the condition that blocks the centripetal movement of the subplasmalemmal actin filaments actually has little effect on pronuclei movement, the process that has been reported to be mainly dependent on the transport system, largely based on microtubules [66,67].

3.8. Effect of DTT Pretreatment of the Eggs on Early Embryonic Development

The eggs restored in NSW after being preincubated in DTT exhibited a certain degree of plasticity in terms of Ca^{2+} signaling, F-actin mobilization, and FE formation during fertilization, albeit a high frequency of polyspermy. To explore this issue a step further, we examined the early phase development of those embryos derived from the monospermic zygotes obtained from each experimental condition. By 2 h and 30 min, the control eggs fertilized in NSW were well into the 8-cell stage. The cleavage also appeared quite synchronous (Figure 7A). In contrast, when the eggs were preincubated and fertilized in the presence of 10 mM DTT, the majority of embryos were at the two-cell stage, suggesting that the cleavages of the zygote had been significantly delayed. Indeed, in some cases, it appeared that the individual daughter cells had difficulties completing cytokinesis, as judged by the partially formed cleavage furrow (Figure 7C, arrow). By 6 h post-fertilization, the control embryos were at the early blastula stage (Figure 7B), but the embryos deriving from the eggs fertilized in the presence of DTT were still at the four-cell stage with some cleavage anomaly (Figure 7D). This problem of cytokinesis seemed to be overcome when the DTT-preincubated eggs were restored and fertilized in NSW. However, 2.5 h after insemination, the embryos were mainly at the four-cell stage (Figure 7E), a clear sign of delay in comparison with the control embryos. Six hours after insemination, the embryos derived from the eggs restored and fertilized in NSW showed varying blastomere sizes, abnormal cleavage, and their leaking from the ruptured evanescent FE (Figure 7F, red arrow).

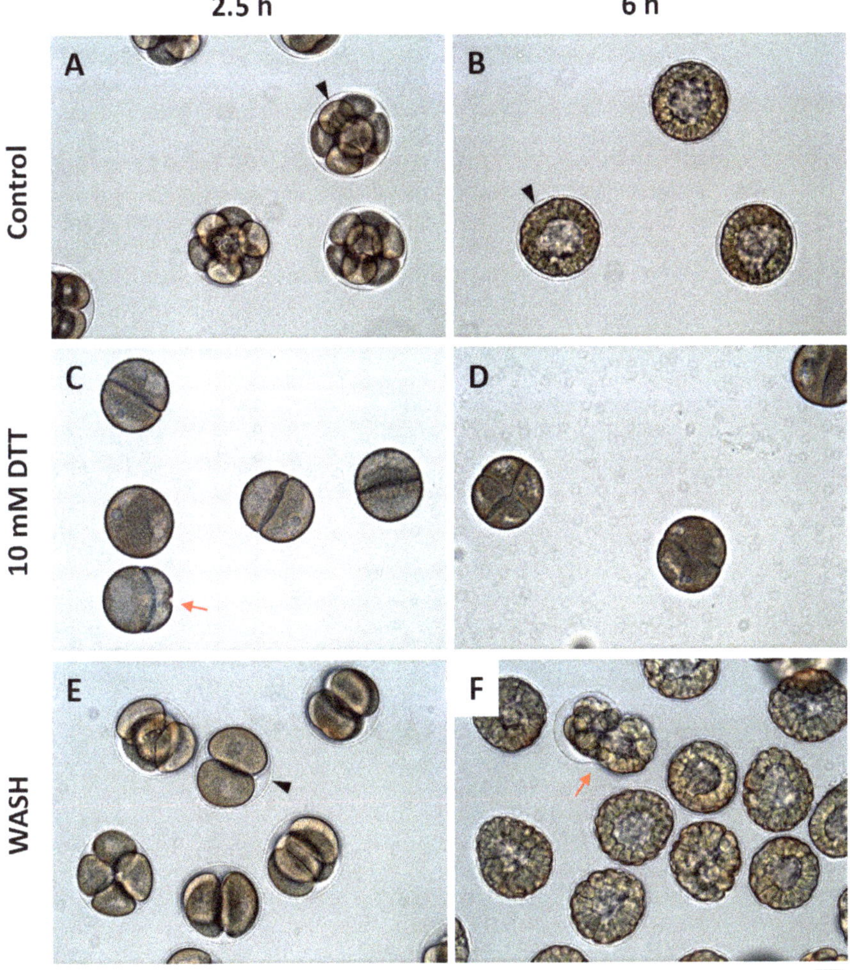

Figure 7. DTT pretreatment of the eggs induces anomalies during early development. (**A–D**) *P. lividus* eggs were incubated in the presence or absence (control) of 10 mM DTT for 10 min and fertilized. (**E–F**) Some of the DTT-pretreated eggs were rinsed and restored in NSW (10 min) and fertilized (WASH). The cleavage and development of the embryos in each condition were monitored by light microscopy. Arrowheads indicate the FE. Note that the FE covering the control embryo is still intact at 6 h (**B**), where it is totally absent in the embryos fertilized in the presence of DTT. The FE is evident again in the embryos derived from the eggs restored in NSW prior to fertilization (**F**). Red arrows indicate signs of leakage in the embryo from the ruptured FE. Scale bar, 10 μm. FE, fertilization envelope. The number of fertilized eggs examined for each set of experiments was 100. Representative results from one of the three independent experiments are shown.

4. Discussion

As a part of the extracellular matrix, the VL is thought to play important roles during fertilization and early embryonic development. In this study, we induced partial disruption of the VL in *P. lividus* eggs through the use of reducing agents that cleave intra- and intermolecular disulfide bonds. The experimental condition (10 mM DTT, 10 min) of the egg pretreatment was mild enough not to cause total removal of the VL, as judged by the images in the TEM and SEM analyses, but introduced subtle ultrastructural changes on the egg surface. The VL was evidently interrupted (Figure 1) and failed to transform into FE at fertilization (Figure 2). This is presumably because the VL was not physically sealed up due to the patchy structure induced by DTT, and thereby the osmotic pressure failed to build up in the perivitelline space, while proteins and other materials extruded from the cortical granules were eventually lost to the external medium [7,41,68].

The lack of VL integrity is likely to have altered the mechanical property of the egg surface because the plasma membrane is intimately ensheathed by the VL. Furthermore, the structure of the microvilli underlying the VL and the plasma membrane was appreciably modified in the eggs pretreated with 10 mM DTT. Another indication of cytoskeletal changes in the outer cytoplasmic region of the eggs fertilized in seawater containing DTT was the lack of cytoplasmic contractility [69], which prevented the formation of the dimple-like structure where the sperm fused with the oolemma (Figure 2A, Supplemental Videos S1 and S2) [42,70]. Moreover, the appearance of an extraordinarily enhanced fertilization cone in these eggs (Figure S1), together with the delayed sperm incorporation (Figure 2E), underline the alteration of the F-actin dynamics necessary to engulf the sperm [71].

The plasticity of *P. lividus* eggs was also manifested when the DTT-pretreated eggs were restored in NSW. While the eggs fertilized in the presence of DTT exhibited some anomalies (e.g., repressed Ca^{2+} response and failed FE elevation), the eggs pretreated with DTT but restored in NSW experienced less alterations in sperm-induced Ca^{2+} signals in terms of CF and traverse time. Likewise, the reversal of the altered Ca^{2+} response at fertilization was also observed in *P. lividus* eggs that were restored in NSW after being exposed to low salinity and other conditions [72,73], as well as in *Arbacia lixula* eggs that were incubated in hypertonic salinity and then inseminated in NSW [74]. In the eggs restored in NSW after DTT pretreatment, however, the physicochemical and structural changes of the cortex were accompanied by polyspermy at fertilization, as well as some anomaly in embryonic development (Figure 7). The centripetal translocation of the subplasmalemmal actin filaments during egg activation has been observed in both sea urchin and starfish [25,28,29,58]. It is conceivable that this orderly translocation of the actin filaments may contribute to cytoplasmic sorting and reorganization of the egg organelles in preparation for the subsequent cleavage. The eggs fertilized in the presence of DTT failed to exhibit the centripetal translocation of F-actin, and these eggs had significant delay and problems in the cleavages. Beyond this, the precise role and physiological significance of the translocation of F-actin in activated eggs are still largely unknown. Nevertheless, the results of our study using the eggs microinjected with AlexaFluor568-phalloidin and exposed to the two different reducing agents suggested that this translocation of cortical actin filaments may be sensitive to the cytoplasmic redox state of the eggs, as judged by its inhibition by DTT, the membrane-permeant reducing agent. In support of this idea, TCEP, which is unable to penetrate the cell membrane, had no such inhibitory effect on the reshuffling of cortical actin filaments (Figure 5). Nevertheless, further investigation is needed to understand to what extent the redox state of the egg cytoplasm is affected by the given conditions of DTT treatment and to identify the major targets of DTT that are accountable for our observations. Indeed, the importance of exquisite control of the redox state in oocytes and embryonic cells is increasingly being appreciated [75]. On the other hand, our observation of the apparently normal migration of female and male pronuclei in the eggs pretreated and fertilized in seawater containing DTT (Figure 5C) suggests that it is not likely that the centripetal movement of the cortical actin filaments makes a significant contribution to the transportation of the egg-engulfed sperm. While the molecular mechanism through which DTT inhibits the translocation of the cortical actin filaments is not known, it has been shown that myosin can form disulfide bonds within its regulatory light chain [76]. Thus, it is tempting to speculate that a DTT-induced shift of the myosin pool to its reduced form might have interfered with the translocation of the cortical actin filaments. In support of this idea, the inner cytoplasm of the eggs fertilized in the presence of DTT showed a conspicuous occurrence of stress fibers during egg activation, which might be a sign of deregulated actin-myosin interaction (Figure 6B). Alternatively, it is noteworthy that actin itself can exist as a dimer linked by disulfide bonds in vitro [77–79]. If such a transitory dimerization of actin molecules utilizing cysteine residues takes place inside the eggs and contributes to the centripetal translocation of microfilaments, it cannot be ruled out that actin itself might be the direct target of DTT.

The findings in this study raise a few more significant points on the cell biology of oocytes and eggs. In conventional practice, to follow the cell surface events at fertilization, eggs were often attached to polylysine-coated plates [80] or deprived of their jelly coat and VL [15]. These so-called "denuded eggs" may better adhere to the solid substrate such as slide glass, and were relatively immobilized during egg activation because of the lack of the FE elevation. This feature was advantageous for morphological analysis and other experiments. As the veil of VL was removed, it was relatively easier to disclose the sperm's interaction with the microvilli in SEM. The ultrastructural images obtained with such methods were strikingly similar to the ones that we obtained with the DTT pretreatment [81]. However, it should be underscored that this is not a natural view and may be quite deviant from reality [82]. Instead, the control eggs visualized with SEM in our study were in their natural state, freely suspended in NSW until they were fixed for morphological analyses. As the DTT-pretreated eggs varied from the ones in natural conditions, as shown here in the analyses of their cell physiology, it bears emphasis that the experimental results obtained from those denuded or modified eggs may deviate from what really happens in natural conditions. This small but important methodological difference should be recognized. Secondly, we note that the significant alteration in the structure of the VL and the defective formation of the FE did not lead to polyspermy in the eggs pretreated with DTT. Combined with the previous findings [29,33,39], this observation again suggests that the FE may not be the decisive factor in preventing polyspermy in echinoderm eggs. Our result is also compatible with the idea that, if preferential binding and penetration sites exist on the VL, such receptors are not sensitive to DTT treatment by allowing the incorporation of the first sperm arriving at the egg surface. Finally, with the evidence of selective labelling at the VL and plasma membrane junction, we suggest that BPA-C8-Cy3 could be used as a convenient fluorescent probe in visualizing the VL of the egg.

5. Conclusions

Quiescent it may seem, a sea urchin egg is at the tight equilibrium for maintaining the homeostasis of intracellular Ca^{2+}, pH, and redox state. Here, when the reducing agent DTT was added to the incubation media, the sea urchin egg displayed altered Ca^{2+} responses and F-actin translocation at fertilization. While the eggs fertilized in this condition were mostly monospermic, their development was severely impaired if DTT was still present in the media. When DTT was removed prior to fertilization, the eggs were overly polyspermic. All of these changes took place while the alteration of the VL was rather moderate; that is, the VL was not completely removed. In view of the fact that TCEP, another reducing agent but membrane-impermeant, did not interfere with the translocation of actin filaments following fertilization, the intracellular redox state is thought to be an important parameter that contributes to egg activation. Compared with the well characterized intracellular Ca^{2+} signaling in fertilized eggs, the roles played by intracellular pH and the redox state have been far less explored, but deserve more intense studies in the future.

Supplementary Materials: The following are available online at https://www.mdpi.com/article/10.3390/cells10123573/s1, Figure S1: Enlarged view of DTT-pre-treated sea urchin egg 1 min after insemination, Video S1: Fertilization of sea urchin control eggs, Video S2: Fertilization of DTT pretreated sea urchin eggs, File S1: Synthesis of the fluorescent molecular probe BPA-C8-Cy3.

Author Contributions: Conceptualization, L.S., N.L., J.-M.L. and J.T.C.; methodology, N.L. and L.S.; resources, S.C., J.-L.S., J.-M.L. and L.S.; formal analysis, N.L., J.T.C. and L.S.; data curation, N.L. and L.S.; visualization N.L., J.T.C. and L.S.; writing—original draft preparation, J.T.C., N.L. and L.S.; writing—review and editing, all authors; supervision, L.S.; project administration, L.S.; funding acquisition, L.S. All authors have read and agreed to the published version of the manuscript.

Funding: This work was financially supported by the research fellowship to N.L. granted by the Stazione Zoologica Anton Dohrn for collaboration with L.S.

Institutional Review Board Statement: Sea urchins *P. lividus* used for the present study were collected according to the Italian legislation (DPR 1639/68, 19 September 1980 and confirmed on 1 October 2000). All the experimental procedures were carried out in accordance with the guidelines of the European Union (Directive 609/86).

Informed Consent Statement: Not applicable.

Data Availability Statement: Not applicable.

Acknowledgments: N.L. was supported by the institutional fund for research fellowship granted by the Stazione Zoologica Anton in Naples (SZN). The authors are grateful to Davide Caramiello for maintaining the sea urchins, and to Giovanni Gragnaniello for preparing the samples for SEM and TEM analyses. The authors also thank the staff of the Advanced Microscopy Center at the SZN for the technical assistance in the observations with the electron microscopes.

Conflicts of Interest: The authors declare no conflict of interest.

References

1. Ward, G.E.; Brokaw, C.J.; Garbers, D.L.; Vacquier, V.D. Chemotaxis of Arbacia punctulata spermatozoa to resact, a peptide from the egg jelly layer. *J. Cell Biol.* **1985**, *101*, 2324–2329. [CrossRef]
2. Kaupp, U.B.; Solzin, J.; Hildebrand, E.; Brown, J.E.; Helbig, A.; Hagen, V.; Beyermann, M.; Pampaloni, F.; Weyand, I. The signal flow and motor response controling chemotaxis of sea urchin sperm. *Nat. Cell Biol.* **2003**, *5*, 109–117. [CrossRef]
3. Ramírez-Gómez, H.V.; Jimenez Sabinina, V.; Velázquez Pérez, M.; Beltran, C.; Carneiro, J.; Wood, C.D.; Tuval, I.; Darszon, A.; Guerrero, A. Sperm chemotaxis is driven by the slope of the chemoattractant concentration field. *eLife* **2020**, *9*, e50532. [CrossRef]
4. Aketa, K. On the sperm-egg bonding as the initial step of fertilization in the sea urchin. *Embryologia (Nagoya)* **1967**, *9*, 238–245. [CrossRef]
5. Aketa, K.; Onitake, K.; Tsuzuki, H. Tryptic disruption of sperm-binding site of sea urchin egg surface. *Exp. Cell Res.* **1972**, *71*, 27–32. [CrossRef]
6. Glabe, C.G.; Vacquier, V.D. Isolation and characterization of the vitelline layer of sea urchin eggs. *J. Cell Biol.* **1977**, *75*, 410–421. [CrossRef] [PubMed]
7. Vacquier, V.D.; Epel, D.; Douglas, L.A. Sea urchin eggs release protease activity at fertilization. *Nature* **1972**, *237*, 34–36. [CrossRef] [PubMed]
8. Vacquier, V.D.; Moy, G.W. Isolation of bindin: The protein responsible for adhesion of sperm to sea urchin eggs. *Proc. Natl. Acad. Sci. USA* **1977**, *74*, 2456–2460. [CrossRef]
9. Chandler, D.E.; Heuser, J. The vitelline layer of the sea urchin egg and its modification during fertilization. A freeze-fracture study using quick-freezing and deep-etching. *J. Cell Biol.* **1980**, *84*, 618–632. [CrossRef]
10. Glabe, C.G.; Lennarz, W.J. Species-specific sperm adhesion in sea urchins. A quantitative investigation of bindin-mediated egg agglutination. *J. Cell Biol.* **1979**, *83*, 595–604. [CrossRef]
11. Vacquier, V.D. Activation of sea urchin spermatozoa during fertilization. *Trends Biochem. Sci.* **1986**, *11*, 77–81. [CrossRef]
12. Kamei, N.; Glabe, C.G. The species-specific egg receptor for sea urchin sperm adhesion is EBR1, a novel ADAMTS protein. *Genes Dev.* **2003**, *17*, 2502–2507. [CrossRef]
13. Vacquier, V.D. The quest for the sea urchin egg receptor for sperm. *Biochem. Biophys. Res. Commun.* **2012**, *425*, 583–587. [CrossRef] [PubMed]
14. Wessel, G.M.; Wada, Y.; Yajima, M.; Kiyomoto, M. Bindin is essential for fertilization in the sea urchin. *Proc. Natl. Acad. Sci. USA* **2021**, *118*, e2109636118. [CrossRef]
15. Epel, D.; Weaver, A.M.; Mazia, D. Methods for removal of the vitelline membrane of sea urchin eggs: I. Use of dithiothreitol (Cleland Reagent). *Exp. Cell Res.* **1970**, *61*, 64–68. [CrossRef]
16. Schatten, G. Motility during fertilization. *Int. Rev. Cytol.* **1982**, *79*, 59–163. [CrossRef]

17. Ohlendieck, K.; Lennarz, W.J. Role of the sea urchin egg receptor for sperm in gamete interactions. *Trends Biochem. Sci.* **1995**, *20*, 29–33. [CrossRef]
18. Santella, L.; Vasilev, F.; Chun, J.T. Fertilization in echinoderms. *Biochem. Biophys. Res. Commun.* **2012**, *425*, 588–594. [CrossRef]
19. Gilkey, J.C.; Jaffe, L.F.; Ridgway, E.B.; Reynolds, G.T. A free calcium wave traverses the activating egg of the medaka, Oryzias latipes. *J. Cell Biol.* **1978**, *76*, 448–466. [CrossRef] [PubMed]
20. Santella, L.; Lim, D.; Moccia, F. Calcium and fertilization: The beginning of life. *Trends Biochem. Sci.* **2004**, *29*, 400–408. [CrossRef] [PubMed]
21. Miyazaki, S. Thirty years of calcium signals at fertilization. *Semin. Cell Dev. Biol.* **2006**, *17*, 233–243. [CrossRef]
22. Whitaker, M. Calcium at fertilization and in early development. *Physiol. Rev.* **2006**, *86*, 25–88. [CrossRef]
23. Stein, P.; Savy, V.; Williams, A.M.; Williams, C.J. Modulators of calcium signalling at fertilization. *Open Biol.* **2020**, *10*, 200118. [CrossRef]
24. Parrington, J.; Davis, L.C.; Galione, A.; Wessel, G. Flipping the switch: How a sperm activates the egg at fertilization. *Dev. Dyn.* **2007**, *236*, 2027–2038. [CrossRef] [PubMed]
25. Terasaki, M. Actin filament translocations in sea urchin eggs. *Cell Motil. Cytoskelet.* **1996**, *34*, 48–56. [CrossRef]
26. Tilney, L.G.; Jaffe, L.A. Actin, microvilli, and the fertilization cone of sea urchin eggs. *J. Cell Biol.* **1980**, *87*, 771–782. [CrossRef]
27. Chun, J.T.; Puppo, A.; Vasilev, F.; Gragnaniello, G.; Garante, E.; Santella, L. The biphasic increase of PIP2 in the fertilized eggs of starfish: New roles in actin polymerization and Ca^{2+} signaling. *PLoS ONE* **2010**, *5*, e14100. [CrossRef]
28. Vasilev, F.; Chun, J.T.; Gragnaniello, G.; Garante, E.; Santella, L. Effects of ionomycin on egg activation and early development in starfish. *PLoS ONE* **2012**, *7*, e39231. [CrossRef]
29. Limatola, N.; Vasilev, F.; Santella, L.; Chun, J.T. Nicotine Induces Polyspermy in Sea Urchin Eggs through a Non-Cholinergic Pathway Modulating Actin Dynamics. *Cells* **2020**, *9*, 63. [CrossRef]
30. Santella, L.; Chun, J.T. Actin, more than just a housekeeping protein at the scene of fertilization. *Sci. China Life Sci.* **2011**, *54*, 733–743. [CrossRef]
31. Santella, L.; Limatola, N.; Chun, J.T. Calcium and actin in the saga of awakening oocytes. *Biochem. Biophys. Res. Commun.* **2015**, *460*, 104–113. [CrossRef] [PubMed]
32. Chun, J.T.; Vasilev, F.; Limatola, N.; Santella, L. Fertilization in Starfish and Sea Urchin: Roles of Actin. *Results Probl. Cell Differ.* **2018**, *65*, 33–47. [CrossRef] [PubMed]
33. Santella, L.; Limatola, N.; Chun, J.T. Cellular and molecular aspects of oocyte maturation and fertilization: A perspective from the actin cytoskeleton. *Zool. Lett.* **2020**, *6*, 5. [CrossRef]
34. Vasilev, F.; Ezhova, Y.; Chun, J.T. Signaling Enzymes and Ion Channels Being Modulated by the Actin Cytoskeleton at the Plasma Membrane. *Int. J. Mol. Sci.* **2021**, *22*, 10366. [CrossRef]
35. Foltz, K.R.; Partin, J.S.; Lennarz, W.J. Sea urchin egg receptor for sperm: Sequence similarity of binding domain and hsp70. *Science* **1993**, *259*, 1421–1425. [CrossRef] [PubMed]
36. Ohlendieck, K.; Partin, J.S.; Lennarz, W.J. The biologically active form of the sea urchin egg receptor for sperm is a disulfide-bonded homo-multimer. *J. Cell Biol.* **1994**, *125*, 817–824. [CrossRef]
37. Jaalouk, D.E.; Lammerding, J. Mechanotransduction gone awry. *Nat. Rev. Mol. Cell Biol.* **2009**, *10*, 63–73. [CrossRef]
38. Bonnans, C.; Chou, J.; Werb, Z. Remodelling the extracellular matrix in development and disease. *Nat. Rev. Mol. Cell Biol.* **2014**, *15*, 786–801. [CrossRef]
39. Chun, J.T.; Limatola, N.; Vasilev, F.; Santella, L. Early events of fertilization in sea urchin eggs are sensitive to actin-binding organic molecules. *Biochem. Biophys. Res. Commun.* **2014**, *450*, 1166–1174. [CrossRef]
40. Eddy, E.M.; Shapiro, B.M. Changes in the topography of the sea urchin egg after fertilization. *J. Cell Biol.* **1976**, *71*, 35–48. [CrossRef]
41. Green, J.D.; Summers, R.G. Formation of the cortical concavity at fertilization in the sea urchin egg. *Dev. Growth Differ.* **1980**, *22*, 821–829. [CrossRef]
42. Limatola, N.; Vasilev, F.; Chun, J.T.; Santella, L. Sodium-mediated fast electrical depolarization does not prevent polyspermic fertilization in Paracentrotus lividus eggs. *Zygote* **2019**, *27*, 241–249. [CrossRef]
43. Hamaguchi, Y.; Mabuchi, I. Effects of phalloidin microinjection and localization of fluorescein-labeled phalloidin in living sand dollar eggs. *Cell Motil.* **1982**, *2*, 103–113. [CrossRef]
44. Wessel, G.M.; Wong, J.L. Cell surface changes in the egg at fertilization. *Mol. Reprod. Dev.* **2009**, *76*, 942–953. [CrossRef] [PubMed]
45. Shen, S.S.; Buck, W.R. Sources of calcium in sea urchin eggs during the fertilization response. *Dev. Biol.* **1993**, *157*, 157–169. [CrossRef] [PubMed]
46. Lange, K. Microvillar ion channels: Cytoskeletal modulation of ion fluxes. *J. Theor. Biol.* **2000**, *206*, 561–584. [CrossRef]
47. Vasilev, F.; Limatola, N.; Chun, J.T.; Santella, L. Contributions of suboolemmal acidic vesicles and microvilli to the intracellular Ca^{2+} increase in the sea urchin eggs at fertilization. *Int. J. Biol. Sci.* **2019**, *15*, 757–775. [CrossRef]
48. Gillot, I.; Ciapa, B.; Payan, P.; Sardet, C. The calcium content of cortical granules and the loss of calcium from sea urchin eggs at fertilization. *Dev. Biol.* **1991**, *146*, 396–405. [CrossRef]
49. Dale, B.; De Belice, L.J.; Taglietti, V. Membrane noise and conductance increase during single spermatozoon-egg interactions. *Nature* **1978**, *275*, 217–219. [CrossRef]

50. Cline, D.J.; Redding, S.E.; Brohawn, S.G.; Psathas, J.N.; Schneider, J.P.; Thorpe, C. New water-soluble phosphines as reductants of peptide and protein disulfide bonds: Reactivity and membrane permeability. *Biochemistry* **2004**, *43*, 15195–15203. [CrossRef]
51. Han, J.C.; Han, G.Y. A procedure for quantitative determination of tris(2-carboxyethyl)phosphine, an odorless reducing agent more stable and effective than dithiothreitol. *Anal. Biochem.* **1994**, *220*, 5–10. [CrossRef] [PubMed]
52. Getz, E.B.; Xiao, M.; Chakrabarty, T.; Cooke, R.; Selvin, P.R. A comparison between the sulfhydryl reductants tris(2-carboxyethyl)phosphine and dithiothreitol for use in protein biochemistry. *Anal. Biochem.* **1999**, *273*, 73–80. [CrossRef] [PubMed]
53. Byrd, W.; Belisle, B.W. Microvillar elongation following parthenogenetic activation of sea urchin eggs. *Exp. Cell Res.* **1985**, *159*, 211–223. [CrossRef]
54. Santella, L.; Limatola, N.; Chun, J.T. *Unpublished Data*, Department of Research Infrastructures for Marine Biological Resources, Stazione Zoologica Anton Dohrn: Napoli, Italy, 2021; manuscript in preparation.
55. Nusco, G.A.; Chun, J.T.; Ercolano, E.; Lim, D.; Gragnaniello, G.; Kyozuka, K.; Santella, L. Modulation of calcium signalling by the actin-binding protein cofilin. *Biochem. Biophys. Res. Commun.* **2006**, *348*, 109–114. [CrossRef]
56. Puppo, A.; Chun, J.T.; Gragnaniello, G.; Garante, E.; Santella, L. Alteration of the cortical actin cytoskeleton deregulates Ca^{2+} signaling, monospermic fertilization, and sperm entry. *PLoS ONE* **2008**, *3*, e3588. [CrossRef]
57. Chun, J.T.; Vasilev, F.; Santella, L. Antibody against the actin-binding protein depactin attenuates Ca^{2+} signaling in starfish eggs. *Biochem. Biophys. Res. Commun.* **2013**, *441*, 301–307. [CrossRef]
58. Limatola, N.; Vasilev, F.; Chun, J.T.; Santella, L. Altered actin cytoskeleton in ageing eggs of starfish affects fertilization process. *Exp. Cell Res.* **2019**, *381*, 179–190. [CrossRef]
59. Fisher, G.W.; Rebhun, L.I. Sea urchin egg cortical granule exocytosis is followed by a burst of membrane retrieval via uptake into coated vesicles. *Dev. Biol.* **1983**, *99*, 456–472. [CrossRef]
60. Smith, R.M.; Baibakov, B.; Ikebuchi, Y.; White, B.H.; Lambert, N.A.; Kaczmarek, L.K.; Vogel, S.S. Exocytotic insertion of calcium channels constrains compensatory endocytosis to sites of exocytosis. *J. Cell Biol.* **2000**, *148*, 755–767. [CrossRef]
61. Terasaki, M. Visualization of exocytosis during sea urchin egg fertilization using confocal microscopy. *J. Cell Sci.* **1995**, *108*, 2293–2300. [CrossRef]
62. Emans, N.; Zimmermann, S.; Fischer, R. Uptake of a fluorescent marker in plant cells is sensitive to brefeldin A and wortmannin. *Plant Cell* **2002**, *14*, 71–86. [CrossRef]
63. Nedeva, I.; Koripelly, G.; Caballero, D.; Chièze, L.; Guichard, B.; Romain, B.; Pencreach, E.; Lehn, J.M.; Carlier, M.F.; Riveline, D. Synthetic polyamines promote rapid lamellipodial growth by regulating actin dynamics. *Nat. Commun.* **2013**, *4*, 2165. [CrossRef] [PubMed]
64. Oriol-Audit, C. Polyamine-induced actin polymerization. *Eur. J. Biochem.* **1978**, *87*, 371–376. [CrossRef]
65. Wallace, H.M. Polyamines: Specific metabolic regulators or multifunctional polycations? *Biochem. Soc. Trans.* **1998**, *26*, 569–571. [CrossRef]
66. Hamaguchi, M.S.; Hiramoto, Y. Fertilization process in the heart-urchin, clypeaster japonicus observed with a differential interference microscope. *Dev. Growth Differ.* **1980**, *22*, 517–530. [CrossRef]
67. Holy, J.; Schatten, G. Spindle pole centrosomes of sea urchin embryos are partially composed of material recruited from maternal stores. *Dev. Biol.* **1991**, *147*, 343–353. [CrossRef]
68. Zarski, D.; Krejszeff, S.; Palińska, K.; Targońska, K.; Kupren, K.; Fontaine, P.; Kestemont, P.; Kucharczyk, D. Cortical reaction as an egg quality indicator in artificial reproduction of pikeperch, Sander lucioperca. *Reprod. Fertil. Dev.* **2012**, *24*, 843–850. [CrossRef] [PubMed]
69. Stack, C.; Lucero, A.J.; Shuster, C.B. Calcium-responsive contractility during fertilization in sea urchin eggs. *Dev. Dyn.* **2006**, *235*, 1042–1052. [CrossRef]
70. Santella, L. Polyspermy-preventing mechanisms in sea urchin eggs: New developments for an old problem. *Biochem. Biophys. Res. Commun.* **2019**, *520*, 695–698. [CrossRef] [PubMed]
71. Cline, C.A.; Schatten, G. Microfilaments during sea urchin fertilization: Fluorescence detection with rhodaminyl phalloidin. *Gamete Res.* **1986**, *14*, 277–291. [CrossRef]
72. Limatola, N.; Chun, J.T.; Santella, L. Effects of salinity and pH of seawater on the reproduction of the sea urchin *Paracentrotus lividus*. *Biol. Bull.* **2020**, *239*, 13–23. [CrossRef]
73. Limatola, N.; Bertocci, I.; Chun, J.T.; Musco, L.; Munari, M.; Caramiello, D.; Danovaro, R.; Santella, L. Oxygen supersaturation mitigates the impact of the regime of contaminated sediment reworking on sea urchin fertilization process. *Mar. Environ. Res.* **2020**, *158*, 104951. [CrossRef]
74. Limatola, N.; Chun, J.T.; Santella, L. Fertilization and development of Arbacia lixula eggs are affected by osmolality conditions. *Biosystems* **2021**, *206*, 104448. [CrossRef] [PubMed]
75. Hardy, M.L.M.; Day, M.L.; Morris, M.B. Redox Regulation and Oxidative Stress in Mammalian Oocytes and Embryos Developed In Vivo and In Vitro. *Int. J. Environ. Res. Public Health* **2021**, *18*, 11374. [CrossRef]
76. Huber, P.J.; Brunner, U.T.; Schaub, M.C. Disulfide formation within the regulatory light chain of skeletal muscle myosin. *Biochemistry* **1989**, *28*, 9116–9123. [CrossRef] [PubMed]
77. Ishiwata, S. Freezing of actin. Reversible oxidation of a sulfhydryl group and structural change. *J. Biochem.* **1976**, *80*, 595–609. [CrossRef]

78. Tang, J.X.; Janmey, P.A.; Stossel, T.P.; Ito, T. Thiol oxidation of actin produces dimers that enhance the elasticity of the F-actin network. *Biophys. J.* **1999**, *76*, 2208–2215. [CrossRef]
79. Grintsevich, E.E.; Phillips, M.; Pavlov, D.; Phan, M.; Reisler, E.; Muhlrad, A. Antiparallel dimer and actin assembly. *Biochemistry* **2010**, *49*, 3919–3927. [CrossRef]
80. Schatten, G.; Mazia, D. The surface events of fertilization: The movements of the spermatozoon through the sea urchin egg surface and the roles of the surface layers. *J. Supramol. Struct.* **1976**, *5*, 343–369. [CrossRef]
81. Schatten, G.; Mazia, D. The penetration of the spermatozoon through the sea urchin egg surface at fertilization. Observations from the outside on whole eggs and from the inside on isolated surfaces. *Exp. Cell Res.* **1976**, *98*, 325–337. [CrossRef]
82. Just, E.E. *The Biology of the Cell Surface*; P. Blakiston's Son & Co.: Philadelphia, PA, USA, 1939.

Review

Regulation of Cell Cycle Progression by Growth Factor-Induced Cell Signaling

Zhixiang Wang 

Department of Medical Genetics, Faculty of Medicine and Dentistry, University of Alberta, Edmonton, AB T6G2H7, Canada; zhixiang.wang@ualberta.ca

Abstract: The cell cycle is the series of events that take place in a cell, which drives it to divide and produce two new daughter cells. The typical cell cycle in eukaryotes is composed of the following phases: G1, S, G2, and M phase. Cell cycle progression is mediated by cyclin-dependent kinases (Cdks) and their regulatory cyclin subunits. However, the driving force of cell cycle progression is growth factor-initiated signaling pathways that control the activity of various Cdk–cyclin complexes. While the mechanism underlying the role of growth factor signaling in G1 phase of cell cycle progression has been largely revealed due to early extensive research, little is known regarding the function and mechanism of growth factor signaling in regulating other phases of the cell cycle, including S, G2, and M phase. In this review, we briefly discuss the process of cell cycle progression through various phases, and we focus on the role of signaling pathways activated by growth factors and their receptor (mostly receptor tyrosine kinases) in regulating cell cycle progression through various phases.

Keywords: cell cycle; growth factors; receptor tyrosine kinases; G1 phase; S phase; G2 phase; M phase; Ras/Erk; PI3K/Akt

Citation: Wang, Z. Regulation of Cell Cycle Progression by Growth Factor-Induced Cell Signaling. *Cells* **2021**, *10*, 3327. https://doi.org/10.3390/cells10123327

Academic Editors: Stephen Yarwood and Ludger Hengst

Received: 9 October 2021
Accepted: 24 November 2021
Published: 26 November 2021

Publisher's Note: MDPI stays neutral with regard to jurisdictional claims in published maps and institutional affiliations.

1. Introduction

The cell cycle, or the cell division cycle, is the series of events that take place in a cell that drive it to divide and produce two new daughter cells. The typical cell cycle in eukaryotes is composed of four phases including the G1, S, G2, and M phase. G1, S, and G2 together are called interphase. M phase is comprised of mitosis, in which the cell's nucleus divides, and cytokinesis, in which the cell's cytoplasm divides to form two daughter cells. Mitosis and cytokinesis are tightly coupled together. Mitosis is further divided into five subphases including prophase, prometaphase, metaphase, anaphase, and telophase (Figure 1). Each phase of cell cycle progression is reliant on the proper completion of the previous cell cycle phase. A cell could also exit from cell cycle to enter G0 phase, a state of quiescence [1,2].

Cell cycle progression is mediated by cyclin-dependent kinases (Cdks) and their regulatory cyclin subunits. Cdks, such as Cdk4/6, Cdk2, and Cdk1 (also known as Cdc2) are serine/threonine kinases with a wide variety of substrates. Cdks are activated mainly by binding to their cyclin partners, whose expressions rise and fall throughout the cell cycle to mediate the temporal activation of each Cdks. Various cell cycle checkpoints exist to ensure that critical processes are engaged prior to progression to the next phase. There are three major cell cycle checkpoints, including the G1/S checkpoint (also referred as restriction point), the G2/M DNA damage checkpoint, and the spindle assembly checkpoint (SAC) [3 5].

Growth factors (GFs) are a group of proteins stimulating the growth of specific tissues. GF plays important roles in the regulation of cell division that drives cell proliferation. Each GF binds to a specific cell–surface receptor. A specific group of GF receptors possesses tyrosine kinases activity and is termed as receptor tyrosine kinases (RTKs). RTK plays most important roles in cell cycle regulation. RTKs are classified into 20 families. The most studied RTKs in terms of cell cycle include epidermal growth factor (EGF) receptor (EGFR) family, insulin receptor family, platelet-derived growth factor (PDGF) receptor (PDGFR) family, and nerve growth factor (NGF) receptor (NGFR). GFs drive cell cycle by activating RTKs and downstream signaling pathways, which regulates cyclin-Cdk complexes [6,7].

In this review, we will briefly discuss the process of cell cycle progression through various phases and will focus on the role of GF/RTK-activated signaling cascades in regulating cell cycle.

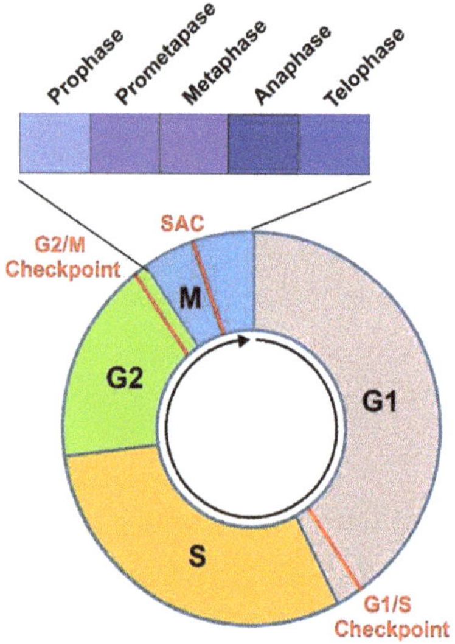

Figure 1. Diagram to illustrate a complete cell cycle progression through four cell cycle phases (G1, S, G2, and M) and three major checkpoints (G1/S, G2/M, and SAC). M phase is further divided into Prophase, Prometaphase, Metaphase, Anaphase, and Telophase.

2. Early History of Cell Cycle Discovery

Cell theory was developed in the middle of the 19th century. This theory has three main components: (1) Every living organism is composed of one or more cells; (2) cells are the basic unit of life for all living organisms, and (3) cells only arise from pre-existing cells. While the first two components were the contribution of Theodor Schwann and Matthias Jakob Schleiden, the last is the contribution of German scientist and physician Rudolf Virchow. His discovery that all cells arise from pre-existing cells is the start point of cell cycle research [8–10] (Figure 2).

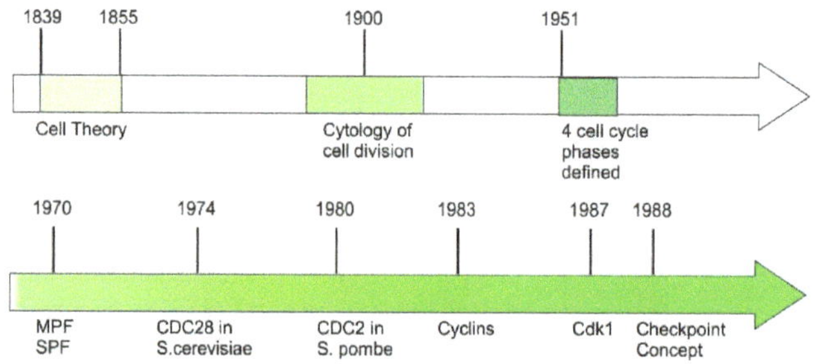

Figure 2. Timeline of major discoveries in the early cell cycle research.

At the turn of the 19th century to 20th century, the cell cycle has been the subject of intense study. The cytology of cell division is described in great detail by microscopists and embryologists, however, the underlying mechanisms driving cell division are mostly unknown. In the late 1970s and 1980s, the advancement of modern molecular biology provided means and knowledge to study the molecular mechanisms regulating cell cycle. Cell biologists, biochemists, and geneticists joined forces and demonstrated that the basic processes and control mechanisms of cell cycle are universal in eukaryotes.

In the late 19th century, early light microscopic studies recognized that cell division follows mitosis, during which cells condensed their chromosomes. Based on his observations of cell division in various stages, German biologist and a founder of cytogenetics Walther Flemming identified the sequence of chromosome movements in mitosis. Flemming's discovery was proven correct decades later by the study of live dividing cells [11]. However, the only observable morphological changes outside of mitosis is the growth of the cell size. Interphase remained a black box and recognized as one phase until the discovery that DNA synthesis occurs only in a short period during interphase [12]. This discovery split interphase into three phases: This DNA synthesis period is termed as S phase, the gap between mitosis and S phase is termed G1 phase, and the gap between S and M phases is termed as G2 phase [13].

Following the recognition of four major cell cycle states G1, S, G2, and M, the focus of cell cycle study shifted to understand the transition between these phases. A major task is to identify the factors driving the transition. In the early 1970s, by fusing cells at different stages of the cell cycle, it was shown that late G2 or M phase cells contained an M phase-promoting factor (MPF) capable of accelerating the onset of mitosis in early G2 cells [14]. It was further shown that S phase cells contains an S phase-promoting factor (SPF) in nuclei, which is able to accelerate S phase [14,15].

While there is no biochemical method available to purify either MPF or SPF at the time, genetic studies of cell cycle related genes are fruitful. At the end of the 1960s, Leland Hartwell realized the possibility of using genetic methods to study cell cycles. He established budding yeast *Saccharomyces cerevisiae* as a highly suitable model system to study cell cycles. In an elegant series of experiments in 1970–1971, he used the temperature sensitive lethal mutants of *S. cerevisiae* to isolate yeast cells with mutated genes, controlling the cell cycle. By this approach, he successfully identified more than one hundred genes which specifically involved in cell cycle control. Among these genes are genes encoding SPF and MPF. Hartwell named these genes Cdc-genes (cell division cycle genes) [16–18]. One particularly important gene identified is Cdc28, which controls the first step of cell cycle progression in G1 phase and was also known as "start".

In the middle of the 1970s, Paul Nurse followed Hartwell's approach to study cell cycle regulation with similar genetic methods but using fission yeast *Schizosaccharomyces pombe* as a model system. Through this research, Paul Nurse discovered the gene Cdc2 in fission yeast. Cdc2 is identical to Cdc28 identified in budding yeast. Nurse found that Cdc2 had a key function in the control of transition from G2 to mitosis during cell cycle [19]. In 1987, Nurse isolated the human version of Cdc2 gene, Cdk1. Cdk1 encodes a protein called cyclin-dependent kinase (Cdk). They found that phosphorylation status of the mammalian Cdc2 protein (p34Cdc2) is closely related to cell cycle progression. It is phosphorylated when cells are stimulated to enter the cell cycle in G1 phase, but dephosphorylated when cells go to quiescence [20,21]. Based on these findings, half a dozen different Cdk molecules have been found in humans.

In the early 1980s, Tim Hunt discovered the first cyclin molecule by studying sea urchins, Arbacia. There are eight very rapid cell divisions during the cleavage in embryos of the sea urchin. To sustain these cell divisions, the continual protein synthesis is required. Hunt found that one protein is always destroyed each time the cells divide. This protein was named cyclin as the level of the protein vary periodically during the cell cycle [22]. In the following years, more cyclins were identified in various species by Hunt and other groups. Moreover, it was discovered that the cyclins bind to the Cdk molecules to regulate the Cdk activity and determine the substrate specificity of Cdks [23].

Another important concept introduced during this period is "Checkpoint". In the late 1980s, by studying the sensitivity of yeast cells to irradiation, Hartwell developed concept of checkpoint [24,25] (Figure 2). He observed that the cell cycle is arrested at certain point when DNA is damaged. This cell cycle checkpoint concept is then expanded as surveillance mechanisms used by the cells to check the integrity, fidelity, and the sequences of the major cell cycle events. The events being monitored include cell size growth, DNA replication, and integrity, and the accurate chromosome segregation [3].

The historical contribution of Leland H. Hartwell, Paul M. Nurse, and R. Timothy (Tim) Hunt earned them 2001 Nobel Prize in Physiology or Medicine for their discovery of "key regulators of the cell cycle".

3. Cell Cycle Progression through Various Phases

The cell cycle consists of G1, S, G2, and M phases. In G1 phase, the cell grows and becomes larger. The cell enters S phase when it reaches a certain size. S phase is the period for DNA-synthesis, during which the cell duplicates its DNA. In the following G2 phase, the cell monitors the completion of DNA-replication and prepares for mitosis. Chromosome segregation and cell division are completed in M phase. The proper cell cycle progression ensures that each of the two daughter cells receives identical chromosome from parent cell. After cell division, the cell cycle is completed, and the cells are back in G1 phase. The duration of the cell cycle varies between 10 and 30 h in most mammalian cells. Cells in the G1 can exit from the cell cycle and enter G0 phase, a state of quiescence.

Cell cycle progression is mainly driven and regulated by two classes of proteins, Cdks and cyclins [26] (Figure 3). In yeast, while several Cdks are expressed, including Cdk1, PHO85, and Kin28, only Cdk1 directly regulates cell cycle progression. Cdk1 is equivalent to p34Cdc2 in *S. pombe* and p34Cdc28 in *S cerevisiae*. By associating with different cell-cycle stage-specific cyclins Cdk1 regulates diverse cell cycle transitions including G1 to S transition and G2 to M transition. The roles of PHO85 and Kin28 in cell-cycle regulation are indirect [26,27]. Higher organisms possess many yeast Cdk1 functional homologues. These functional homologues are phase-specific Cdks. Each phase-specific Cdk acts in a specific cell phase to perform the function of Cdk1 in yeast. Approximately 20 Cdk-related proteins are discovered, which leads to the concept that cell cycle events in higher eukaryotic cells are regulated by complex combinations of Cdks and cyclins in various cell cycle phases. For Cdk/cyclin complexes, cyclins confer substrate specificity and determine the regulatory consequence of the substrates such as activation, inactivation, and localization. Based on this hypothesis, the classical model of cell cycle regulation is established through extensive research in eukaryotic cells.

According to this model, Cdk4 and/or Cdk6 form complexes with D-type cyclins, which activates Cdk4/6 and initiates phosphorylation of the retinoblastoma protein (Rb) family in early G1 phase [28,29]. Rb phosphorylation stimulates the release of transcription factor E2F, which then stimulates the transcription of early E2F responsive genes required for the progression of the cell cycle [30,31]. Early E2F responsive genes include A- and E-type cyclins [28,32]. In the late G1 phase, cyclin E binds to and activate Cdk2, which leads to the full Rb phosphorylation and the further activation of E2F mediated transcription [28,29]. Together, the above events drive the passage of the cell through the restriction point at the boundary of the G1/S phase and initiate the S phase. At the onset of the S phase, A-type cyclins are synthesized and form complex with Cdk2, which phosphorylates proteins involved in DNA replication and drive the cell progression to G2 phase [33,34]. At the late G2 phase, Cdk1/cyclin A is formed and activated, which is required for the G2/M transition and the initiation of prophase [35]. Finally, Cdk1/cyclin B complexes are formed in M phase and drive the completion of mitosis [36] (Figure 3).

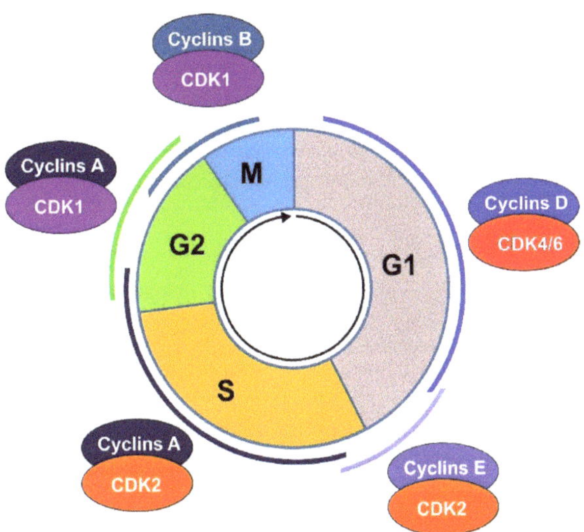

Figure 3. Regulation of cell cycle progression by Cdks and Cyclins.

3.1. G1 Phase

Cells enter G1 either from the preceding M phase or from G0 phase. The transition of cells between G0 and G1 phase is determined by extracellular mitogenic signals [37,38]. G1 phase is the growth phase. The biosynthetic activities of the cell are slowed down considerably in M phase; however, it resumes at a high rate in G1 phase. In G1 phase, the cells synthesize many proteins, amplify organelles including ribosomes and mitochondria, and grow in size. The duration of cell cycle phases varies considerably in different types of cells. For a typical proliferating human cell, if we assume the total cycle time is 24 h, the duration of G1 phase is approximately 11 h, S phase duration last 8 h, G2 phase last 4 h, and the duration of M phase is approximately 1 h.

During G1 phase, diverse signals, including environmental cues, stress, and metabolic cues intervene to influence cell's developmental program. These signals are integrated and interpreted by the cells. Based on these inputs, the cell decides whether to self-renew, differentiate, or die; however, to enter S phase for starting its renewal, all cells must fulfill one essential requirement: activation of Cdks [38,39].

3.2. S Phase

S phase is marked by DNA synthesis. In S phase, each chromosome consists of two sister chromatids following replication to double the amount of DNA. However, S phase also marked with low activities of gene expression and protein synthesis. A noticeable exception is the production of histone. Most histones are produced in the S phase [40].

It is suggested that an intra-S phase checkpoint exists to control S phase progression. Intra-S phase checkpoint turns off Cdk2 in response to DNA damage and other replication stress, which blocks origin firing to avoid replication of damaged DNA [41]. S phase to G2 phase transition is regulated by the active checkpoint kinase ATR (ataxia-telangiectasia and Rad3-related) [42].

3.3. G2 Phase

The cell enters G2 phase after successful completion of S phase. G2 phase ends with the onset of mitosis. The major task of cells in G2 phase is to prepare itself for mitosis. G2 phase is marked by significant protein/lipid synthesis and cell growth [43]. While it is known that protein synthesis inhibitor arrests cells at G2 phase, a recent study suggests that this may be due to the inhibition of p38, and the protein synthesis is not absolutely required for mitosis entry [44]. Interestingly, some cell types, including certain cancer cells and *Xenopus* embryos, lack the G2 phase. Cell cycle proceeds directly from S phase to M phase. It is hypothesized that cell size controls the growth in G2 phase, however, this is only demonstrated in fission yeast [45]. Another process that occurs during G2 phase is to repair DNA double-strand breaks. During and after DNA replication, DNA double-strand breaks accumulate in the cell and need to be repaired before cell can move to pass G2/M checkpoint [24,46,47].

3.4. Mitosis and Cytokinesis

M phase is comprised of mitosis, in which the cell's nucleus divides, and cytokinesis, in which the cell's cytoplasm divides to form two daughter cells. Mitosis is further divided into prophase, prometaphase, metaphase, anaphase, and telophase (Figure 1).

Prophase is characterized with chromatin/chromosome condensation, centrosome separation, and nuclear membrane breakdown. The migration of centrosome to two opposite poles is important for the later formation of the bipolar mitotic spindle apparatus. A recent detailed study shows that the interphase organization is rapidly lost in prophase by a condensin-dependent manner [48]. Observations with a microscope indicate that chromosomes become recognizable as linearly organized structures in early prophase [49]. Sister chromatids are mixed in early prophase, but they are separated in late prophase. Each chromatid is shown as an array of loops radiating from an axial core that contains topoisomerase II alpha and condensin complexes [50]. The rise of cyclin B-Cdk1 activity is a defining molecular event of prophase [51].

Prometaphase starts from the nuclear envelope breakdown, which marks the end of prophase, ends when chromosome alignment at the spindle equator completes, which defines the beginning of metaphase. For faithful chromosome segregation, it is essential to establish a metaphase plate in which all chromosomes aligned at the cell equator attach to mitotic spindle microtubules. The achievement of this configuration depends on the precise coordination of several mitotic events including nuclear envelope breakdown, connection between chromosome kinetochores, and microtubules of the mitotic spindle assembly, and the congression of all chromosomes to the spindle equator. A kinetochore is a disc-shaped protein structure in duplicated chromatids [52]. During prometaphase the chromatids shorten and become thicker [49] and ultimately form fully condensed metaphase chromosomes [53].

Metaphase starts when the duplicated chromosomes are aligned along the metaphase plate in the middle of the cell. During metaphase, the sister chromatids are pulled back and forth by the kinetochore microtubules until they align along the equatorial plane. The chromosome segregation process is monitored by SAC pathway to ensures that all kinetochores are attached to microtubules of the opposite poles before segregation proceeds after metaphase-to-anaphase transition. Once all the chromosomes are properly aligned and the kinetochores are correctly attached, the cohesion between sister chromatids is dissolved, leading to the migration of the separated chromatids towards opposite sides of the cell by the pulling force of spindle microtubules. The cell now enters the anaphase [54].

Anaphase involves two mechanistically distinct steps, the shortening of kinetochore microtubules and the spindle elongation in the midzone. The shortening of kinetochore microtubules causes the migration of each chromatid towards its respective pole. The disjointed sister chromatids are further separated through spindle elongation in the midzone. These two steps may be temporally divided in some organisms while occurring simultaneously in other organisms. These two steps are called anaphase A and anaphase B, respectively [55]. In human mitotic cells, anaphase B usually starts 30–50 s later than the start of the anaphase A [46,56]. During anaphase, the spindle elongates 8 μm and additional 3 μm in telophase [46,55].

Telophase follows anaphase and starts at the onset of the chromosome recondensation and the nuclear envelope reformation [47]. During telophase the duplicated chromosomes in the nucleus of a parent cell separate into two identical daughter cells. A nuclear membrane forms around each set of chromosomes to divide the nuclear DNA from the cytoplasm. Simultaneously, the chromosome decondensation begins [55].

Cytokinesis results the physical separation of the cytoplasm of a mother cell into two daughter cells [57,58]. The segregation of chromosomes and cytoplasm needs to be tightly coordinated to generate offspring with the right complement of chromosomes [59]. Cell cytokinesis is initiated in anaphase, when lower Cdk1 activity causes the reorganization of the mitotic spindle and the stabilization of microtubules. The assembly of the central spindle is the key early event, which provides the template for the midbody and contributes to division plane specification. The division plane is positioned between the two sets of segregated chromosomes. The precise position of the plane is critical to prevent segregation errors. Cytokinetic furrow ingression of the attached plasma membrane is then initiated by the contraction of the actomyosin ring, which partitions the cytoplasm into two domains of emerging daughter cells. The last step of cytokinesis is abscission [60]. Abscission is the physical separation of the plasma membrane of the two daughter cells. During abscission, cells remove the cytoskeletal structures from the intercellular bridge, followed by constriction of the cell cortex, and finally the division of the plasma membrane [61,62].

4. The Regulation of Cell Cycle by GF-Initiated Signaling Pathways

GFs play vital role in driving cell proliferation by activating RTK and the downstream signaling cascades. The aberrant activity of these signaling cascades frequently leads to cancer [63,64]. Cell proliferation can only be realized through cell division and early studies have demonstrated that GF stimulation is the driving force of cell cycle initiation and progression. However, after extensive research in the late 1990s and early 2000s, little research directly studies the role of GF/RTK in the regulation of cell cycle. While the role of GF receptor in G1 phase has been well studied, very little is known regarding their role in other phases of cell cycle.

GFs regulate diverse functions of the cells, and the effects of most GFs are mediated by RTKs (Figure 4). Signal transduction starts when a GF binds to its receptor at the cell surface, which stimulates the dimerization of RTK and the activation of its kinase. Two monomers of the RTK dimer phosphorylate each other to fully activate RTK and the phosphorylated Tyr residues in the C-terminus of the receptor become the binding sites for recruiting multiple downstream signaling molecules. The formation of the RTK-signaling protein complex initiates the activation of multiple signaling pathways including Ras/Erk, PI3K/Akt, Src/Jak/Stat, and PLC-γ1. These signaling pathways interact each other to form a signaling network. Signals from various signaling pathways are eventually integrated at the level of transcription. There are several major transcriptional nodes: (1) The p53 node mediates cytostasis and apoptosis in response to high level of mitogenic signals or DNA damage; (2) The SMAD node mediate the expression of cytostatic (arrest of cell growth and multiplication) and apoptotic factors in response to TGF-β; (3) The FOXO node also mediates cytostasis and apoptosis, but in response to oxidative stress and starvation; and (4) ID and Myc node that suppress Cdk inhibitors to favor cell proliferation. By controlling the transcription of specific genes, GF-initiated cell signaling regulates diverse cellular functions including cell migration, cell survival, cell cycle progression, and differentiation (Figure 4) [7,64].

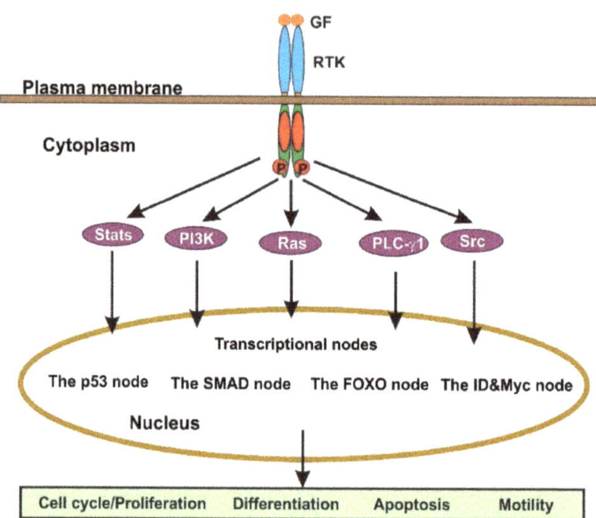

Figure 4. Diagram to illustrate signaling pathways initiated by GF through their membrane receptors, mostly RTKs.

4.1. EGFR-Mediated Signaling Pathways

As a prototypical receptor of all RTKs, EGFR signaling network has been extensively studied and very well understood [7,64]. Here, we are using EGFR as an example to illustrate the signaling pathways of GF/RTK. Following its identification in the 1970s [65], EGFR is shown to possess intrinsic kinase activity. The full-length receptor is cloned in 1984 [66]. EGFR is a single polypeptide chain transmembrane glycoprotein. The following is a brief description of the two major signaling pathways activated by EGFR that is most relevant to cell cycle progression.

4.1.1. Ras/Erk Pathway

A major signaling pathway downstream of EGFR is the Ras/Erk pathway. The phosphorylated EGFR interacts with SHC/Grb2, which recruits Sos to cell membrane to activate Ras. Activated Ras then stimulates the activation of Raf. Mitogen-activated protein kinase (MEK) is phosphorylated and activated by Raf, which activates Erk. Erk activation stimulates the activation and nuclear translocation of RSK. In the nucleus, RSK stimulates the activation of transcription factors including c-Fos and SRF. On the other hand, following its activation, Erk also translocates into the nucleus where it stimulates c-Fos and Elk1 (Figure 5A) [67–73].

4.1.2. PI3K/Akt Pathway

EGFR stimulates PI3K either by binding to its p85 subunit directly or indirectly by activating Ras [74,75] (Figure 5B). The function of PI3K is to generate phosphatidylinositol-3,4,5-trisphosphate (PIP3) by phosphorylating phosphatidylinositol-4,5-bisphosphate (PIP2). However, this process is reversed by the phosphatase and tensin homologue deleted on chromosome 10 (PTEN). PTEN acts as a direct antagonist of PI3K and provides important negative control over the PI3K pathway. It is also reported that PI3K directly interacts with PTEN to stimulate PTEN activity [76,77]. PIP3 interacts with pleckstrin homology (PH) domains to recruit PH-domain containing proteins to the plasma membrane. Among the recruited PH-domain containing protein is the serine threonine kinase Akt. In the plasma membrane, phosphoinositide-dependent kinase 1 (PDK-1) phosphorylates Akt Thr 308 to partially activate Akt. The following phosphorylation of Akt in Ser 473 by the rapamycin complex 2 (mTORC2) results the full activation of Akt. Activated Akt activates multiple downstream substrates, including the forkhead box O transcription factors (FoxO), the BCL2-associated agonist of cell death (BAD), and glycogen synthase kinase 3 (GSK3), to promote cell cycle entry and cell survival. Akt also activates the small G-protein ras homologue enriched in the brain (Rheb), leading to the activation of mTORC1. mTORC1 phosphorylates the eukaryotic translation initiation factor 4E binding protein 1 (4EBP1) and the p70S6 kinase (S6K1), which promotes protein translation and protects the cell from apoptosis [78–80].

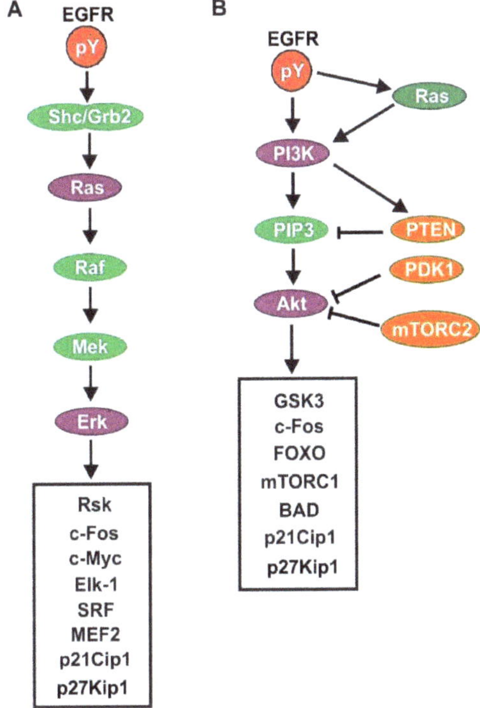

Figure 5. Activation of Ras/Erk (**A**) and PI3K/Akt (**B**) signaling pathways downstream of EGFR.

4.2. GF-Activated Ras/Erk Pathway Activation in Cell Cycle Progression

An earlier study indicates that Erk is rapidly phosphorylated in response to GFs including EGF and PDGF [81]. The first evidence regarding the role of Erk in cell proliferation comes from the Erk-inhibition experiments. Inhibition of Erk by various means block the GF-induced cell proliferation, which indicates that GF activation of p42mapk and p44mapk is an absolute requirement for triggering the proliferative response [82]. However, Erk1/2 activation is not sufficient to drive cells into S phase [83–85].

Accumulated evidence indicates that Erk activation promotes G1 progression with multiple mechanisms [86] (Figure 6). One mechanism is by induction of cyclin D and assembly of cyclin D–Cdk4 complexes. Expression of activated Ras in different cell types is sufficient to induce the accumulation of cyclin D1 [87–89]. As a downstream signaling protein of Ras, Erk activation is both necessary and sufficient for transcriptional induction of the Cyclin D1 gene [90]. In order for cells to enter S phase, sustained Erk activation is essential to maintain the level of cyclin D1 in G1 phase [91]. The precise mechanism that connects Erk signaling to Cyclin D1 transcription is not clear. However, it is known that the Cyclin D1 promoter contains a functional AP-1 binding site. It is also shown that Erk signaling stimulates the expression of c-Fos and c-Jun, the critical AP-1 components in response to GFs [91,92]. Sustained Erk activation also stabilizes AP-1 proteins by stimulating the phosphorylation of their C-terminal residues [86]. The Erk pathway also regulates the cellular cyclin D1 expression level through the post-transcriptional regulation [86,93]. There is also evidence to support the role of Erk in the induction of cyclin D2 and cyclin D3 [94,95].

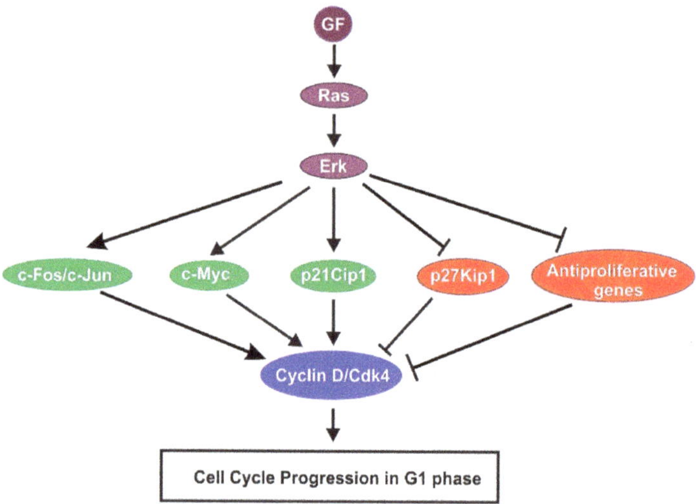

Figure 6. Regulation of cell cycle progression in G1 phase by GF-induced activation of Ras/Erk signaling pathway.

The second mechanism is to stabilize transcription factor c-Myc. C-Myc regulates cell cycle progression and apoptosis [96]. Activation of Erk1 by GF strongly enhances the stability of c-Myc protein by direct phosphorylating c-Myc at Ser 62 [97]. C-Myc regulates many proteins that function directly in cell cycle control. These proteins include Cdk4 [98], cyclin D2 [99], Cdc25A [100], and p21 [99]. It is shown that both Erk activation and c-Myc expression are required to drive cells from G0 to late G1 phase [85].

Another mechanism is through the regulation of p21Cip1 and p27Kip1 expression. While p21Cip1 was initially thought to be a cyclin-dependent kinase inhibitor (CKI), further study indicates that p21Cip1 plays multiple roles in regulating Cdk activity. p21Cip1 interact with both Cdk4–cyclin D and Cdk6–cyclin D complexes under physiological conditions, which induces their kinase activation from early G1 to middle S phase [28]. In response to GFs p21Cip1 is transiently accumulated in early G1 phase by an Erk-dependent and p53-independent mechanism [101,102]. A transient activation of Erk is sufficient to induce the expression of p21Cip1 [103], which contributes to the stabilization of cyclin D/Cdk4 complexes in G1 [104]. On the other hand, p21Cip1 also inhibits the activation of Cdk2–cyclin A and Cdk2–cyclin E complexes from late G1 to S phase [105]. Thus, p21Cip1 regulates cell cycle progression by controlling the activation of various Cdks.

Ras/Erk signaling pathway also mediates the GF-induced downregulation of p27Kip1 [84,106–108]. It is possible that Erk1/2 downregulates p27Kip1 by stimulating the activation of cyclin D/Cdk4/6 complexes, which induces the degradation of p27Kip1 at the G1/S transition. It is also suggested that Erk signaling downregulates p27Kip1 expression by a Skp2- and Cdk2-independent mechanism [109]. Finally, Erk signaling also contributes indirectly to p27Kip1 regulation by stimulating the synthesis of autocrine GFs [86].

The fourth mechanism is to downregulate antiproliferative genes. As revealed by a gene profiling analysis, 173 antiproliferative genes are downregulated by an Erk-dependent mechanism during G1 phase [110]. It is further shown that to maintain decreased expression levels of antiproliferative genes, continuous activation of Erk throughout G1 is required.

While the activation of the Ras–Erk pathway stimulates cell cycle progression in G1 phase in general, it is interesting to note that some studies suggest that too strong Erk activation results reversible or permanent cell cycle arrest [86] by stimulating the expression of the Cdk inhibitor p21Cip1 [111–114].

4.3. GF-Activated PI3K/Akt Pathway Activation in Cell Cycle Progression

GFs also regulate cell cycle progression through the activation of PI3K/Akt pathway. It was shown in the early 1990s that PI3K mediated the mitogenic signals of PDGFR [115,116]. In the following years, the role of PI3K in mediating PDGF-induced cell cycle progression were revealed [117,118]. The activation of Akt by PI3K or inhibition by PTEN has been shown to play critical role in GF-induced cell cycle progression [119–121]. These studies firmly established roles of PI3K in GF induced cell cycle progression. Extensive research during this period also revealed various mechanisms by which PI3K/Akt pathway regulate in GF-induced cell cycle progression (Figure 7).

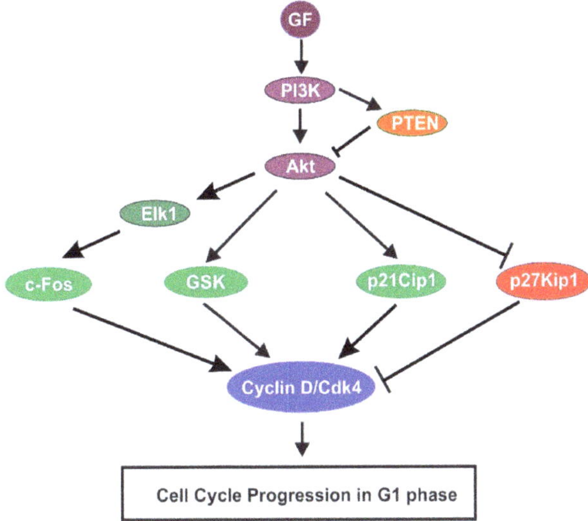

Figure 7. Regulation of cell cycle progression in G1 phase by GF-induced activation of PI3K/Akt signaling pathway.

Stimulation of p21Cip1 is an important mechanism underlying PI3K/Akt regulation of cell cycle progression. The PI3K/Akt signaling is required for the accumulation of p21Cip1 in human ovarian carcinoma cells. While expression of a constitutively active mutant Akt increases the expression of p21Cip1, expression of dominant negative Akt decreases p21Cip1 expression [122]. Further research about the mechanisms underlying the PI3K/Akt pathway in regulating p21Cip1 expression indicates that Akt directly phosphorylates T145 and S146 near the c-terminus of p21Cip1. These phosphorylations stimulate DNA synthesis and Cdk activity, increasing cellular proliferation [123]. Akt also phosphorylates p21Cip1 at S146, which enhances p21Cip1 protein stability and significantly prolongs the half-life of p21Cip1. High level of p21Cip1 induces assembly and activation of p21Cip1/Cdk4/cyclin D and p21Cip1/Cdk6/cyclin D complexes [124].

The inhibition of p27Kip1 is another mechanism. The PI3K/Akt pathway decreases the expression of p27KIP1. Expression of dominant negative Akt caused transcriptional induction of p27Kip1 in mesenchymal cells, inhibiting both Cdk2 activity and DNA synthesis [125]. It is shown simultaneously by three groups that Akt directly phosphorylates p27Kip1 on T157, which leads to the retention of p27Kip1 in the cytoplasm. Cytoplasm-localized p27Kip1 cannot bind to nuclear Cdk2, thus cannot inhibit Cdk-2 [126–128].

The third mechanism is to stimulate cyclin D1 by inhibiting GSK-3. GSK-3 is constitutively active in unstimulated cells and phosphorylates numerous proteins including cyclin D, which triggers the degradation of Cyclin D through the ubiquitin-dependent proteolysis pathway, thus maintains cyclin D in an inactive state by reducing its expression level. Phosphorylation of GSK-3 by PI3K/Akt pathway causes the inactivation of GSK-3, thus promoting the accumulation and activation of cyclin D [129–132].

The fourth mechanism is the regulation of c-Fos. Expression of a constitutively active PI3K p110 mutant in NIH-3T3 cells induced c-Fos transcription [125]. Moreover, expression of a dominant negative Akt mutant rat mesenchymal cells decreases c-Fos transcription. As discussed above, c-Fos transcription stimulates cell proliferation. It is further shown that PI3K/Akt-induced c-Fos transcription is mediated by Elk-1 [116].

It is notable that the role of PI3K/Akt in the regulation of the cell cycle is heavily influenced by PTEN [133]. As a lipid phosphatase, PTEN antagonizes the function of phosphoinositide 3-kinase (PI3K) by converting phosphatidylinositol-3, 4, 5-trisphosphate (PIP3) to phosphatidylinositol-4, 5-biphosphate (PIP2) [134]. Thus, PTEN inactivates Akt and suppresses cell cycle progression [133].

4.4. Co-Regulation of Cell Cycle Progression by PI3K/Akt Pathway and Ras/Erk Pathway

Upon GF stimulation, the PI3K/Akt and Ras/Erk pathways co-regulate cell cycle progression by interacting closely [135]. Ras/Erk pathway and PI3K/Akt pathway interact at multiple point to coordinately regulate cell cycle progression. Many crosstalk mechanisms have been revealed between these two pathways. These include cross-activation, cross-inhibition, and pathway convergence on substrates [136] (Figure 8).

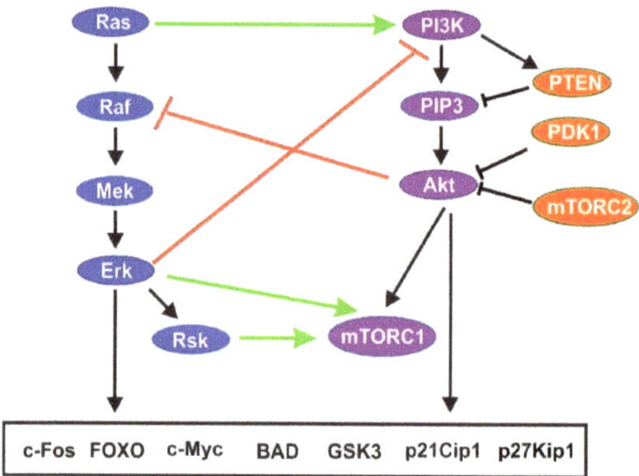

Figure 8. Interplay between Ras/Erk pathway and PI3K/Akt pathway in terms of cell cycle regulation. Green arrow indicates stimulation and red line indicates inhibition. Bottom box listed cell cycle related protein regulated by both signaling pathways.

First, Ras itself is an upstream activator of PI3K, thus activation of Ras leads to the activation of both Erk and PI3K [137]. In addition, Raf may activate Akt in hematopoietic cells [131].

The second mechanism is through mutual inhibition. Ras/Erk signaling may negatively regulate PI3K/Akt signaling and vice versa [138,139]. It is known that activated Erk phosphorylates GAB1, which inhibits GAB1-mediated recruitment of PI3K to the EGFR for activation [140]. It is shown that Runx2 expression relieves Erk-mediated negative regulation of EGFR and Akt [141]. It is also reported that MEK suppresses PI3K signaling by promoting membrane localization of the phosphatase PTEN [142]. Cross-inhibition between Akt and Raf by IGF1 stimulation is also reported [143]. Akt inhibits Raf1 by phosphorylating Raf N-terminal inhibitory sites, which leads to the inhibition of Erk [144,145].

The most striking feature of the crosstalk is that both Ras/Erk pathway and PI3K/Akt pathway regulate same proteins involved in the regulation of cell cycle. As discussed above, both pathways regulate p21Cip1, p27Kip1, GSK-3, Cyclin D, and c-Fos, which control cell cycle progression [86,131].

5. Two Waves of GF Signaling Drives Cell Cycle in G1 Phase

Cell cycle progression is controlled at multiple stages. During the early time of cell cycle research, a key regulatory step identified was the restriction point in cultured cells. GF is needed to drive the cell cycle to pass the restriction point, and once the restriction point is passed, GF is no longer required for the continuation of cell cycle progression [146]. What we know now is that the restriction point is G1/S checkpoint, which represents commitment to subsequent DNA replication (S phase). As discussed above, at a restriction point a signaling activity threshold must be reached to allow the continuation of the cell cycle progression. Mitogenic signals generated from GFs/RTKs are sufficient to drive the cells pass the restriction point.

The earliest studies of the restriction point in mammalian cells indicated that different GFs function differently in regulating cell cycle [147,148]. PDGF was shown as a competence factor, and EGF and IGF-I were defined as progression factors. However, later research shows that different GF may function similarly in driving cell cycle progression [85,149]. As revealed by live imaging in a study using dual-reporter cell model, when compared with serum-free medium, insulin-like GF-I (IGF-I) enhances cell cycle entry by more than 5-fold. Similar results are also obtained when GFs such as EGF, PDGF-AA, and PDGF-BB are used [150]. Another early observation is that the continuous presence of the GF for at least 6–8 h is required for cell cycle progression to pass restrict point [151,152]. Later research finds that the continuous presence of GF could be substituted with two short pulses of GF, one at early G1 phase and one at late G1 phase [6,153]. It is revealed that even with the continuous presence of GFs, the downstream signaling proteins including Ras/Erk and PI3K show two waves of activation (Figure 9).

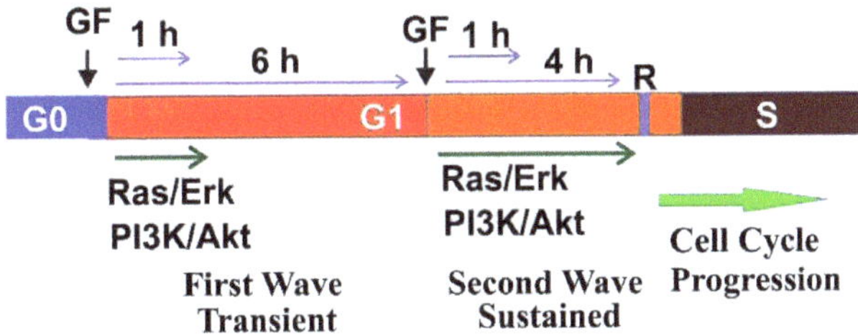

Figure 9. Two waves of GF-induced cell signaling in G1 phase drive the cell cycle progression to pass the restriction point (R).

In the continuous presence of GFs, Ras shows two waves of activation, an early transient activation and a more sustained activation approximately 4 h later [154]. Interestingly, it has been shown by earlier study that Ras activity is needed for cell cycle progression in two distinct times during the G1 phase [155,156], which corresponds to the two waves of Ras activation. In addition to functioning as a molecular switch for re-entry into the cell cycle from the G0 to G1 phase, Ras functions late in G1 to induce p27kip1 downregulation and cell cycle progression to pass the restriction point in response to EGF [157].

Similarly, GF also stimulates two waves of PI3K activation during G1 phase. It is shown that inhibition of PI3K catalytic subunit p110α (but not p110β) several hours after PDGF stimulation suppresses PDGF-dependent DNA synthesis [158]. In the mid-G1 phase, treatment with PI3K inhibitors inhibits GF-dependent cell-cycle progression and the inhibition is due to impaired degradation of p27Kip1 [159]. Indeed, a second, prolonged wave of accumulation of PI3K products is detected 3–7 h after PDGF stimulation [117]. By using an inhibition–rescue approach, this research further shows that, unlike the initial wave, the second wave of PI3K activation is required for PDGF-dependent DNA synthesis.

While prolonged presence of EGF is required to stimulate cell proliferation, we show that two short pulses of EGF are also sufficient to stimulate cell proliferation. We also show that two pulses of endosomal EGFR signaling are sufficient to stimulate cell proliferation [153,160]. The first pulse of EGFR signaling stimulates the exit from G0 into G1 phase. The second pulse, required 4–8 h later, drives the transition of cells from late G1 into S phase [153]. We also show that two waves of PDGFR signaling from endosome are sufficient to drive cell cycle progression [161].

The effects of the two short pulse of GF exposure on cell signaling network is further studied by a comprehensive transcriptomic and proteomic analyses. It is revealed that three processes are responsible for regulating restriction point crossing. In addition to activating essential metabolic enzymes, the first pulse also relieves p53-related restraining processes. The study also shows that the second pulse eliminates the suppressive action of p53 by activating the PI3K/Akt pathway. Finally, the second pulse uses the Erk-EGR1 threshold mechanism to digitize the graded external signal into an all-or-nothing decision-making obligation into the S phase [149]. Recent research shows that Ras signals mainly through Erk in G1, however, it also functions through PI3K/Akt to induce Cyclin D, driving S-phase entry [162].

Together, the findings to date highlight the importance of GF signaling for the progression of the cell cycle. These studies also demonstrated the presence of two action period for GF signaling in G1 phase. The first is in the early stage of G1 phase, which is critical for the G0/G1 transition to commit the cells to enter cell cycle, instead of going into a quiescent state. The second is right before the R-point, which is critical to drive the cells to pass the G1/S checkpoint and to start DNA synthesis.

6. GF Signaling in S, G2, and M Phase of Cell Cycle

The study regarding the role of GF-mediated signaling in other cell cycle phases are limited and controversial. The initial studies in the early 1990s to establish the role of GF-mediated cell signaling in cell cycle regulation suggests that GF-activated signaling cascades especially Ras/Erk and PI3K/Akt signaling pathways are required for the cell cycle progression from G1 to S phase by passing through restriction point [86,131]. The role of GF-activated signaling pathways in other phases only began to emerge in the late 1990s to early 2000s.

While it is proposed that GFs only regulate cell cycle progression during G0 to S interval [85], sporadic publications have emerged and shown the involvement of GF-mediated cell signaling in S and G2 phase [163–165]. It is shown that the presence of hepatocyte GF in S phase induces G2 delay by sustaining the activation of Erk [164]. On the other hand, it is reported that activation of EGFR by EGF in S phase induces centrosome separation, which promotes mitotic progression and cell survival [166]. The data regarding the role of EGFR activation in G2 phase are also limited and controversial. While all data indicate that EGF stimulates the activation of MEK-Erk signaling pathways [165,167], the findings regarding the effects of activated MEK-Erk pathway are quite different.

It is reported that activation of MEK-Erk signaling pathway by okadaic acid is required for entry into M phase and cell survival [168] and activation of both PI3K/Akt and MEK/Erk pathways by FBS is required for mitotic entry [169]. However, other studies indicate that activation of MEK-Erk signaling pathway by EGFR or phorbol 12-myristate 13-acetate (PMA) in G2 phase delays the M phase entry through p21Cip1 [166]. Activation of Erk signaling by EGF or PMA also induces the G2 phase delay and the blocked exit from G2 checkpoint by destabilizing Cdc25B [170]. By following live cells in which ERK1/2 activity was inhibited through late G2 and mitosis, it is shown that, for a timely entry into mitosis, Erk activity in early G2 is essential. However, Erk activity from late G2 through mitosis does not directly affect cell cycle progression [171].

The role of GF-induced cell signaling in mitosis has been less studied and poorly understood. It is reported that while the level of EGFR expression is the same between M phase and interphase, EGFR-mediated cell signaling pathways is tightly suppressed in M phase as EGF fails to bind to EGFR with high affinity to induce EGFR dimerization [170]. It is reported that EGFR, PLC-γ1, GTPase-activating protein, and Erk2 are less phosphorylated in M phase than in interphase [168]. A further study shows that Cdc2 inhibits EGF-induced Erk activation in M phase [169]. These studies argue that inhibition of GF signaling in M phase shelters the cell from extracellular signals during cell division, which helps the cell to preserve the precious energy needed for mitotic structural changes [168–170].

However, we recently showed that in M phase, EGFR is both expressed at the same level and activated to the same level by EGF as in interphase [171]. We further show that, in mitosis, EGFR is phosphorylated at all the major tyrosine residues in the C-terminus to the level similar to that in interphase, suggesting that EGFR is fully activated. However, the fully activated EGFR regulates downstream signaling pathways differently from interphase. It selectively activates some downstream signaling pathways while avoid others. Two major differences include the activation of Akt2, not Akt1, and inability to activate Erk despite of strong activation of Ras [172]. The activation level of other signaling proteins including PLC-γ1, PI3K, Cbl, and Src are all similar between mitosis and interphase. While EGF promotes cell survival in mitosis, it does not alter mitosis progression significantly. The only effect observed is the longer mitosis with the inhibition of EGFR by inhibitor AG1478 [172,173].

By using a cell line that is defective in EGFR downregulation, and thus maintains sustained EGFR signaling, it is shown that EGF-induced activation of EGFR signaling is required in G2 phase to drive the transition of cell cycle from G2 to M phase [174].

A novel method combining quantitative time-lapse fluorescence microscopy and microinjection is developed to examine cell cycle progression without cell synchronization [175–177]. By using this method, series research analyzes the role of Ras in the regulation of cyclin D expression and cell cycle progression. The results show that cyclin D1 is induced in a Ras-dependent manner in asynchronous NIH3T3 cells from S to G2 transition during cell cycle. Interestingly, the expression of cyclin D1 is Ras independent during the next G1 phase once induced in G2 phase. It is further shown that the Ras-dependent induction of cyclin D1 in the S/G2 transition is mediated by post-transcriptional mechanisms [178]. While endogenous Ras is active in all cell cycle phases, cyclin D1 is only induced during G2 phase in cycling cells, which indicates that the function of Ras is regulated by cell cycle phase [179]. Constitutively activated mutant Ras accelerates the cell cycle transition through G2/M and renders the G2/M checkpoint and SAC ineffective [180].

The downstream signaling proteins of Ras have also been shown to be involved in the cell cycle phase rather than G0–G1. For example, the MAPK kinase 1 (MAPKK1) activity in synchronized NIH 3T3 cells affects the kinetics of the cell cycle progression through both the G1 and G2 phases. Inhibition of MAPKK1 by dominant negative mutant is also found to delay progression of cells through G2. Moreover, inhibition of MAPKK1 in cells synchronized to S phase arrests the cell in G2 phase, which demonstrates a role for MAPKK1 in G2/M transition [181]. It is interesting to note that some studies suggest that too much Erk activity at the G2/M transition blocks entry in mitosis. Cells lacking VHR arrest at the G1-S and G2-M transitions of the cell cycle, which is dependent on the hyperactivation of Jnk and Erk. Moreover, this arrest is reversed by Jnk and Erk inhibition [182]. Hyperactivation of the Erk pathway due to expression of activated Ras or Raf mutants arrest cell cycle progression by promoting the accumulation of cyclin-dependent kinase inhibitors [86]. Wentilactone B (WB), a tetranorditerpenoid derivative induces G2/M phase arrest in human hepatoma SMMC-7721 cells via the Ras/Raf/Erk and Ras/Raf/JNK signaling pathways [183].

Inhibition of the PI3K/Akt pathway with PTEN or a PI-3 kinase inhibitor results in the cell cycle arrest in G2 phase and overexpression of constitutively active Akt kinase relieves this inhibition [184–186]. A PI3K inhibitor shortens the IR-induced G2 arrest [187]. These findings suggest a role of PI3K/Akt pathway G2 phase of cell cycle. Further study suggests that Chk1 kinase mediates the function of Akt in G2 phase. Akt kinase activity increase in G2/M, which coincides with the fall in Chk1 kinase activity [186]. Inhibition of PI3K by inhibitor LY294002 increase Chk1 kinase activity by decreasing Akt activity. Moreover, constitutively active Akt inhibits hydroxyurea-induced activation of Chk1 activation, which relieving DNA damage-induced G2 arrest [186]. Multiple mechanisms underly the ability of Akt to inhibit Chk1. These mechanisms include phosphorylation, ubiquitination, and reduced nuclear localization [188]. It is also reported that blocking PI3K/Akt signaling prolongs progression through S/G2 [189].

By using a Time-Resolved Single-Cell Imaging method, EGF-induced signal processing in individual cells is quantitated over time, which reveals the dynamic contribution of various signaling pathways [190]. It is shown that both PI3K and Erk activity are required for initial cell cycle entry, however, only PI3K activity regulates the duration of S phase. Importantly, if PI3k activity is blocked for 10–20 h after EGF treatment, the durations of S and G2 phase is increased dramatically. A likely underlying mechanism is the phosphorylation and subcellular translocation of Cdk2 by Akt during S phase [190].

7. Conclusions

Mitogenic signals of GFs are mediated by their transmembrane receptors, mostly RTKs. RTK activation initiates the activation of multiple downstream signaling cascades. Both Ras/Erk signaling pathway and PI3K/Akt signaling pathway play major role in regulating cell cycle progression. While the mechanism underlying the role of GF signaling in G1 phase of cell cycle progression has been largely revealed due to early extensive research, little is known regarding the function and mechanism of GF signaling in regulating other phases of cell cycle including S, G2, and M phase. Accumulated results so far suggest that GF signaling my regulate cell cycle progression throughout the cell cycle, but further research is needed to sustain these findings and to uncover the underlying molecular mechanisms (Figure 10).

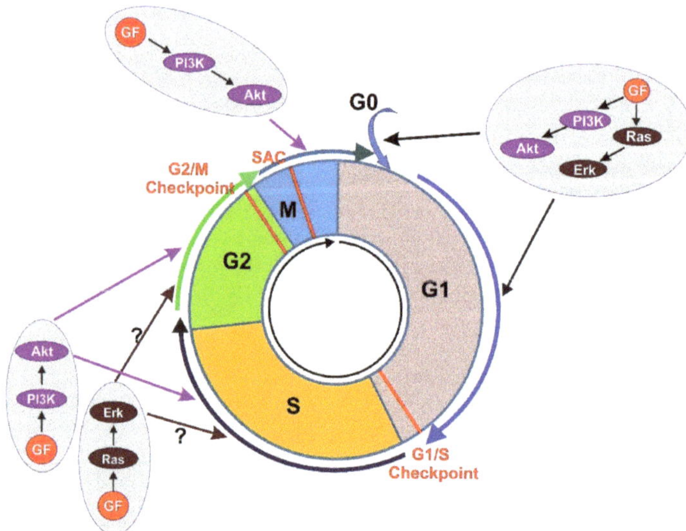

Figure 10. Regulation of cell cycle progression by GF-induced cell signaling in various phases of cell cycle. Question mark indicates that the data are limited and controversial.

Author Contributions: Z.W. is the sole author of this review. The author has read and agreed to the published version of the manuscript.

Funding: Our research was funded in part by grants from the Canadian Institutes of Health Research (CIHR) to Z.W.

Acknowledgments: This work was supported by CIHR to Z.W.

Conflicts of Interest: The author declares no conflict of interest.

References

1. Panda, S.K.; Ray, S.; Nayak, S.; Behera, S.; Bhanja, S.; Acharya, V. A review on cell cycle checkpoints in relation to cancer. *J. Med. Sci.* **2019**, *5*, 88–95. [CrossRef]
2. Panagopoulos, A.; Altmeyer, M. The hammer and the dance of cell cycle control. *Trends Biochem. Sci.* **2021**, *46*, 301–314. [CrossRef] [PubMed]
3. Barnum, K.J.; O'Connell, M.J. Cell cycle regulation by checkpoints. *Methods Mol. Biol.* **2014**, *1170*, 29–40. [CrossRef]
4. Satyanarayana, A.; Kaldis, P. Mammalian cell-cycle regulation: Several cdks, numerous cyclins and diverse compensatory mechanisms. *Oncogene* **2009**, *28*, 2925–2939. [CrossRef] [PubMed]
5. Gao, X.; Leone, G.W.; Wang, H. Cyclin d-cdk4/6 functions in cancer. *Adv. Cancer Res.* **2020**, *148*, 147–169. [CrossRef]
6. Jones, S.M.; Kazlauskas, A. Connecting signaling and cell cycle progression in growth factor-stimulated cells. *Oncogene* **2000**, *20*, 5558–5567. [CrossRef] [PubMed]
7. Wee, P.; Wang, Z. Cell cycle synchronization of hela cells to assay egfr pathway activation. *Methods Mol. Biol.* **2017**, *1652*, 167–181. [CrossRef]
8. Wolpert, L. Evolution of the cell theory. *Philos. Trans. R. Soc. Lond. B Biol. Sci.* **1995**, *349*, 227–233. [CrossRef]
9. Müller-Wille, S. Cell theory, specificity, and reproduction, 1837–1870. *Stud. Hist. Philos. Biol. Biomed. Sci.* **2010**, *41*, 225–231. [CrossRef] [PubMed]
10. Ribatti, D. Rudolf virchow, the founder of cellular pathology. *Rom. J. Morphol. Embryol.* **2019**, *60*, 1381–1382.
11. Paweletz, N. Walther flemming: Pioneer of mitosis research. *Nat. Rev. Mol. Cell Biol.* **2001**, *2*, 72–75. [CrossRef]
12. Howard, A.; Pelc, S.R. Synthesis of nucleoprotein in bean root cells. *Nature* **1951**, *167*, 599–600. [CrossRef]
13. Nasmyth, K. Viewpoint: Putting the cell cycle in order. *Science* **1996**, *274*, 1643–1645. [CrossRef]
14. Johnson, R.T.; Rao, P.N. Nucleo-cytoplasmic interactions in the acheivement of nuclear synchrony in DNA synthesis and mitosis in multinucleate cells. *Biol. Rev. Camb. Philos. Soc.* **1971**, *46*, 97–155. [CrossRef] [PubMed]
15. Rao, P.N.; Johnson, R.T. Mammalian cell fusion: Studies on the regulation of DNA synthesis and mitosis. *Nature* **1970**, *225*, 159–164. [CrossRef]
16. Hartwell, L.H.; Culotti, J.; Reid, B. Genetic control of the cell-division cycle in yeast. I. Detection of mutants. *Proc. Natl. Acad. Sci. USA* **1970**, *66*, 352–359. [CrossRef]
17. Hartwell, L.H.; Culotti, J.; Pringle, J.R.; Reid, B.J. Genetic control of the cell division cycle in yeast. *Science* **1974**, *183*, 46–51. [CrossRef] [PubMed]
18. Hartwell, L.H. Twenty-five years of cell cycle genetics. *Genetics* **1991**, *129*, 975–980. [CrossRef] [PubMed]
19. Nurse, P.; Thuriaux, P. Regulatory genes controlling mitosis in the fission yeast schizosaccharomyces pombe. *Genetics* **1980**, *96*, 627–637. [CrossRef]
20. Lee, M.G.; Nurse, P. Complementation used to clone a human homologue of the fission yeast cell cycle control gene cdc2. *Nature* **1987**, *327*, 31–35. [CrossRef] [PubMed]
21. Lee, M.G.; Norbury, C.J.; Spurr, N.K.; Nurse, P. Regulated expression and phosphorylation of a possible mammalian cell-cycle control protein. *Nature* **1988**, *333*, 676–679. [CrossRef]
22. Evans, T.; Rosenthal, E.T.; Youngblom, J.; Distel, D.; Hunt, T. Cyclin: A protein specified by maternal mrna in sea urchin eggs that is destroyed at each cleavage division. *Cell* **1983**, *33*, 389–396. [CrossRef]
23. Pryor, P.R.; Jackson, L.; Gray, S.R.; Edeling, M.A.; Thompson, A.; Sanderson, C.M.; Evans, P.R.; Owen, D.J.; Luzio, J.P. Molecular basis for the sorting of the snare vamp7 into endocytic clathrin-coated vesicles by the arfgap hrb. *Cell* **2008**, *134*, 817–827. [CrossRef]
24. Weinert, T.A.; Hartwell, L.H. The rad9 gene controls the cell cycle response to DNA damage in saccharomyces cerevisiae. *Science* **1988**, *241*, 317–322. [CrossRef]
25. Hartwell, L.H.; Weinert, T.A. Checkpoints: Controls that ensure the order of cell cycle events. *Science* **1989**, *246*, 629–634. [CrossRef] [PubMed]
26. Morgan, D.O. Cyclin-dependent kinases: Engines, clocks, and microprocessors. *Annu. Rev. Cell Dev. Biol.* **1997**, *13*, 261–291. [CrossRef] [PubMed]
27. Huang, D.; Friesen, H.; Andrews, B. Pho85, a multifunctional cyclin-dependent protein kinase in budding yeast. *Mol. Microbiol.* **2007**, *66*, 303–314. [CrossRef]
28. Sherr, C.J.; Roberts, J.M. Cdk inhibitors: Positive and negative regulators of g1-phase progression. *Genes Dev.* **1999**, *13*, 1501–1512. [CrossRef] [PubMed]

29. Sherr, C.J.; Roberts, J.M. Living with or without cyclins and cyclin-dependent kinases. *Genes Dev.* **2004**, *18*, 2699–2711. [CrossRef]
30. Weinberg, R.A. The retinoblastoma protein and cell cycle control. *Cell* **1995**, *81*, 323–330. [CrossRef]
31. Dyson, N. The regulation of e2f by prb-family proteins. *Genes Dev.* **1998**, *12*, 2245–2262. [CrossRef]
32. Lundberg, A.S.; Weinberg, R.A. Functional inactivation of the retinoblastoma protein requires sequential modification by at least two distinct cyclin-cdk complexes. *Mol. Cell Biol.* **1998**, *18*, 753–761. [CrossRef]
33. Petersen, B.O.; Lukas, J.; Sørensen, C.S.; Bartek, J.; Helin, K. Phosphorylation of mammalian cdc6 by cyclin a/cdk2 regulates its subcellular localization. *EMBO J.* **1999**, *18*, 396–410. [CrossRef]
34. Coverley, D.; Pelizon, C.; Trewick, S.; Laskey, R.A. Chromatin-bound cdc6 persists in s and g2 phases in human cells, while soluble cdc6 is destroyed in a cyclin a-cdk2 dependent process. *J. Cell Sci.* **2000**, *113*, 1929–1938. [CrossRef] [PubMed]
35. Furuno, N.; den Elzen, N.; Pines, J. Human cyclin a is required for mitosis until mid prophase. *J. Cell Biol.* **1999**, *147*, 295–306. [CrossRef] [PubMed]
36. Riabowol, K.; Draetta, G.; Brizuela, L.; Vandre, D.; Beach, D. The cdc2 kinase is a nuclear protein that is essential for mitosis in mammalian cells. *Cell* **1989**, *57*, 393–401. [CrossRef]
37. Duan, L.; Raja, S.M.; Chen, G.; Virmani, S.; Williams, S.H.; Clubb, R.J.; Mukhopadhyay, C.; Rainey, M.A.; Ying, G.; Dimri, M.; et al. Negative regulation of egfr-vav2 signaling axis by cbl ubiquitin ligase controls egf receptor-mediated epithelial cell adherens junction dynamics and cell migration. *J. Biol. Chem.* **2011**, *286*, 620–633. [CrossRef] [PubMed]
38. Limas, J.C.; Cook, J.G. Preparation for DNA replication: The key to a successful s phase. *FEBS Lett.* **2019**, *593*, 2853–2867. [CrossRef]
39. Massague, J. G1 cell-cycle control and cancer. *Nature* **2004**, *432*, 298–306. [CrossRef]
40. Nelson, D.M.; Ye, X.; Hall, C.; Santos, H.; Ma, T.; Kao, G.D.; Yen, T.J.; Harper, J.W.; Adams, P.D. Coupling of DNA synthesis and histone synthesis in s phase independent of cyclin/cdk2 activity. *Mol. Cell Biol.* **2002**, *22*, 7459–7472. [CrossRef]
41. Ciardo, D.; Goldar, A.; Marheineke, K. On the interplay of the DNA replication program and the intra-s phase checkpoint pathway. *Genes* **2019**, *10*, 94. [CrossRef] [PubMed]
42. Saldivar, J.C.; Hamperl, S.; Bocek, M.J.; Chung, M.; Bass, T.E.; Cisneros-Soberanis, F.; Samejima, K.; Xie, L.; Paulson, J.R.; Earnshaw, W.C.; et al. An intrinsic s/g(2) checkpoint enforced by atr. *Science* **2018**, *361*, 806–810. [CrossRef]
43. Hannen, R.; Selmansberger, M.; Hauswald, M.; Pagenstecher, A.; Nist, A.; Stiewe, T.; Acker, T.; Carl, B.; Nimsky, C.; Bartsch, J.W. Comparative transcriptomic analysis of temozolomide resistant primary gbm stem-like cells and recurrent gbm identifies up-regulation of the carbonic anhydrase ca2 gene as resistance factor. *Cancers* **2019**, *11*, 921. [CrossRef]
44. Lockhead, S.; Moskaleva, A.; Kamenz, J.; Chen, Y.; Kang, M.; Reddy, A.R.; Santos, S.D.M.; Ferrell, J.E., Jr. The apparent requirement for protein synthesis during g2 phase is due to checkpoint activation. *Cell Rep.* **2020**, *32*, 107901. [CrossRef] [PubMed]
45. Moseley, J.B.; Mayeux, A.; Paoletti, A.; Nurse, P. A spatial gradient coordinates cell size and mitotic entry in fission yeast. *Nature* **2009**, *459*, 857–860. [CrossRef] [PubMed]
46. Vukušić, K.; Buđa, R.; Tolić, I.M. Force-generating mechanisms of anaphase in human cells. *J. Cell Sci.* **2019**, *132*, jcs231985. [CrossRef]
47. Afonso, O.; Matos, I.; Pereira, A.J.; Aguiar, P.; Lampson, M.A.; Maiato, H. Feedback control of chromosome separation by a midzone aurora b gradient. *Science* **2014**, *345*, 332–336. [CrossRef]
48. Gibcus, J.H.; Samejima, K.; Goloborodko, A.; Samejima, I.; Naumova, N.; Nuebler, J.; Kanemaki, M.T.; Xie, L.; Paulson, J.R.; Earnshaw, W.C.; et al. A pathway for mitotic chromosome formation. *Science* **2018**, *359*. [CrossRef] [PubMed]
49. Liang, Z.; Zickler, D.; Prentiss, M.; Chang, F.S.; Witz, G.; Maeshima, K.; Kleckner, N. Chromosomes progress to metaphase in multiple discrete steps via global compaction/expansion cycles. *Cell* **2015**, *161*, 1124–1137. [CrossRef]
50. Samejima, K.; Samejima, I.; Vagnarelli, P.; Ogawa, H.; Vargiu, G.; Kelly, D.A.; Alves, F.d.; Kerr, A.; Green, L.C.; Hudson, D.F.; et al. Mitotic chromosomes are compacted laterally by kif4 and condensin and axially by topoisomerase iiα. *J. Cell Biol.* **2012**, *199*, 755–770. [CrossRef] [PubMed]
51. Gavet, O.; Pines, J. Progressive activation of cyclinb1-cdk1 coordinates entry to mitosis. *Dev. Cell* **2010**, *18*, 533–543. [CrossRef]
52. Santaguida, S.; Musacchio, A. The life and miracles of kinetochores. *EMBO J.* **2009**, *28*, 2511–2531. [CrossRef]
53. Paulson, J.R.; Laemmli, U.K. The structure of histone-depleted metaphase chromosomes. *Cell* **1977**, *12*, 817–828. [CrossRef]
54. Dhatchinamoorthy, K.; Mattingly, M.; Gerton, J.L. Regulation of kinetochore configuration during mitosis. *Curr. Genet.* **2018**, *64*, 1197–1203. [CrossRef]
55. Vukušić, K.; Tolić, I.M. Anaphase b: Long-standing models meet new concepts. *Semin. Cell Dev. Biol.* **2021**, *117*, 127–139. [CrossRef]
56. Su, K.C.; Barry, Z.; Schweizer, N.; Maiato, H.; Bathe, M.; Cheeseman, I.M. A regulatory switch alters chromosome motions at the metaphase-to-anaphase transition. *Cell Rep.* **2016**, *17*, 1728–1738. [CrossRef]
57. Green, R.A.; Paluch, E.; Oegema, K. Cytokinesis in animal cells. *Annu. Rev. Cell Dev. Biol.* **2012**, *28*, 29–58. [CrossRef]
58. Fededa, J.P.; Gerlich, D.W. Molecular control of animal cell cytokinesis. *Nat. Cell Biol.* **2012**, *14*, 440–447. [CrossRef]
59. Lens, S.M.A.; Medema, R.H. Cytokinesis defects and cancer. *Nat. Rev. Cancer* **2019**, *19*, 32–45. [CrossRef]
60. Mierzwa, B.; Gerlich, D.W. Cytokinetic abscission: Molecular mechanisms and temporal control. *Dev. Cell* **2014**, *31*, 525–538. [CrossRef]

61. Gromley, A.; Yeaman, C.; Rosa, J.; Redick, S.; Chen, C.T.; Mirabelle, S.; Guha, M.; Sillibourne, J.; Doxsey, S.J. Centriolin anchoring of exocyst and snare complexes at the midbody is required for secretory-vesicle-mediated abscission. *Cell* **2005**, *123*, 75–87. [CrossRef]
62. Schiel, J.A.; Park, K.; Morphew, M.K.; Reid, E.; Hoenger, A.; Prekeris, R. Endocytic membrane fusion and buckling-induced microtubule severing mediate cell abscission. *J. Cell Sci.* **2011**, *124*, 1411–1424. [CrossRef]
63. Wee, P.; Wang, Z. Epidermal growth factor receptor cell proliferation signaling pathways. *Cancers* **2017**, *9*, 52. [CrossRef]
64. Wang, Z. Erbb receptors and cancer. *Methods Mol. Biol.* **2017**, *1652*, 3–35. [CrossRef]
65. Carpenter, G.; Lembach, K.J.; Morrison, M.M.; Cohen, S. Characterization of the binding of 125-i-labeled epidermal growth factor to human fibroblasts. *J. Biol. Chem.* **1975**, *250*, 4297–4304. [CrossRef]
66. Ullrich, A.; Coussens, L.; Hayflick, J.S.; Dull, T.J.; Gray, A.; Tam, A.W.; Lee, J.; Yarden, Y.; Libermann, T.A.; Schlessinger, J. Human epidermal growth factor receptor cdna sequence and aberrant expression of the amplified gene in a431 epidermoid carcinoma cells. *Nature* **1984**, *309*, 418–425. [CrossRef]
67. Yarden, Y.; Sliwkowski, M.X. Untangling the erbb signalling network. *Nat. Rev. Mol. Cell Biol.* **2001**, *2*, 127–137. [CrossRef]
68. Pearson, G.; Robinson, F.; Gibson, T.B.; Xu, B.E.; Karandikar, M.; Berman, K.; Cobb, M.H. Mitogen-activated protein (map) kinase pathways: Regulation and physiological functions. *Endocr. Rev.* **2001**, *22*, 153–183.
69. Avruch, J. Map kinase pathways: The first twenty years. *Biochim. Biophys. Acta-Mol. Cell Res.* **2007**, *1773*, 1150–1160. [CrossRef]
70. Marshall, C.J. Map kinase kinase kinase, map kinase kinase and map kinase. *Curr. Opin. Genet. Dev.* **1994**, *4*, 82–89. [CrossRef]
71. Marshall, C.J. Cell signalling. Raf gets it together. *Nature* **1996**, *383*, 127–128. [CrossRef]
72. Pawson, T. Protein modules and signalling networks. *Nature* **1995**, *373*, 573–580. [CrossRef] [PubMed]
73. Wu, P.; Wee, P.; Jiang, J.; Chen, X.; Wang, Z. Differential regulation of transcription factors by location-specific egf receptor signaling via a spatio-temporal interplay of erk activation. *PLoS ONE* **2012**, *7*, e41354. [CrossRef] [PubMed]
74. Downward, J. Ras signalling and apoptosis. *Curr. Opin. Genet. Dev.* **1998**, *8*, 49–54. [CrossRef]
75. Castellano, E.; Downward, J. Ras interaction with pi3k: More than just another effector pathway. *Genes Cancer* **2011**, *2*, 261–274. [CrossRef]
76. Taniguchi, C.M.; Tran, T.T.; Kondo, T.; Luo, J.; Ueki, K.; Cantley, L.C.; Kahn, C.R. Phosphoinositide 3-kinase regulatory subunit p85alpha suppresses insulin action via positive regulation of pten. *Proc. Natl. Acad. Sci. USA* **2006**, *103*, 12093–12097. [CrossRef]
77. Chagpar, R.B.; Links, P.H.; Pastor, M.C.; Furber, L.A.; Hawrysh, A.D.; Chamberlain, M.D.; Anderson, D.H. Direct positive regulation of pten by the p85 subunit of phosphatidylinositol 3-kinase. *Proc. Natl. Acad. Sci. USA* **2010**, *107*, 5471–5476. [CrossRef]
78. Burgering, B.M.; Coffer, P.J. Protein kinase b (c-akt) in phosphatidylinositol-3-oh kinase signal transduction. *Nature* **1995**, *376*, 599–602. [CrossRef] [PubMed]
79. Downward, J. Mechanisms and consequences of activation of protein kinase b/akt. *Curr. Opin. Cell Biol.* **1998**, *10*, 262–267. [CrossRef]
80. Okano, J.; Gaslightwala, I.; Birnbaum, M.J.; Rustgi, A.K.; Nakagawa, H. Akt/protein kinase b isoforms are differentially regulated by epidermal growth factor stimulation. *J. Biol. Chem.* **2000**, *275*, 30934–30942. [CrossRef]
81. Nakamura, K.D.; Martinez, R.; Weber, M.J. Tyrosine phosphorylation of specific proteins after mitogen stimulation of chicken embryo fibroblasts. *Mol. Cell Biol.* **1983**, *3*, 380–390. [CrossRef]
82. Pagès, G.; Lenormand, P.; L'Allemain, G.; Chambard, J.C.; Meloche, S.; Pouysségur, J. Mitogen-activated protein kinases p42mapk and p44mapk are required for fibroblast proliferation. *Proc. Natl. Acad. Sci. USA* **1993**, *90*, 8319–8323. [CrossRef]
83. Cheng, M.; Sexl, V.; Sherr, C.J.; Roussel, M.F. Assembly of cyclin d-dependent kinase and titration of p27kip1 regulated by mitogen-activated protein kinase kinase (mek1). *Proc. Natl. Acad. Sci. USA* **1998**, *95*, 1091–1096. [CrossRef]
84. Treinies, I.; Paterson, H.F.; Hooper, S.; Wilson, R.; Marshall, C.J. Activated mek stimulates expression of ap-1 components independently of phosphatidylinositol 3-kinase (pi3-kinase) but requires a pi3-kinase signal to stimulate DNA synthesis. *Mol. Cell Biol.* **1999**, *19*, 321–329. [CrossRef]
85. Jones, S.M.; Kazlauskas, A. Growth factor-dependent signaling and cell cycle progression. *FEBS Lett.* **2001**, *490*, 110–116. [CrossRef]
86. Meloche, S.; Pouysségur, J. The erk1/2 mitogen-activated protein kinase pathway as a master regulator of the g1- to s-phase transition. *Oncogene* **2007**, *26*, 3227–3239. [CrossRef]
87. Filmus, J.; Robles, A.I.; Shi, W.; Wong, M.J.; Colombo, L.L.; Conti, C.J. Induction of cyclin d1 overexpression by activated ras. *Oncogene* **1994**, *9*, 3627–3633. [PubMed]
88. Arber, N.; Sutter, T.; Miyake, M.; Kahn, S.M.; Venkatraj, V.S.; Sobrino, A.; Warburton, D.; Holt, P.R.; Weinstein, I.B. Increased expression of cyclin d1 and the rb tumor suppressor gene in c-k-ras transformed rat enterocytes. *Oncogene* **1996**, *12*, 1903–1908. [PubMed]
89. Winston, J.T.; Coats, S.R.; Wang, Y.Z.; Pledger, W.J. Regulation of the cell cycle machinery by oncogenic ras. *Oncogene* **1996**, *12*, 127–134. [PubMed]
90. Lavoie, J.N.; L'Allemain, G.; Brunet, A.; Müller, R.; Pouysségur, J. Cyclin d1 expression is regulated positively by the p42/p44mapk and negatively by the p38/hogmapk pathway. *J. Biol. Chem.* **1996**, *271*, 20608–20616. [CrossRef] [PubMed]
91. Balmanno, K.; Cook, S.J. Sustained map kinase activation is required for the expression of cyclin d1, p21cip1 and a subset of ap-1 proteins in ccl39 cells. *Oncogene* **1999**, *18*, 3085–3097. [CrossRef] [PubMed]

92. Hill, C.S.; Treisman, R. Transcriptional regulation by extracellular signals: Mechanisms and specificity. *Cell* **1995**, *80*, 199–211. [CrossRef]
93. Strudwick, S.; Borden, K.L. The emerging roles of translation factor eif4e in the nucleus. *Differentiation* **2002**, *70*, 10–22. [CrossRef]
94. Dey, A.; She, H.; Kim, L.; Boruch, A.; Guris, D.L.; Carlberg, K.; Sebti, S.M.; Woodley, D.T.; Imamoto, A.; Li, W. Colony-stimulating factor-1 receptor utilizes multiple signaling pathways to induce cyclin d2 expression. *Mol. Biol. Cell* **2000**, *11*, 3835–3848. [CrossRef]
95. Tetsu, O.; McCormick, F. Proliferation of cancer cells despite cdk2 inhibition. *Cancer Cell* **2003**, *3*, 233–245. [CrossRef]
96. Pelengaris, S.; Khan, M.; Evan, G. C-myc: More than just a matter of life and death. *Nat. Rev. Cancer* **2002**, *2*, 764–776. [CrossRef]
97. Sears, R.; Nuckolls, F.; Haura, E.; Taya, Y.; Tamai, K.; Nevins, J.R. Multiple ras-dependent phosphorylation pathways regulate myc protein stability. *Genes Dev.* **2000**, *14*, 2501–2514. [CrossRef]
98. Hermeking, H.; Rago, C.; Schuhmacher, M.; Li, Q.; Barrett, J.F.; Obaya, A.J.; O'Connell, B.C.; Mateyak, M.K.; Tam, W.; Kohlhuber, F.; et al. Identification of cdk4 as a target of c-myc. *Proc. Natl. Acad. Sci. USA* **2000**, *97*, 2229–2234. [CrossRef]
99. Coller, H.A.; Grandori, C.; Tamayo, P.; Colbert, T.; Lander, E.S.; Eisenman, R.N.; Golub, T.R. Expression analysis with oligonucleotide microarrays reveals that myc regulates genes involved in growth, cell cycle, signaling, and adhesion. *Proc. Natl. Acad. Sci. USA* **2000**, *97*, 3260–3265. [CrossRef]
100. Galaktionov, K.; Chen, X.; Beach, D. Cdc25 cell-cycle phosphatase as a target of c-myc. *Nature* **1996**, *382*, 511–517. [CrossRef]
101. Li, Y.; Jenkins, C.W.; Nichols, M.A.; Xiong, Y. Cell cycle expression and p53 regulation of the cyclin-dependent kinase inhibitor p21. *Oncogene* **1994**, *9*, 2261–2268.
102. Liu, Y.; Martindale, J.L.; Gorospe, M.; Holbrook, N.J. Regulation of p21waf1/cip1 expression through mitogen-activated protein kinase signaling pathway. *Cancer Res.* **1996**, *56*, 31–35. [PubMed]
103. Bottazzi, M.E.; Zhu, X.; Böhmer, R.M.; Assoian, R.K. Regulation of p21(cip1) expression by growth factors and the extracellular matrix reveals a role for transient erk activity in g1 phase. *J. Cell Biol.* **1999**, *146*, 1255–1264. [CrossRef] [PubMed]
104. LaBaer, J.; Garrett, M.D.; Stevenson, L.F.; Slingerland, J.M.; Sandhu, C.; Chou, H.S.; Fattaey, A.; Harlow, E. New functional activities for the p21 family of cdk inhibitors. *Genes Dev.* **1997**, *11*, 847–862. [CrossRef]
105. Rank, K.B.; Evans, D.B.; Sharma, S.K. The n-terminal domains of cyclin-dependent kinase inhibitory proteins block the phosphorylation of cdk2/cyclin e by the cdk-activating kinase. *Biochem. Biophys. Res. Commun.* **2000**, *271*, 469–473. [CrossRef] [PubMed]
106. Kawada, M.; Yamagoe, S.; Murakami, Y.; Suzuki, K.; Mizuno, S.; Uehara, Y. Induction of p27kip1 degradation and anchorage independence by ras through the map kinase signaling pathway. *Oncogene* **1997**, *15*, 629–637. [CrossRef]
107. Lenferink, A.E.; Simpson, J.F.; Shawver, L.K.; Coffey, R.J.; Forbes, J.T.; Arteaga, C.L. Blockade of the epidermal growth factor receptor tyrosine kinase suppresses tumorigenesis in mmtv/neu + mmtv/tgf-alpha bigenic mice. *Proc. Natl. Acad. Sci. USA* **2000**, *97*, 9609–9614. [CrossRef]
108. Gysin, S.; Lee, S.H.; Dean, N.M.; McMahon, M. Pharmacologic inhibition of RAF→MEK→ERK signaling elicits pancreatic cancer cell cycle arrest through induced expression of p27kip1. *Cancer Res.* **2005**, *65*, 4870–4880. [CrossRef]
109. Mirza, A.M.; Gysin, S.; Malek, N.; Nakayama, K.; Roberts, J.M.; McMahon, M. Cooperative regulation of the cell division cycle by the protein kinases raf and akt. *Mol. Cell Biol.* **2004**, *24*, 10868–10881. [CrossRef] [PubMed]
110. Yamamoto, T.; Ebisuya, M.; Ashida, F.; Okamoto, K.; Yonehara, S.; Nishida, E. Continuous erk activation downregulates antiproliferative genes throughout g1 phase to allow cell-cycle progression. *Curr. Biol.* **2006**, *16*, 1171–1182. [CrossRef]
111. Sewing, A.; Wiseman, B.; Lloyd, A.C.; Land, H. High-intensity raf signal causes cell cycle arrest mediated by p21cip1. *Mol. Cell Biol.* **1997**, *17*, 5588–5597. [CrossRef]
112. Woods, D.; Parry, D.; Cherwinski, H.; Bosch, E.; Lees, E.; McMahon, M. Raf-induced proliferation or cell cycle arrest is determined by the level of raf activity with arrest mediated by p21cip1. *Mol. Cell Biol.* **1997**, *17*, 5598–5611. [CrossRef] [PubMed]
113. Kerkhoff, E.; Rapp, U.R. High-intensity raf signals convert mitotic cell cycling into cellular growth. *Cancer Res.* **1998**, *58*, 1636–1640.
114. Tombes, R.M.; Auer, K.L.; Mikkelsen, R.; Valerie, K.; Wymann, M.P.; Marshall, C.J.; McMahon, M.; Dent, P. The mitogen-activated protein (map) kinase cascade can either stimulate or inhibit DNA synthesis in primary cultures of rat hepatocytes depending upon whether its activation is acute/phasic or chronic. *Biochem. J.* **1998**, *330*, 1451–1460. [CrossRef] [PubMed]
115. Valius, M.; Bazenet, C.; Kazlauskas, A. Tyrosines 1021 and 1009 are phosphorylation sites in the carboxy terminus of the platelet-derived growth factor receptor beta subunit and are required for binding of phospholipase c gamma and a 64-kilodalton protein, respectively. *Mol.Cell Biol.* **1993**, *13*, 133–143. [PubMed]
116. Choudhury, G.G.; Karamitsos, C.; Hernandez, J.; Gentilini, A.; Bardgette, J.; Abboud, H.E. Pi-3-kinase and mapk regulate mesangial cell proliferation and migration in response to pdgf. *Am. J. Physiol.* **1997**, *273*, 931–938. [CrossRef]
117. Jones, S.M.; Klinghoffer, R.; Prestwich, G.D.; Toker, A.; Kazlauskas, A. Pdgf induces an early and a late wave of pi 3-kinase activity, and only the late wave is required for progression through g1. *Curr. Biol.* **1999**, *9*, 512–521. [CrossRef]
118. Gille, H.; Downward, J. Multiple ras effector pathways contribute to g(1) cell cycle progression [in process citation]. *J. Biol. Chem.* **1999**, *274*, 22033–22040. [CrossRef]
119. Sun, H.; Lesche, R.; Li, D.M.; Liliental, J.; Zhang, H.; Gao, J.; Gavrilova, N.; Mueller, B.; Liu, X.; Wu, H. Pten modulates cell cycle progression and cell survival by regulating phosphatidylinositol 3,4,5,-trisphosphate and akt/protein kinase b signaling pathway. *Proc. Natl. Acad. Sci. USA* **1999**, *96*, 6199–6204. [CrossRef]

120. Medema, R.H.; Kops, G.J.; Bos, J.L.; Burgering, B.M. Afx-like forkhead transcription factors mediate cell-cycle regulation by ras and pkb through p27kip1. *Nature* **2000**, *404*, 782–787. [CrossRef]
121. Choudhury, G.G. Akt serine threonine kinase regulates platelet-derived growth factor-induced DNA synthesis in glomerular mesangial cells: Regulation of c-fos and p27(kip1) gene expression. *J. Biol. Chem.* **2001**, *276*, 35636–35643. [CrossRef]
122. Mitsuuchi, Y.; Johnson, S.W.; Selvakumaran, M.; Williams, S.J.; Hamilton, T.C.; Testa, J.R. The phosphatidylinositol 3-kinase/akt signal transduction pathway plays a critical role in the expression of p21waf1/cip1/sdi1 induced by cisplatin and paclitaxel. *Cancer Res.* **2000**, *60*, 5390–5394.
123. Rössig, L.; Jadidi, A.S.; Urbich, C.; Badorff, C.; Zeiher, A.M.; Dimmeler, S. Akt-dependent phosphorylation of p21(cip1) regulates pcna binding and proliferation of endothelial cells. *Mol. Cell Biol.* **2001**, *21*, 5644–5657. [CrossRef] [PubMed]
124. Li, Y.; Dowbenko, D.; Lasky, L.A. Akt/pkb phosphorylation of p21cip/waf1 enhances protein stability of p21cip/waf1 and promotes cell survival. *J. Biol. Chem.* **2002**, *277*, 11352–11361. [CrossRef]
125. Hu, Q.; Klippel, A.; Muslin, A.J.; Fantl, W.J.; Williams, L.T. Ras-dependent induction of cellular responses by constitutively active phosphatidylinositol-3 kinase. *Science* **1995**, *268*, 100–102. [CrossRef]
126. Viglietto, G.; Motti, M.L.; Bruni, P.; Melillo, R.M.; D'Alessio, A.; Califano, D.; Vinci, F.; Chiappetta, G.; Tsichlis, P.; Bellacosa, A.; et al. Cytoplasmic relocalization and inhibition of the cyclin-dependent kinase inhibitor p27(kip1) by pkb/akt-mediated phosphorylation in breast cancer. *Nat. Med.* **2002**, *8*, 1136–1144. [CrossRef] [PubMed]
127. Shin, I.; Yakes, F.M.; Rojo, F.; Shin, N.Y.; Bakin, A.V.; Baselga, J.; Arteaga, C.L. Pkb/akt mediates cell-cycle progression by phosphorylation of p27(kip1) at threonine 157 and modulation of its cellular localization. *Nat. Med.* **2002**, *8*, 1145–1152. [CrossRef]
128. Liang, J.; Zubovitz, J.; Petrocelli, T.; Kotchetkov, R.; Connor, M.K.; Han, K.; Lee, J.H.; Ciarallo, S.; Catzavelos, C.; Beniston, R.; et al. Pkb/akt phosphorylates p27, impairs nuclear import of p27 and opposes p27-mediated g1 arrest. *Nat. Med.* **2002**, *8*, 1153–1160. [CrossRef] [PubMed]
129. Diehl, J.A.; Cheng, M.; Roussel, M.F.; Sherr, C.J. Glycogen synthase kinase-3beta regulates cyclin d1 proteolysis and subcellular localization. *Genes Dev.* **1998**, *12*, 3499–3511. [CrossRef]
130. Van Weeren, P.C.; de Bruyn, K.M.; de Vries-Smits, A.M.; van Lint, J.; Burgering, B.M. Essential role for protein kinase b (pkb) in insulin-induced glycogen synthase kinase 3 inactivation. Characterization of dominant-negative mutant of pkb. *J. Biol. Chem.* **1998**, *273*, 13150–13156. [CrossRef]
131. Chang, F.; Lee, J.T.; Navolanic, P.M.; Steelman, L.S.; Shelton, J.G.; Blalock, W.L.; Franklin, R.A.; McCubrey, J.A. Involvement of pi3k/akt pathway in cell cycle progression, apoptosis, and neoplastic transformation: A target for cancer chemotherapy. *Leukemia* **2003**, *17*, 590–603. [CrossRef]
132. Dong, J.; Peng, J.; Zhang, H.; Mondesire, W.H.; Jian, W.; Mills, G.B.; Hung, M.C.; Meric-Bernstam, F. Role of glycogen synthase kinase 3beta in rapamycin-mediated cell cycle regulation and chemosensitivity. *Cancer Res.* **2005**, *65*, 1961–1972. [CrossRef]
133. Brandmaier, A.; Hou, S.Q.; Shen, W.H. Cell cycle control by pten. *J. Mol. Biol.* **2017**, *429*, 2265–2277. [CrossRef]
134. Maehama, T.; Dixon, J.E. The tumor suppressor, pten/mmac1, dephosphorylates the lipid second messenger, phosphatidylinositol 3,4,5-trisphosphate. *J. Biol. Chem.* **1998**, *273*, 13375–13378. [CrossRef]
135. Chen, J.Y.; Lin, J.R.; Cimprich, K.A.; Meyer, T. A two-dimensional erk-akt signaling code for an ngf-triggered cell-fate decision. *Mol. Cell* **2012**, *45*, 196–209. [CrossRef]
136. Mendoza, M.C.; Er, E.E.; Blenis, J. The ras-erk and pi3k-mtor pathways: Cross-talk and compensation. *Trends Biochem. Sci.* **2011**, *36*, 320–328. [CrossRef] [PubMed]
137. Kodaki, T.; Woscholski, R.; Hallberg, B.; Rodriguez-Viciana, P.; Downward, J.; Parker, P.J. The activation of phosphatidylinositol 3-kinase by ras. *Curr. Biol.* **1994**, *4*, 798–806. [CrossRef]
138. Yu, C.F.; Liu, Z.X.; Cantley, L.G. Erk negatively regulates the epidermal growth factor-mediated interaction of gab1 and the phosphatidylinositol 3-kinase. *J. Biol. Chem.* **2002**, *277*, 19382–19388. [CrossRef] [PubMed]
139. Hoeflich, K.P.; O'Brien, C.; Boyd, Z.; Cavet, G.; Guerrero, S.; Jung, K.; Januario, T.; Savage, H.; Punnoose, E.; Truong, T.; et al. In vivo antitumor activity of mek and phosphatidylinositol 3-kinase inhibitors in basal-like breast cancer models. *Clin. Cancer Res.* **2009**, *15*, 4649–4664. [CrossRef]
140. Lehr, S.; Kotzka, J.; Avci, H.; Sickmann, A.; Meyer, H.E.; Herkner, A.; Muller-Wieland, D. Identification of major erk-related phosphorylation sites in gab1. *Biochemistry* **2004**, *43*, 12133–12140. [CrossRef] [PubMed]
141. Tandon, M.; Chen, Z.; Pratap, J. Role of runx2 in crosstalk between mek/erk and pi3k/akt signaling in mcf-10a cells. *J. Cell Biochem.* **2014**, *115*, 2208–2217. [CrossRef] [PubMed]
142. Zmajkovicova, K.; Jesenberger, V.; Catalanotti, F.; Baumgartner, C.; Reyes, G.; Baccarini, M. Mek1 is required for pten membrane recruitment, akt regulation, and the maintenance of peripheral tolerance. *Mol. Cell* **2013**, *50*, 43–55. [CrossRef]
143. Moelling, K.; Schad, K.; Bosse, M.; Zimmermann, S.; Schweneker, M. Regulation of raf-akt cross-talk. *J. Biol. Chem.* **2002**, *277*, 31099–31106. [CrossRef]
144. Zimmermann, S.; Moelling, K. Phosphorylation and regulation of raf by akt (protein kinase b). *Science* **1999**, *286*, 1741–1744. [CrossRef] [PubMed]
145. Cheung, M.; Sharma, A.; Madhunapantula, S.V.; Robertson, G.P. Akt3 and mutant v600e b-raf cooperate to promote early melanoma development. *Cancer Res.* **2008**, *68*, 3429–3439. [CrossRef] [PubMed]
146. Pardee, A.B. A restriction point for control of normal animal cell proliferation. *Proc. Natl. Acad. Sci. USA* **1974**, *71*, 1286–1290. [CrossRef]

147. Pledger, W.J.; Stiles, C.D.; Antoniades, H.N.; Scher, C.D. Induction of DNA synthesis in balb/c 3t3 cells by serum components: Reevaluation of the commitment process. *Proc. Natl. Acad. Sci. USA* **1977**, *74*, 4481–4485. [CrossRef] [PubMed]
148. Pledger, W.J.; Stiles, C.D.; Antoniades, H.N.; Scher, C.D. An ordered sequence of events is required before balb/c-3t3 cells become committed to DNA synthesis. *Proc. Natl. Acad. Sci. USA* **1978**, *75*, 2839–2843. [CrossRef]
149. Zwang, Y.; Sas-Chen, A.; Drier, Y.; Shay, T.; Avraham, R.; Lauriola, M.; Shema, E.; Lidor-Nili, E.; Jacob-Hirsch, J.; Amariglio, N.; et al. Two phases of mitogenic signaling unveil roles for p53 and egr1 in elimination of inconsistent growth signals. *Mol. Cell* **2011**, *42*, 524–535. [CrossRef]
150. Gross, S.M.; Rotwein, P. Unraveling growth factor signaling and cell cycle progression in individual fibroblasts. *J. Biol. Chem.* **2016**, *291*, 14628–14638. [CrossRef]
151. Stiles, C.D.; Capone, G.T.; Scher, C.D.; Antoniades, H.N.; van Wyk, J.J.; Pledger, W.J. Dual control of cell growth by somatomedins and platelet-derived growth factor. *Proc. Natl. Acad. Sci. USA* **1979**, *76*, 1279–1283. [CrossRef]
152. Bennett, A.M.; Hausdorff, S.F.; O'Reilly, A.M.; Freeman, R.M.; Neel, B.G. Multiple requirements for shptp2 in epidermal growth factor- mediated cell cycle progression. *Mol.Cell Biol.* **1996**, *16*, 1189–1202. [CrossRef] [PubMed]
153. Pennock, S.; Wang, Z. Stimulation of cell proliferation by endosomal epidermal growth factor receptor as revealed through two distinct phases of signaling. *Mol.Cell Biol.* **2003**, *23*, 5803–5815. [CrossRef] [PubMed]
154. Taylor, S.J.; Shalloway, D. Cell cycle-dependent activation of ras. *Curr. Biol.* **1996**, *6*, 1621–1627. [CrossRef]
155. Mulcahy, L.S.; Smith, M.R.; Stacey, D.W. Requirement for ras proto-oncogene function during serum- stimulated growth of nih 3t3 cells. *Nature* **1985**, *313*, 241–243. [CrossRef]
156. Dobrowolski, S.; Harter, M.; Stacey, D.W. Cellular ras activity is required for passage through multiple points of the g0/g1 phase in balb/c 3t3 cells. *Mol. Cell Biol.* **1994**, *14*, 5441–5449. [CrossRef]
157. Takuwa, N.; Takuwa, Y. Ras activity late in g1 phase required for p27kip1 downregulation, passage through the restriction point, and entry into s phase in growth factor-stimulated nih 3t3 fibroblasts. *Mol. Cell Biol.* **1997**, *17*, 5348–5358. [CrossRef]
158. Roche, S.; Koegl, M.; Courtneidge, S.A. The phosphatidylinositol 3-kinase alpha is required for DNA synthesis induced by some, but not all, growth factors. *Proc. Natl. Acad. Sci. USA* **1994**, *91*, 9185–9189. [CrossRef] [PubMed]
159. Casagrande, F.; Bacqueville, D.; Pillaire, M.J.; Malecaze, F.; Manenti, S.; Breton-Douillon, M.; Darbon, J.M. G1 phase arrest by the phosphatidylinositol 3-kinase inhibitor ly 294002 is correlated to up-regulation of p27kip1 and inhibition of g1 cdks in choroidal melanoma cells. *FEBS Lett.* **1998**, *422*, 385–390. [CrossRef]
160. Wang, Y.; Pennock, S.; Chen, X.; Wang, Z. Endosomal signaling of epidermal growth factor receptor stimulates signal transduction pathways leading to cell survival. *Mol.Cell Biol.* **2002**, *22*, 7279–7290. [CrossRef]
161. Wang, Y.; Pennock, S.D.; Chen, X.; Kazlauskas, A.; Wang, Z. Platelet-derived growth factor receptor-mediated signal transduction from endosomes. *J. Biol. Chem.* **2004**, *279*, 8038–8046. [CrossRef]
162. Vasjari, L.; Bresan, S.; Biskup, C.; Pai, G.; Rubio, I. Ras signals principally via erk in g1 but cooperates with pi3k/akt for cyclin d induction and s-phase entry. *Cell Cycle* **2019**, *18*, 204–225. [CrossRef] [PubMed]
163. Nam, H.J.; Kim, S.; Lee, M.W.; Lee, B.S.; Hara, T.; Saya, H.; Cho, H.; Lee, J.H. The erk-rsk1 activation by growth factors at g2 phase delays cell cycle progression and reduces mitotic aberrations. *Cell Signal.* **2008**, *20*, 1349–1358. [CrossRef] [PubMed]
164. Park, Y.Y.; Nam, H.J.; Lee, J.H. Hepatocyte growth factor at s phase induces g2 delay through sustained erk activation. *Biochem. Biophys. Res. Commun.* **2007**, *356*, 300–305. [CrossRef]
165. Dangi, S.; Chen, F.M.; Shapiro, P. Activation of extracellular signal-regulated kinase (erk) in g(2) phase delays mitotic entry through p21(cip1). *Cell Prolif.* **2006**, *39*, 261–279. [CrossRef]
166. Mardin, B.R.; Isokane, M.; Cosenza, M.R.; Kramer, A.; Ellenberg, J.; Fry, A.M.; Schiebel, E. Egf-induced centrosome separation promotes mitotic progression and cell survival. *Dev. Cell* **2013**, *25*, 229–240. [CrossRef]
167. Astuti, P.; Pike, T.; Widberg, C.; Payne, E.; Harding, A.; Hancock, J.; Gabrielli, B. Mapk pathway activation delays g2/m progression by destabilizing cdc25b. *J. Biol. Chem.* **2009**, *284*, 33781–33788. [CrossRef]
168. Gomez-Cambronero, J. P42-map kinase is activated in egf-stimulated interphase but not in metaphase-arrested hela cells. *FEBS Lett.* **1999**, *443*, 126–130. [CrossRef]
169. Dangi, S.; Shapiro, P. Cdc2-mediated inhibition of epidermal growth factor activation of the extracellular signal-regulated kinase pathway during mitosis. *J. Biol. Chem.* **2005**, *280*, 24524–24531. [CrossRef] [PubMed]
170. Kiyokawa, N.; Lee, E.K.; Karunagaran, D.; Lin, S.Y.; Hung, M.C. Mitosis-specific negative regulation of epidermal growth factor receptor, triggered by a decrease in ligand binding and dimerization, can be overcome by overexpression of receptor. *J. Biol. Chem.* **1997**, *272*, 18656–18665. [CrossRef]
171. Liu, L.; Shi, H.; Chen, X.; Wang, Z. Regulation of egf-stimulated egf receptor endocytosis during m phase. *Traffic* **2011**, *12*, 201–217. [CrossRef]
172. Wee, P.; Shi, H.; Jiang, J.; Wang, Y.; Wang, Z. Egf stimulates the activation of egf receptors and the selective activation of major signaling pathways during mitosis. *Cell Signal.* **2015**, *27*, 638–651. [CrossRef]
173. Bruns, A.F.; Herbert, S.P.; Odell, A.F.; Jopling, H.M.; Hooper, N.M.; Zachary, I.C.; Walker, J.H.; Ponnambalam, S. Ligand-stimulated vegfr2 signaling is regulated by co-ordinated trafficking and proteolysis. *Traffic* **2010**, *11*, 161–174. [CrossRef]
174. Walker, F.; Zhang, H.H.; Burgess, A.W. Identification of a novel egf-sensitive cell cycle checkpoint. *Exp. Cell Res.* **2007**, *313*, 511–526. [CrossRef]

175. Hitomi, M.; Stacey, D.W. Cellular ras and cyclin d1 are required during different cell cycle periods in cycling nih 3t3 cells. *Mol. Cell Biol.* **1999**, *19*, 4623–4632. [CrossRef]
176. Stacey, D.W.; Hitomi, M.; Chen, G. Influence of cell cycle and oncogene activity upon topoisomerase iialpha expression and drug toxicity. *Mol. Cell Biol.* **2000**, *20*, 9127–9137. [CrossRef]
177. Hitomi, M.; Stacey, D.W. Ras-dependent cell cycle commitment during g2 phase. *FEBS Lett.* **2001**, *490*, 123–131. [CrossRef]
178. Guo, Y.; Stacey, D.W.; Hitomi, M. Post-transcriptional regulation of cyclin d1 expression during g2 phase. *Oncogene* **2002**, *21*, 7545–7556. [CrossRef]
179. Sa, G.; Hitomi, M.; Harwalkar, J.; Stacey, A.W.; Gc, G.C.; Stacey, D.W. Ras is active throughout the cell cycle, but is able to induce cyclin d1 only during g2 phase. *Cell Cycle* **2002**, *1*, 50–58. [CrossRef]
180. Knauf, J.A.; Ouyang, B.; Knudsen, E.S.; Fukasawa, K.; Babcock, G.; Fagin, J.A. Oncogenic ras induces accelerated transition through g2/m and promotes defects in the g2 DNA damage and mitotic spindle checkpoints. *J. Biol. Chem.* **2006**, *281*, 3800–3809. [CrossRef]
181. Wright, J.H.; Munar, E.; Jameson, D.R.; Andreassen, P.R.; Margolis, R.L.; Seger, R.; Krebs, E.G. Mitogen-activated protein kinase kinase activity is required for the g(2)/m transition of the cell cycle in mammalian fibroblasts. *Proc. Natl. Acad. Sci. USA* **1999**, *96*, 11335–11340. [CrossRef]
182. Rahmouni, S.; Cerignoli, F.; Alonso, A.; Tsutji, T.; Henkens, R.; Zhu, C.; Louis-dit-Sully, C.; Moutschen, M.; Jiang, W.; Mustelin, T. Loss of the vhr dual-specific phosphatase causes cell-cycle arrest and senescence. *Nat. Cell Biol.* **2006**, *8*, 524–531. [CrossRef]
183. Zhang, Z.; Miao, L.; Lv, C.; Sun, H.; Wei, S.; Wang, B.; Huang, C.; Jiao, B. Wentilactone b induces g2/m phase arrest and apoptosis via the ras/raf/mapk signaling pathway in human hepatoma smmc-7721 cells. *Cell Death Dis.* **2013**, *4*, e657. [CrossRef]
184. Kandel, E.S.; Skeen, J.; Majewski, N.; di Cristofano, A.; Pandolfi, P.P.; Feliciano, C.S.; Gartel, A.; Hay, N. Activation of akt/protein kinase b overcomes a g(2)/m cell cycle checkpoint induced by DNA damage. *Mol. Cell Biol.* **2002**, *22*, 7831–7841. [CrossRef]
185. Saito, Y.; Swanson, X.; Mhashilkar, A.M.; Oida, Y.; Schrock, R.; Branch, C.D.; Chada, S.; Zumstein, L.; Ramesh, R. Adenovirus-mediated transfer of the pten gene inhibits human colorectal cancer growth in vitro and in vivo. *Gene Ther.* **2003**, *10*, 1961–1969. [CrossRef]
186. Shtivelman, E.; Sussman, J.; Stokoe, D. A role for pi 3-kinase and pkb activity in the g2/m phase of the cell cycle. *Curr. Biol.* **2002**, *12*, 919–924. [CrossRef]
187. Blasina, A.; de Weyer, I.V.; Laus, M.C.; Luyten, W.H.; Parker, A.E.; McGowan, C.H. A human homologue of the checkpoint kinase cds1 directly inhibits cdc25 phosphatase. *Curr. Biol.* **1999**, *9*, 1–10. [CrossRef]
188. Puc, J.; Keniry, M.; Li, H.S.; Pandita, T.K.; Choudhury, A.D.; Memeo, L.; Mansukhani, M.; Murty, V.V.; Gaciong, Z.; Meek, S.E.; et al. Lack of pten sequesters chk1 and initiates genetic instability. *Cancer Cell* **2005**, *7*, 193–204. [CrossRef]
189. Dangi, S.; Cha, H.; Shapiro, P. Requirement for phosphatidylinositol-3 kinase activity during progression through s-phase and entry into mitosis. *Cell. Signal.* **2003**, *15*, 667–675. [CrossRef]
190. Benary, M.; Bohn, S.; Lüthen, M.; Nolis, I.K.; Blüthgen, N.; Loewer, A. Disentangling pro-mitotic signaling during cell cycle progression using time-resolved single-cell imaging. *Cell Rep.* **2020**, *31*, 107514. [CrossRef]

Review

Cell Cycle, Telomeres, and Telomerase in *Leishmania* spp.: What Do We Know So Far?

Luiz H. C. Assis [1,†], Débora Andrade-Silva [1,†], Mark E. Shiburah [1], Beatriz C. D. de Oliveira [1], Stephany C. Paiva [1], Bryan E. Abuchery [2], Yete G. Ferri [2], Veronica S. Fontes [1], Leilane S. de Oliveira [1], Marcelo S. da Silva [2,*] and Maria Isabel N. Cano [1,*]

[1] Telomeres Laboratory, Department of Chemical and Biological Sciences, Biosciences Institute, São Paulo State University (UNESP), Botucatu 18618-689, Brazil; lhc.assis@unesp.br (L.H.C.A.); debora.dede@gmail.com (D.A.-S.); me.shiburah@unesp.br (M.E.S.); bcd.oliveira@unesp.br (B.C.D.d.O.); stephany.paiva@unesp.br (S.C.P.); vs.fontes@unesp.br (V.S.F.); leilane.oliveira@unesp.br (L.S.d.O.)

[2] DNA Replication and Repair Laboratory (DRRL), Department of Chemical and Biological Sciences, Biosciences Institute, São Paulo State University (UNESP), Botucatu 18618-689, Brazil; bryan.abuchery@unesp.br (B.E.A.); yete.gambarini@unesp.br (Y.G.F.)

* Correspondence: marcelo.santos-silva@unesp.br (M.S.d.S.); maria.in.cano@unesp.br (M.I.N.C.)

† Both authors contributed equally to the paper.

Abstract: Leishmaniases belong to the inglorious group of neglected tropical diseases, presenting different degrees of manifestations severity. It is caused by the transmission of more than 20 species of parasites of the *Leishmania* genus. Nevertheless, the disease remains on the priority list for developing new treatments, since it affects millions in a vast geographical area, especially low-income people. Molecular biology studies are pioneers in parasitic research with the aim of discovering potential targets for drug development. Among them are the telomeres, DNA–protein structures that play an important role in the long term in cell cycle/survival. Telomeres are the physical ends of eukaryotic chromosomes. Due to their multiple interactions with different proteins that confer a likewise complex dynamic, they have emerged as objects of interest in many medical studies, including studies on leishmaniases. This review aims to gather information and elucidate what we know about the phenomena behind *Leishmania* spp. telomere maintenance and how it impacts the parasite's cell cycle.

Keywords: *Leishmania* spp.; leishmaniases; cell cycle; telomeres; telomerase

Citation: Assis, L.H.C.; Andrade-Silva, D.; Shiburah, M.E.; de Oliveira, B.C.D.; Paiva, S.C.; Abuchery, B.E.; Ferri, Y.G.; Fontes, V.S.; de Oliveira, L.S.; da Silva, M.S.; et al. Cell Cycle, Telomeres, and Telomerase in *Leishmania* spp.: What Do We Know So Far? *Cells* **2021**, *10*, 3195. https://doi.org/10.3390/cells10113195

Academic Editor: Zhixiang Wang

Received: 16 October 2021
Accepted: 14 November 2021
Published: 16 November 2021

Publisher's Note: MDPI stays neutral with regard to jurisdictional claims in published maps and institutional affiliations.

1. Introduction

Leishmaniases are among the poverty-related endemic diseases. They are well-known to cause a wide spectrum of clinical manifestations and their harsh incidences in East Africa, the Indian subcontinent, and Latin America, where approximately one million new diagnostics are expected yearly [1]. The disease is vector-induced and caused by more than twenty species of the *Leishmania* genus, protozoan parasites that belong to the Trypanosomatidae family [1]. The invertebrate host is a phlebotomine insect that is infected during a blood meal with amastigote forms. Inside the insect digestive system, amastigotes transform into procyclic promastigotes, which are noninfective but highly proliferative forms. Procyclics eventually migrate to the proboscis and differentiate into infective metacyclic promastigote forms [2,3]. The transmission to humans occurs when a preinfected insect (female phlebotomines) regurgitates during its blood meal infective but nonproliferative forms (metacyclic promastigotes) into the mammalian host skin. Afterward, metacyclics are phagocytosed by neutrophils or macrophages, and further inside the phagolysosomes, the parasites undergo a series of morphogenetic modifications, leading to the formation of amastigotes. The amastigotes then multiply and reach the bloodstream, causing the initiation of clinical manifestations. A new cycle of infection can occur when infected macrophages are ingested by other female phlebotomines (Figure 1).

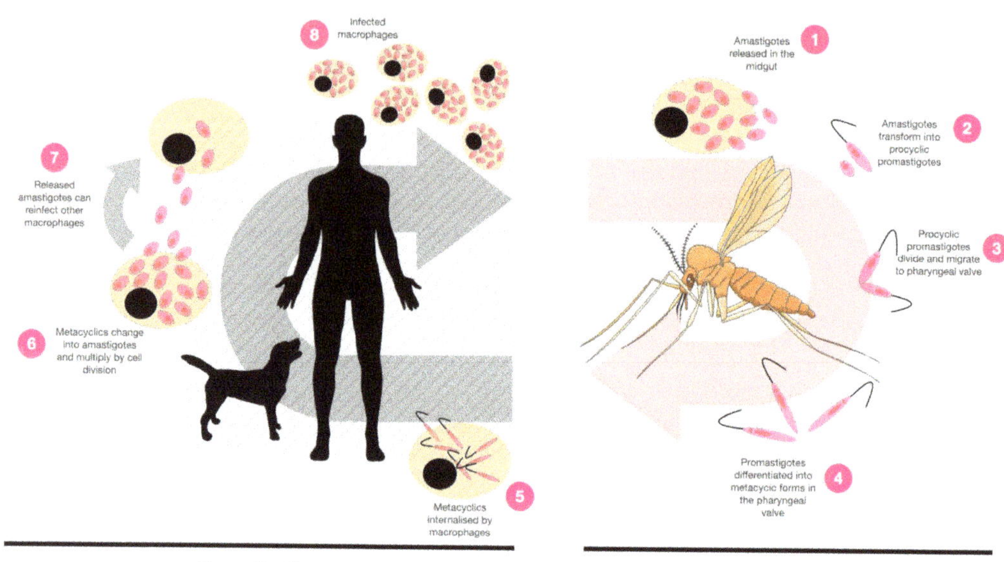

Figure 1. Schematic illustration of the *Leishmania* spp. infective cycle showing different parasite stages of development. Phlebotomine female sandflies get infected with amastigote forms during bloodmeals (1). Amastigotes transform into procyclic promastigotes, which will proliferate inside the invertebrate midgut (2 and 3). Promastigotes then migrate to the sandflies' pharyngeal valve (4) and differentiate into metacyclic forms. The metacyclic forms are transferred to the mammalian hosts' bloodstream during a new blood meal and infect macrophages and other cells from the mononuclear phagocyte system (5). Inside the macrophages, there is a metacyclic change into amastigote forms, which multiply, lyse the macrophage, and reinfect new macrophages (6 and 7). In a new cycle of infection, the infected macrophages are ingested by new phlebotomines (8). The silhouettes of man and dog and the sandfly clipart are free for use and were withdrawn from the websites HiClipart (https://www.hiclipart.com, accessed on 8 November 2021) and Gratispng (https://www.gratispng.com, accessed on 8 November 2021), respectively.

Different species of *Leishmania* associated with the host biology and vector factors can lead to different clinical symptoms, ranging from light and self-cured cutaneous lesions (e.g., *Leishmania major*) to life-threatening visceral complications (e.g., *Leishmania infantum*) [2]. The disease is endemic in more than 60 countries, with East Africa and the Indian subcontinent remarkably impacted. Although a significant decrease in the number of cases has been reported in the past years, mainly due to efforts in vector control, leishmaniasis is still on the top priority list for developing new treatments. Such a scenario is typified, because toxic antiparasitic drugs (e.g., antimonial pentavalent and amphotericin B) are still in use nowadays. Additionally, other therapeutic approaches such as vaccination have been shown to be not adequate in eradicating the disease [4]. Thus, there is urgency for new specific and efficient treatments against leishmaniasis. For this purpose, several studies on *Leishmania* spp. have been conducted to elucidate the different aspects of parasite biology, including genetic studies and full sequencing of the parasite genome [4–6]. Among these studies, trypanosomatid telomere biology has awakened great interest in the scientific community, as telomeres are essential for genome stability and cell cycle progression. Furthermore, some components involved in parasite telomere maintenance, such as the telomerase RNA component and members of the shelterin-like telomeric complex (e.g., TRF, RPA-1, and Rbp38), are unique and present parasite-specific features relative to their hosts [7–14]. Therefore, new advances in understanding and exploring the parasite telomeric environment may reveal how these and other unknown parasite factors could be used as potential and specific targets for drug development [8–19].

Telomeres are the physical ends of eukaryotic linear chromosomes. Structurally, they are an association between tandemly repeated noncoding DNA sequences and nucleoproteins forming complexes at the end of chromosomes [20]. Telomeric DNA is composed of a double-stranded sequence (one of them rich in cytosine, the other in guanine) and a G-rich single strand that protrudes toward the ends of the chromosome, known as the 3′ overhang [20]. In humans and other vertebrates, the repeated telomeric sequence is 5′-TTAGGG-3′, approximately 3–15 kb in length and associated with a six-telomeric protein complex called shelterin [21–25]. These arrangements of DNA and proteins influence cell homeostasis. They are crucial to cell cycle maintenance, being decisive in important cellular processes such as cell aging, genome integrity, and maintenance of the nuclear arrangement [26–28]. It is known that, lengthwise, telomeres tend to shorten after each cell division due to the inability of DNA polymerase to complete replications in the lagging strand of linear chromosomes [29], culminating in the progressive loss of telomere repeats. Thus, telomeres act as a molecular clock. If its size reaches critical levels, it can lead to early/unprogrammed cell senescence or even activate local DNA damage repair, eventually triggering a mitotic catastrophe [14,30–32]. In most organisms, including *Leishmania* spp., telomeric DNA is elongated by telomerase. This specialized reverse transcriptase forms a ribonucleoprotein (RPN) complex [31], whose function is strictly regulated throughout the cell cycle [17,21,32]. In that sense, telomerase, telomeres, and their associated proteins are now recognized as potential drug targets [33]. In the face of the critical medical relevance of leishmaniases, peculiarities associated with the cell cycle, telomeres dynamics, and telomerase have reached the mainstream studies on parasite biology. This review aims to shed light on what we know about the phenomena behind telomere maintenance and how it impacts parasites' cell cycle and survival.

2. *Leishmania* spp. Cell Cycle

The cell cycle comprises a repeating series of events encompassing growth, DNA synthesis, and cell division. For most eukaryotes (including *Leishmania* spp.), the cell cycle follows a single pattern of organization consisting of the following phases: G1 (gap 1), S (DNA synthesis), G2 (gap 2), M (mitosis), and C (cytokinesis). In addition, the G0 state is sometimes included as part of the cell cycle. However, it is noteworthy that cells in the G0 state are not stimulated to proliferate [34]. In other words, cells in the G0 state are not yet committed to genome replication and cell division. Interestingly, the infective forms of most parasitic microorganisms are at the G0 state (e.g., metacyclic promastigotes of *Leishmania* spp.) [35,36] (Figure 2), which makes us wonder if, in these organisms, proliferation and infection are mutually exclusive events.

In *Leishmania* spp., as in most trypanosomatids, the G1 phase corresponds to the larger proportion of the cell cycle, while the other phases vary slightly in duration [42,43]. No studies have evidenced the metabolic changes and main events during the G1 phase in *Leishmania* spp. However, compared with the other trypanosomatids, we can infer that, in the G1 phase, there is an increase in the transcription rate and intense protein synthesis of the factors related to DNA replication [44,45]. Moreover, in the G1 phase occurs the establishment of a divergent prereplication protein complex at specific sites on the chromosomes called replication origins, which can give rise to a replication bubble [45].

The firing of replication origins starts the S phase, which briefly consists of the reliable replication of DNA molecules. Usually, eukaryotes have many replication origins per chromosome. However, in *Leishmania* spp., the number of replication origins per chromosome is an issue that generates debate. Six years ago, Marques et al., (2015) [46] used a marker frequency analysis coupled with deep sequencing (MFA-seq) to evidence that *L. major* replicates each of its chromosomes during the S phase using a single replication origin. However, in a more recent study, da Silva et al., (2020) [47] used mathematical equations to reveal that it is improbable that *L. major* uses only a single origin per chromosome during the S phase. The explanation for these discrepancies is that the MFA-seq approach is not sensitive enough to identify all replication origins [47]. Nevertheless, further sensitive assays are needed to establish how many origins are used per chromosome during the S phase in *Leishmania* spp.

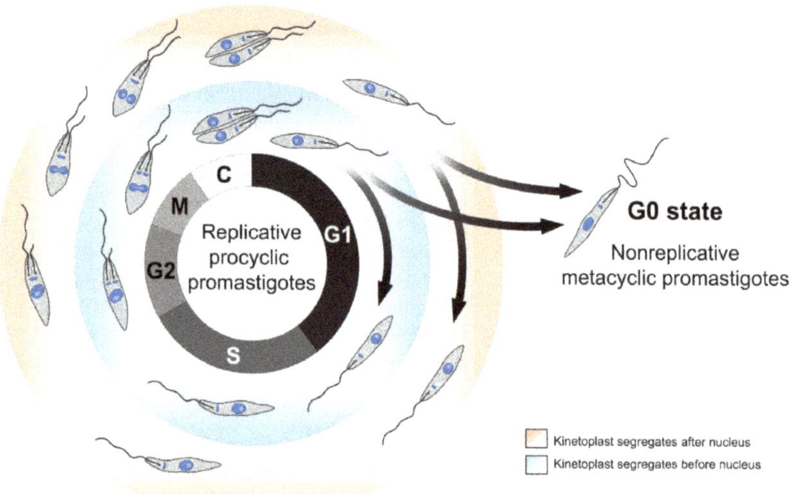

Figure 2. Scheme showing the two distinct patterns of kinetoplast segregation relative to the nucleus in promastigotes of *Leishmania* spp. throughout the cell cycle. Most *Leishmania* species presents one of the two kinetoplast segregation patterns presented: kinetoplast segregates after the nucleus (light red) or kinetoplast segregates before the nucleus (light blue). For instance, *L. mexicana* segregates its kinetoplast predominantly after the nucleus [37], while *L. major* and *L. tarentolae* do the opposite [38,39]. However, *L. donovani* and *L. amazonensis* exhibit these two patterns distributed in the population [40,41].

In model eukaryotes, the G2 phase is characterized by the duplication of centrioles and other cytoplasmic organelles [48]. In trypanosomatids, homologs of the proteins described in model eukaryotes involved with centriole biogenesis are associated with the basal body and flagellum biogenesis [49,50]. Furthermore, based on studies with other organisms [51], we can infer that, in the G2 phase, *Leishmania* spp. increase the rate of transcription and resumption of intense protein synthesis, which are necessary for the completion of cell division. This entire process results in an increase in cell volume and size [37,40].

During the M phase, *Leishmania* spp. and all other trypanosomatids do not disassemble their nuclear envelope and perform a closed mitosis process, which is organized by spindle pole body-like structures [52]. Moreover, due to the absence of the N-terminal portion and globular domain of histone H1 and the absence of phosphorylation on serine 10 of histone H3 (H3S10), *Leishmania* spp. and other trypanosomatids are unable to condensate their chromosomes into 30-nm fibers [53,54].

In mammals, a curious feature worth being highlighted is that mitosis and cytokinesis overlap, since cytokinesis begins before the mitotic chromosome segregation is complete. However, *Leishmania* spp. seem not to strictly follow this premise. Once mitosis ends, *Leishmania* spp. undergo a rapid remodeling in shape, first growing in length and then in width prior to cytokinesis, which ends cell division [52,55]. Although it is challenging, all these peculiarities related to the cell cycle phases may provide new routes toward the search for suitable targets for parasite cell cycle interventions aiming at their elimination.

A set of events that deserve being highlighted during the *Leishmania* spp. cell cycles refers to the replication and segregation of the kinetoplast. The coordination of these events throughout the *Leishmania* cell cycle does not follow those equivalents in model eukaryotes, where mitochondrial DNA replicates at any cell cycle stage [56,57]. The nuclear and kinetoplast S phase occurs almost simultaneously, but the effective segregation of these organelles can occur at different periods according to the species analyzed. Many studies have established a pattern of segregation for the kinetoplast relative to the nucleus in some species of *Leishmania* [37–41].

One of these studies characterized the main morphological events of the cell cycle of *L. mexicana* promastigotes [37]. Wheeler et al., (2011) [37] described the cell cycle phase durations and established a duplication order for the kinetoplast and nucleus. Two years later, da Silva et al., (2013) [40] characterized the length of the *L. amazonensis* cell cycle phases and evidenced that the promastigote forms present two distinct modes of kinetoplast segregation relative to the nucleus, following a distinct temporal order in different proportions of the cells. Using DAPI staining and EdU (5-ethynyl-2′-deoxyuridine) to monitor, respectively, the segregation of DNA-containing organelles and DNA replication, the authors found a curious feature: 65% of the dividing promastigotes duplicate the kinetoplast before the nucleus, and the remaining 35% do the opposite or duplicate both organelles concomitantly. This finding corroborates another study carried out with *L. donovani* promastigotes, where the authors found that about 80% of the cells divide the nucleus before kinetoplast, and the remaining 20% do the opposite [41]. In other words, *L. donovani* and *L. amazonensis* exhibit a nonfixed pattern of nucleus and kinetoplast segregation. In fact, when we compare the cell cycle among different *Leishmania* spp., the order and timing of the kinetoplast and nucleus division are not consensual and cannot be generalized [37–41].

For instance, *L. mexicana*, *L. major*, and *L. tarentolae* exhibit fixed patterns of kinetoplast and nucleus segregation. However, *L. mexicana* segregates its kinetoplast predominantly after the nucleus [37], while *L. major* and *L. tarentolae* do the opposite [39,41] (Figure 2). A possible explanation for these different behaviors resides in the fact that, although belonging to the same genus, these parasites show considerable phylogenetic distance [58]. In other words, this phylogenetic divergence may reflect possible species-specific differences relative to kinetoplast segregation, suggesting that some *Leishmania* spp. have less stringent control over the order of division of their DNA-containing organelles (nucleus and kinetoplast). More studies are needed to uncover the potential players involved in controlling cell division and organelle segregation, since some of them could be explored for precise interventions related to the parasite cell cycle.

3. *Leishmania* spp. Telomeres

In *Leishmania* spp., telomeres constitute short noncoding repetitive sequences of 5′ TTAGGG 3′ [16,59], except for *Leishmania braziliensis*, which, in addition to the conventional sequence, is also observed the presence of 5′-CCTAACCCGTGGA-3′ sequences at the ends of some chromosomes [60]. While, in *L. amazonensis*, the 3′ G overhang has an approximate size of 12 nt long (5′-GTTAGGGTTAGG-3′), in *L. donovani* and *L. major*, the 3′G overhang is a 9-nt sequence composed of 5′ GGTTAGGGT 3′ [61]. The studies performed by Genest & Borst [62] described that the length of the *L. tarentolae* and *L. major* telomeres increase over time by approximately 1 bp per population doubling. Similar results were obtained by Oliveira et al., (2021) [15], where telomeres of the *L. amazonensis* procyclic promastigotes increased during continuous passages. It was also observed that amastigote telomeres (Figure 1, numbers 1, 6, and 7) are shorter compared to procyclic (Figure 1, number 3) and metacyclic promastigotes (Figure 1, numbers 4 and 5).

Near telomeres, there are the subtelomeric sequences, which, in *Leishmania*, are composed of 100-bp-long conserved telomere-associated sequences (LCTAS) that contain two conserved domains, the conserved sequence block 1 (CSB1) and conserved sequence block 2 (CSB2) [60]. These sequences have a high degree of conservation among the different *Leishmania* species, which may indicate their importance in chromosome segregation or as binding sites for telomeric proteins [60]. However, despite the CSB sequences and depending on the species, LCTAS also contains some nonconserved sequences and a high degree of polymorphism [59,60,63,64]. Furthermore, telomeres are commonly associated with proteins that interact directly with both telomeric strands and other telomeric proteins that can influence the telomere size by regulating the telomerase access, providing chromosome stability and protecting telomeres from degradation [65]. So far, within the *Leishmania* genus, the *L. amazonensis* telomeric complex (Figure 3) is the best-characterized.

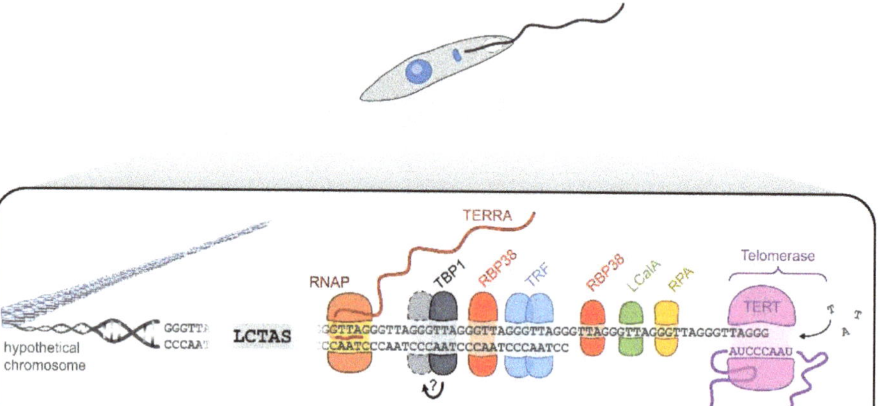

Figure 3. *Leishmania* spp. telomeres are formed by subtelomeric regions (LCTAS); a double-stranded region; a single-stranded protrusion (3′-G overhang); and associated proteins, such as TBP-1, TRF, Rbp38, RPA-1, and LCalA. Of note, it is not clear if TBP1 binds DNA as a monomer or dimer. The telomeres are elongated by telomerase, a ribonucleoprotein composed minimally by TERT and TER. From the C-rich strand of the telomeric region, ncRNAs called TERRAs are transcribed and appear to be involved in telomere length regulation. Adapted from da Silva et al., 2012 [66].

Lira et al., (2007) [67] described a 45-kDa telomere-binding protein named LaTBP1 as a *Leishmania* double-stranded DNA-binding protein that interacts with GT-rich and telomeric DNA sequences. LaTBP1 has at least a central Myb-like DNA-binding domain containing a conserved hydrophobic cavity involved in DNA binding, a feature conserved among the proteins that bind the telomeric double-stranded DNA [67]. However, it is unclear if LaTBP1 binds DNA as a monomer like RAP1 (Repressor Activator Protein 1) in mammals or as a dimer like the TRFs. It is worth highlighting that the LaTBP1 Myb-like DNA-binding domain is related closer to the two centrally located RAP1 Myb domains than the single C-terminal TRF (Telomere Repeat-binding Factor) Myb domain [67].

Later, a TRF homolog was described in *L. amazonensis* (LaTRF) by da Silva et al., 2010 [68]. This protein presents 82.5 kDa, shares structural features with mammalian TRF1 and TRF2, and is highly similar to *Trypanosoma brucei* TRF. It also contains a Myb-like DNA-binding domain that allows it to bind double-stranded telomeric DNA. Apparently, LaTRF is little abundant, and it is localized in the nucleus [68].

Some proteins that bind single-stranded telomeric DNA were identified in *L. amazonensis* [69]. RNA-binding protein 38 (LaRbp38) is one of those. It is a 38-kDa protein with a noncanonical binding site with an affinity to both single- and double-stranded G-rich telomeric DNA and GT-rich kinetoplast DNA (kDNA) [11,70]. LaRbp38 has dual cellular localization, and in synchronized *Leishmania* promastigotes, it shuttles from the mitochondria to the nucleus at the late S and G2 phases via importin alpha [11]. LaRbp38 was first identified in a telomerase-positive extract together with RPA-1, LCalA, and DNA polymerase alpha [69]. The LaRbp38 ability to bind telomeres and kDNA, and the fact that this protein is involved with kDNA stability and replication, suggest that it may also be involved with telomere elongation. However, further assays are necessary to check the veracity of this hypothesis.

Replication protein A1 (LaRPA-1) was another protein pulled down with LaRbp38 [71]. LaRPA-1 is a nuclear protein of 51 kDa that presents a single N-terminal OB-fold domain. This protein binds the telomeric G-rich ssDNA, protecting it from $3'$-$5'$ exonucleolytic degradation. Thus, LaRPA1 is considered a potential trypanosomatid telomere end-binding protein (TEBP) that shares structural and functional features with other TEBPs described in model eukaryotes. It binds at least one telomeric repeat using an OB-fold domain, protects telomeres from exonuclease degradation, and unfolds the telomeric G-quadruplex [12,71,72]. The G-quadruplex structure is known to impair telomerase activity in humans [73,74]. Although not tested against *Leishmania* spp., drugs that specifically bind these structures could be a new source for antiparasitic therapies.

LCalA (MW 16 kDa) was the third protein identified in the parasite telomerase-positive extracts [75]. It is the first reported protozoa calmodulin-like nuclear protein that binds in vivo to the G-rich telomeric strand and the $3'$ G overhang. The binding of LCalA to telomeres is calcium-dependent. Biophysical assays showed structural changes in LCalA in the presence of calcium ions, increasing the affinity of this protein to telomeres. Additionally, LCalA partially colocalizes with telomeres throughout the parasite's cell cycle. LCalA resembles human KIP1 protein, since both share with calmodulins two EF hand domains and the affinity for calcium. Furthermore, KIP1 was shown to be involved with telomere homeostasis by its interactions with the telomerase TERT component and with shelterin member TRF2 through the EF hand domains. Whether LCalA is a functional homolog of KIP1 is a question that remains open.

Long noncoding RNAs called TERRA (telomeric repeat-containing RNA) were also described at *Leishmania* telomeres. TERRA is transcribed from the C-rich subtelomeric strand towards the end of chromosomes. Its main function is to regulate the telomere length [76]. In *L. major*, TERRA was shown to be polyadenylated and processed by trans-splicing [77]. It was also observed that the number of TERRA transcripts was higher in the infective forms of the parasite (metacyclic promastigotes) relative to the procyclic promastigote and amastigote forms [77]. Moreover, Morea et al., (2021) [77] showed that TERRA could form R-loops, suggesting that *L. major* TERRA is engaged with various cellular processes, including telomere maintenance and the regulation of telomere lengths.

Another interesting finding related to *Leishmania* telomeres is the presence of base J (β-d-glucosyl-hydroxymethyluracil). Base J is a modified thymine that can interfere with RNA polymerase II function, and its presence abrogates DNA cleavage by restriction endonucleases [78]. This hypermodified DNA base was first described in the bloodstream form of *T. brucei* [79,80]. Later, base J was characterized in many other trypanosomatids, and in most of them, except the *Leishmania* spp., it was found spread out in different chromosome regions [81,82]. In *Leishmania* spp., ~98% of base J is found in telomeres [62], suggesting that it can be related to telomere function [83]. Furthermore, the recent results of Morea et al. (2021) [77] suggest that differences in the telomeric base J levels may control TERRA transcription in the *L. major* developmental forms and during continuous in vitro passages.

4. *Leishmania* spp. Telomerase

4.1. Structure and Function

Telomerase is the ribonucleoprotein responsible for telomere elongation. The enzyme adds $5'$-TTAGGG-$3'$ repetitions at the single-stranded $3'$G overhangs of telomeres [83]. Thus, telomerase is thought to be essential to the regulation of telomere lengths and maintenance of genomic stability. Telomerase presents two main components, protein Telomerase Reverse Transcriptase (TERT) and Telomerase RNA (TER), minimally required for enzyme activity in vitro [84,85]. TERT is the catalytic component, and TER contains the template used by TERT to synthesize telomeric repeats. These two components form a complex with accessory proteins and are necessary for in vivo biogenesis, enzyme activity, and nucleotide addition processivity [13,32,86]. *Leishmania* spp. telomerase activity was first reported by Cano et al., (1999) [87], and enzyme purification and biochemical characterization were further described by Giardini et al., 2011 [88]. Parasite enzyme activity was detected in all three parasite life forms presenting the canonical properties of other telomerases. However, the catalysis was shown to be temperature and life-stage dependent [15]. The genes encoding the *Leishmania* spp. TERT was characterized by Giardini et al., (2006) [89], and the TER component was described later by Vasconcelos et al., 2014 [7].

In trypanosomatids, as in most eukaryotes, the TERT component is structurally composed of four domains [13]: The Telomerase Essential N-terminal (TEN), Telomerase RNA-Binding Domain (TRBD), Reverse Transcriptase domain (RT), and the C-terminal extension (CTE). The TEN domain is involved in telomerase recruitment to the telomeres and enzyme processivity [90]. The TRBD interacts with TER and is connected to the TEN domain by an unstructured linker and creates the RNA-binding pocket that binds single stranded and paired RNA [91]. Both domains can also interact with proteins that stabilize the complex and help to recruit telomerase to telomeres and regulate enzyme activity [92–97]. The RT is the catalytic core of the enzyme, which interacts with TER through the pseudoknot region [98], and it is involved in the interactions with hybrid RNA–DNA. The CTE domain stabilizes the RNA–DNA duplex, and differently from the other three domains, it is less conserved among different species [13]. *Leishmania* TERT preserves all the canonical domains found in other TERTs but shows some amino acid substitutions that are specific to the genus [89]. The knockout of *Leishmania* TERT seems to be very harmful to the parasite, because it induces a gradual decrease in cell density in the culture, apparent during G1/G0 cell cycle arrest, morphological alterations, and telomere shortening (unpublished data).

The RNA component TER varies in length and sequence, presenting a conserved secondary structure in most eukaryotes. Variations of the TER size and sequence are observed among different organisms, and they are more prominent than TERT [99], which is conserved even among different taxa. In *Leishmania* spp., TER (LeishTER) is about ~2100 nucleotides long, and the mature molecule modified by trans-splicing contains a 5′ cap, a spliced leader sequence (SL), and a 3′ polyA tail. It also presents a C/D box snoRNA domain found in other TER [7]. LeishTER is expressed at similar levels in its procyclic and metacyclic promastigote forms. The mature molecule coimmunoprecipitates and colocalizes with the TERT component in a cell cycle-dependent manner. Its secondary structure prediction shows the template sequence (5′→3′) in a single-stranded form localized near the 5′ end of the RNA molecule. LeishTER also presents a conserved TBE (Template Boundary Element) motif C[U/C]GUCA in helix II, which is responsible for interacting with the TERT TRB domain [7]. Its double knockout led to partial cell cycle arrest and increased apoptosis in procyclic promastigotes. TER KO also triggers a progressive telomere shortening during continuous parasite passages (unpublished data). A similar effect was observed in *T. brucei* TER knockouts [100].

4.2. Biogenesis and Mechanisms

The biogenesis of the telomerase ribonucleoprotein complex is an intricate process involving the assembly of proteins and TER with the subcellular localization of the ribonucleoprotein. Besides TERT and accessory proteins, other proteins interact transiently with telomerase. These other proteins are important players in maturation, stability, and subcellular localization [101,102].

The telomerase biogenesis starts with the transcription of TERT and TER mRNAs by RNA polymerase II. While TERT mRNA is addressed to the cytoplasm to be translated, the TER remains in the nucleus and is assembled with the accessory proteins at the 3′end—specifically, at the C/D box motif [7]. Unfortunately, no 3′end-binding protein has yet been identified in *Leishmania* spp. Still, considering the other trypanosomatids, it is known that the C/D box motif of *T. brucei* TER interacts with SNU13, a protein known to interact with the C/D box in other eukaryotes. Together with SNU13, in humans, three other proteins recognize and bind to this RNA motif: fibrillarin, NOP56, and NOP58. However, the evidence for homologs of these proteins is still missing in these parasites.

In the case of TERT mRNA, in most organisms, after translation, the protein migrates from the cytoplasm to the nucleus, where the telomerase is assembled. In humans, the main proteins involved in TERT traffic are HSP90, its cochaperone p23, and two AAA$^+$ ATPases (Pontin and Reptin) [101–103]. However, a complete understanding of the TERT traffic in *Leishmania* spp. remains to be determined. In these parasites, the HSP90 ortholog is the protein HSP83. Recently, Oliveira et al., (2021) [15] revealed an interaction with the TERT component and the importance of HSP83 for telomerase activity and telomere maintenance in *L. amazonensis*. These findings suggest that *Leishmania* spp. TERT traffic to the nucleus may resemble the human telomerase process. Furthermore, in this study, the HSP90 inhibitor 17AAG affected both the *L. amazonensis* telomere length and telomerase activity using a thousand-times less of the drug needed to inhibit HSP90 in host cells [104].

Considering the assemblance of human telomerase as an example, the location of telomerase in Cajal bodies is the key to ribonucleoprotein trafficking and recruitment to the telomeres [101,102]. This subcellular localization is derived by the CAB box sequence at the 3′end of human TER and its interaction with the TCAB1 protein. Otherwise, in the context of *Leishmania*, nothing is known about the existence of Cajal bodies [105]. The closest information is related to the identification of the MTAP protein, a TCAB1 protein ortholog of *T. brucei* [106] that can interact with TER [107]. Therefore, the biogenesis of telomerase in *Leishmania* spp. still has many gaps to be replenished.

Once the telomerase RNP complex is assembled, it is recruited preferentially to the shortest telomeres. The ability to add more than one repeat at one telomere is called repeat addition processivity (RAP) [108]. In humans, telomerase activity is regulated by shelterin subcomplex POT1-TPP1, with TPP1 being the key player [109–111]. These two proteins are important to decrease the dissociation rate of telomerase and help the translocation step. In *Leishmania* spp., RPA-1 seems to play the same role as POT1 at telomeres [12]. However, no ortholog of TPP1 was described in the parasite. The identification of other parasite telomerase cofactors and the characterization of this multistep process are under investigation.

Interestingly, in mammals, it is known that telomerase is possibly involved in other roles not related to telomere maintenance, such as cellular proliferation, mitochondrial activity, and gene expression regulation [112]. In *Leishmania* spp., for example, Campelo et al., (2015) [113] studied the TERT component of *Leishmania major* and observed that, besides the nucleus, it could also be found in its unique mitochondrion. In addition, parasites exposed to hydrogen peroxide showed increased TERT levels, especially in the mitochondria compartment. However, it is important to remember that parasite mitochondrion contains circular DNA (kDNA), which does not have telomeres. Therefore, it is important to demonstrate if parasite telomerase presents extra telomeric functions, since it could increase the druggable potential of this important ribonucleoprotein.

Our belief in telomerase as a good target for antiparasitic drug development is also related to its essential role in cell homeostasis and genome stability. We must also consider the structural differences between the telomerase complex (TERT and TER) of mammals and *Leishmania* spp. [7,89] and the fact that telomerase is inactive in most mammalian somatic cells. Thus, it is reasonable to hypothesize that impairing the infective parasite forms to elongate telomeres would directly affect parasites without harming host cells.

4.3. Phylogenetic Context of Leishmania spp. Telomerases among Other Pathogenic Trypanosomatids

Our very little knowledge about telomerases in *Leishmania* spp. reflects the challenges involved in the genome manipulation of these parasites [114,115]. However, the recent advent of genome editing using CRISPR-Cas9 and its variations should help resolve some of these functional questions that are still lingering, as has been done for other genes and in other species [116–118]. In addition, the available evidence on the subject matter shows some differences in the TERT between members of the trypanosomatid family and may be due to the enzyme's usefulness to the parasite [89,119].

Considering the importance of a high proliferative rate for the cell cycle of these parasites, the question of whether the parasite depends on an enzyme for longevity, as has been reported in other cells, is still unanswered. The knockdown of *T. brucei* telomerase induced telomere shortening without effects on the parasite growth [119]. However, the existent differences between trypanosomatids cannot be confidently extrapolated. In *Leishmania* spp., for example, the absence of the TERT and the TER minimal telomerase components seems to be very harmful to the cells (see Section 4.1).

Between the members of the *Leishmania* genus, Giardini et al., (2006) [89] found that the TERT component of *L. donovani*, *L. amazonensis*, *L. major*, and *L. braziliensis* has a nucleotide identity exceeding 90%, while sharing a little similarity with *T. cruzi* and *T. brucei*. Concerning amino acids, their sequence analysis shows over 70% identity between the *Leishmania* spp. and under 40% similarity with *T. cruzi* and *T. brucei* [89]. The *Leishmania* TER component also does not share nucleotide sequence similarities with other trypanosome TERs. They only share similarities in their secondary structures [7].

In other eukaryotes, the TERT component is also involved with noncanonical and extratelomeric functions, such as cell proliferation and apoptosis [120–124]. However, these functions may more likely result from evolutional adaptation events, which sometimes result in the loss and gain of certain protein domains with organism-specific functions. Hopefully, with the new approaches based on the CRISPR/Cas system, the many gaps of the *Leishmania* spp. telomerase will begin to be covered, and more details will probably be available to elucidate the telomere biology of these parasites.

5. Conclusions Remarks

The telomeres dynamics throughout the cell cycle are an essential phenomenon for all eukaryotic cells. Additionally, for most eukaryotes, the holoenzyme telomerase is the key behind these dynamics. When we expand this fact for single-cell parasites, we assume that understanding the differences and similarities between the pathogen and the host is an essential pathway for specific and successful treatment development. Here, we covered the knowledge available so far on *Leishmania* spp. cell cycle and telomere homeostasis, unveiling the remaining gaps and the advances reached in the last years. Among the impressive progress on the biology of these parasites are the facts that they remain in the G0 state during the infective stages and the remarkable divergence of their telomeric shelterin-like complex relative to mammals. However, these aspects are only a few examples of how this subject can be future linked to leishmaniases treatment and how far scientists are from a deeper knowledge of these peculiar eukaryote parasites.

Author Contributions: Conceptualization, M.I.N.C. and M.S.d.S.; writing—original draft preparation, L.H.C.A., D.A.-S., B.C.D.d.O., M.E.S., S.C.P., V.S.F., L.S.d.O., B.E.A. and Y.G.F.; writing—review and editing, L.H.C.A., D.A.-S., M.S.d.S. and M.I.N.C.; supervision, M.I.N.C. and M.S.d.S.; project administration, M.I.N.C. and M.S.d.S.; and funding acquisition, M.I.N.C. and M.S.d.S. All authors have read and agreed to the published version of the manuscript.

Funding: This work was supported by the São Paulo Research Foundation (FAPESP) under grants 2018/04375-2 (M.I.N.C.), and 2019/10753-2 and 2020/10277-3 (M.S.d.S.). L.H.C.A. and D.A.-S. are postdoctoral fellows from FAPESP (grants 2021/04253-7 and 2021/05523-8, respectively). B.C.D.d.O., and M.E.S. are Ph.D. students from FAPESP (grants 2019/25985-6 and 2020/00316-1, respectively). B.E.A. and Y.G.F. are M.Sc. students from FAPESP (grants 2020/16480-5 and 2020/16465-6, respectively). S.C.P. is a M.Sc. student from CAPES (Coordenação de Aperfeiçoamento de Pessoal de Nível Superior—Brazil), and V.S.F. is an undergraduate student (FAPESP grant 2020/08162-3).

Institutional Review Board Statement: Not applicable.

Informed Consent Statement: Not applicable.

Data Availability Statement: Not applicable since this work is a review of published data.

Conflicts of Interest: The authors declare no conflict of interests, and the funders had no role in the design of the study; in the collection, analyses, or interpretation of the data; in the writing of the manuscript; or in the decision to publish the results.

References

1. Cantanhêde, L.M.; Mata-Somarribas, C.; Chourabi, K.; Pereira da Silva, G.; Dias das Chagas, B.; de Oliveira, R.; Pereira, L.; Côrtes Boité, M.; Cupolillo, E. The Maze Pathway of Coevolution: A Critical Review over the Leishmania and Its Endosymbiotic History. *Genes* **2021**, *12*, 657. [CrossRef]
2. Burza, S.; Croft, S.L.; Boelaert, M. Leishmaniasis. *Lancet* **2018**, *392*, 951–970. [CrossRef]
3. Serafim, T.D.; Coutinho-Abreu, I.V.; Dey, R.; Kissinger, R.; Valenzuela, J.G.; Oliveira, F.; Kamhawi, S. Leishmaniasis: The Act of Transmission. *Trends Parasitol.* **2021**, *37*, 976–987. [CrossRef] [PubMed]
4. Ghorbani, M.; Farhoudi, R. Leishmaniasis in Humans: Drug or Vaccine Therapy? *Drug Des. Dev. Ther.* **2017**, *12*, 25–40. [CrossRef]
5. Cruz, A.K.; Freitas-Castro, F. Genome and Transcriptome Analyses of *Leishmania* spp.: Opening Pandora's Box. *Curr. Opin. Microbiol.* **2019**, *52*, 64–69. [CrossRef]
6. Lander, N.; Chiurillo, M.A. State-of-the-art CRISPR/Cas9 Technology for Genome Editing in Trypanosomatids. *J. Eukaryot. Microbiol.* **2019**, *66*, 981–991. [CrossRef]
7. Vasconcelos, E.J.R.; Nunes, V.S.; da Silva, M.S.; Segatto, M.; Myler, P.J.; Cano, M.I.N. The Putative Leishmania Telomerase RNA (LeishTER) Undergoes Trans-Splicing and Contains a Conserved Template Sequence. *PLoS ONE* **2014**, *9*, e112061. [CrossRef] [PubMed]
8. Sandin, S.; Rhodes, D. Telomerase Structure. *Curr. Opin. Struct. Biol.* **2014**, *25*, 104–110. [CrossRef] [PubMed]
9. Podlevsky, J.D.; Chen, J.J.-L. Evolutionary Perspectives of Telomerase RNA Structure and Function. *RNA Biol.* **2016**, *13*, 720–732. [CrossRef] [PubMed]
10. Li, B.; Espinal, A.; Cross, G.A.M. Trypanosome Telomeres Are Protected by a Homologue of Mammalian TRF2. *Mol. Cell. Biol.* **2005**, *25*, 5011–5021. [CrossRef]
11. Fernandes, C.A.H.; Perez, A.M.; Barros, A.C.; Dreyer, T.R.; da Silva, M.S.; Morea, E.G.O.; Fontes, M.R.M.; Cano, M.I.N. Dual Cellular Localization of the Leishmania Amazonensis Rbp38 (LaRbp38) Explains Its Affinity for Telomeric and Mitochondrial DNA. *Biochimie* **2019**, *162*, 15–25. [CrossRef]
12. Fernandes, C.A.H.; Morea, E.G.O.; dos Santos, G.A.; da Silva, V.L.; Vieira, M.R.; Viviescas, M.A.; Chatain, J.; Vadel, A.; Saintomé, C.; Fontes, M.R.M.; et al. A Multi-Approach Analysis Highlights the Relevance of RPA-1 as a Telomere End-Binding Protein (TEBP) in Leishmania Amazonensis. *Biochim. Biophys. Acta BBA-Gen. Subj.* **2020**, *1864*, 129607. [CrossRef]
13. Dey, A.; Chakrabarti, K. Current Perspectives of Telomerase Structure and Function in Eukaryotes with Emerging Views on Telomerase in Human Parasites. *Int. J. Mol. Sci.* **2018**, *19*, 333. [CrossRef]
14. Lai, A.G.; Pouchkina-Stantcheva, N.; Di Donfrancesco, A.; Kildisiute, G.; Sahu, S.; Aboobaker, A.A. The Protein Subunit of Telomerase Displays Patterns of Dynamic Evolution and Conservation across Different Metazoan Taxa. *BMC Evol. Biol.* **2017**, *17*, 107. [CrossRef] [PubMed]
15. Oliveira, B.C.D.; Shiburah, M.E.; Paiva, S.C.; Vieira, M.R.; Morea, E.G.O.; da Silva, M.S.; Alves, C.S.; Segatto, M.; Gutierrez-Rodrigues, F.; Borges, J.C.; et al. Possible Involvement of Hsp90 in the Regulation of Telomere Length and Telomerase Activity During Leishmania amazonensis Developmental Cycle and Population Proliferation. *Front. Cell Dev. Biol.* **2021**, in press. [CrossRef]
16. Cano, M.I.N. Telomere Biology of Trypanosomatids: More Questions than Answers. *Trends Parasitol.* **2001**, *17*, 425–429. [CrossRef]
17. Lira, C.B.B.; Giardini, M.A.; Neto, J.L.S.; Conte, F.F.; Cano, M.I.N. Telomere Biology of Trypanosomatids: Beginning to Answer Some Questions. *Trends Parasitol.* **2007**, *23*, 357–362. [CrossRef]
18. Damasceno, J.D.; Silva, G.L.; Tschudi, C.; Tosi, L.R. Evidence for Regulated Expression of Telomeric Repeat-Containing RNAs (TERRA) in Parasitic Trypanosomatids. *Mem. Inst. Oswaldo Cruz* **2017**, *112*, 572–576. [CrossRef] [PubMed]
19. Poláková, E.; Záhonová, K.; Albanaz, A.T.S.; Butenko, A.; Lukeš, J.; Yurchenko, V. Diverse Telomeres in Trypanosomatids. *Parasitology* **2021**, *148*, 1254–1270. [CrossRef]
20. Blackburn, E.H. Telomeres: Structure and Synthesis. *J. Biol. Chem.* **1990**, *265*, 5919–5921. [CrossRef]
21. Turner, K.; Vasu, V.; Griffin, D. Telomere Biology and Human Phenotype. *Cells* **2019**, *8*, 73. [CrossRef] [PubMed]
22. Greider, C.W. Telomere Length Regulation. *Annu. Rev. Biochem.* **1996**, *65*, 337–365. [CrossRef]
23. Barbé-Tuana, F.; Grun, L.K.; Pierdoná, V.; de Oliveira, B.C.D.; Paiva, S.C.; Shiburah, M.E.; da Silva, V.L.; Morea, E.G.O.; Fontes, V.S.; Cano, M.I.N. Human Chromosome Telomeres. In *Human Genome Structure, Function and Clinical Considerations*; Haddad, L.A., Ed.; Springer: Cham, Switzerland, 2021; pp. 207–243, ISBN 978-3-030-73150-2.
24. Pierce, B.A. *Genetics: A Conceptual Approach*; Editora Guanabara Koogan: Rio de Janeiro, Brazil, 2016; v. 5.
25. Greider, C.W. Regulating Telomere Length from the Inside Out: The Replication Fork Model. *Genes Dev.* **2016**, *30*, 1483–1491. [CrossRef] [PubMed]
26. Wright, W.E.; Tesmer, V.M.; Huffman, K.E.; Levene, S.D.; Shay, J.W. Normal Human Chromosomes Have Long G-Rich Telomeric Overhangs at One End. *Genes Dev.* **1997**, *11*, 2801–2809. [CrossRef]
27. Watson, J.D. Origin of Concatemeric T7DNA. *Nat. New Biol.* **1972**, *239*, 197–201. [CrossRef]
28. Hayflick, L. The Limited in vitro Lifetime of Human Diploid Cell Strains. *Exp. Cell Res.* **1965**, *37*, 614–636. [CrossRef]

29. Harley, C.B.; Futcher, A.B.; Greider, C.W. Telomeres Shorten during Ageing of Human Fibroblasts. *Nature* **1990**, *345*, 458–460. [CrossRef] [PubMed]
30. Liu, J.; Wang, L.; Wang, Z.; Liu, J.-P. Roles of Telomere Biology in Cell Senescence, Replicative and Chronological Ageing. *Cells* **2019**, *8*, 54. [CrossRef]
31. Greider, C.W.; Blackburn, E.H. The Telomere Terminal Transferase of Tetrahymena Is a Ribonucleoprotein Enzyme with Two Kinds of Primer Specificity. *Cell* **1987**, *51*, 887–898. [CrossRef]
32. Giardini, M.A.; Segatto, M.; da Silva, M.S.; Nunes, V.S.; Cano, M.I.N. Telomere and Telomerase Biology. In *Progress in Molecular Biology and Translational Science*; Elsevier: Amsterdam, The Netherlands, 2014; Volume 125, pp. 1–40. [CrossRef]
33. Berei, J.; Eckburg, A.; Miliavski, E.; Anderson, A.D.; Miller, R.; Dein, J.; Giuffre, A.M.; Tang, D.; Deb, S.; Racherla, K.S.; et al. Potential Telomere-related Pharmacological Targets. *Curr. Top. Med. Chem.* **2019**, *20*, 458–484. [CrossRef]
34. Terzi, M.Y.; Izmirli, M.; Gogebakan, B. The Cell Fate: Senescence or Quiescence. *Mol. Biol. Rep.* **2016**, *43*, 1213–1220. [CrossRef] [PubMed]
35. Gossage, S.M.; Rogers, M.E.; Bates, P.A. Two Separate Growth Phases during the Development of Leishmania in Sand Flies: Implications for understanding the life cycle. *Int. J. Parasitol.* **2003**, *33*, 1027–1034. [CrossRef]
36. Rittershaus, E.S.C.; Baek, S.H.; Sassetti, C.M. The Normalcy of Dormancy: Common Themes in Microbial Quiescence. *Cell Host Microbe* **2013**, *13*, 643–651. [CrossRef] [PubMed]
37. Wheeler, R.J.; Gluenz, E.; Gull, K. The Cell Cycle of Leishmania: Morphogenetic Events and Their Implications for Parasite Biology. *Mol. Microbiol.* **2011**, *79*, 647–662. [CrossRef]
38. Ambit, A.; Woods, K.L.; Cull, B.; Coombs, G.H.; Mottram, J.C. Morphological Events during the Cell Cycle of Leishmania major. *Eukaryot. Cell* **2011**, *10*, 1429–1438. [CrossRef]
39. Simpson, L.; Braly, P. Synchronization of *Leishmania tarentolae* by Hydroxyurea. *J. Protozool.* **1970**, *17*, 511–517. [CrossRef]
40. da Silva, M.S.; Monteiro, J.P.; Nunes, V.S.; Vasconcelos, E.J.; Perez, A.M.; Freitas-Júnior, L.H.; Elias, M.C.; Cano, M.I.N. *Leishmania amazonensis* Promastigotes Present Two Distinct Modes of Nucleus and Kinetoplast Segregation during Cell Cycle. *PLoS ONE* **2013**, *8*, e81397. [CrossRef]
41. Minocha, N.; Kumar, D.; Rajanala, K.; Saha, S. Kinetoplast morphology and segregation pattern as a marker for cell cycle progression in Leishmania donovani. *J. Eukaryot. Microbiol.* **2011**, *58*, 249–253. [CrossRef]
42. da Silva, M.S.; Muñoz, P.A.M.; Armelin, H.A.; Elias, M.C. Differences in The Detection of BrdU/EdU Incorporation Assays Alter the Calculation for G1, S, and G2 Phases of the Cell Cycle in Trypanosomatids. *J. Eukaryot. Microbiol.* **2017**, *64*, 756–770. [CrossRef]
43. Archer, S.K.; Inchaustegui, D.; Queiroz, R.; Clayton, C. The Cell Cycle Regulated Transcriptome of Trypanosoma brucei. *PLoS ONE* **2011**, *6*, e18425. [CrossRef] [PubMed]
44. da Silva, M.S.; Cayres-Silva, G.R.; Vitarelli, M.O.; Marin, P.A.; Hiraiwa, P.M.; Araújo, C.B.; Scholl, B.B.; Ávila, A.R.; McCulloch, R.; Reis, M.S.; et al. Transcription Activity Contributes to The Firing of Non-constitutive Origins in African trypanosomes Helping to Maintain Robustness in S-phase Duration. *Sci. Rep.* **2019**, *9*, 18512. [CrossRef] [PubMed]
45. da Silva, M.S.; Pavani, R.S.; Damasceno, J.D.; Marques, C.A.; McCulloch, R.; Tosi, L.R.O.; Elias, M.C. Nuclear DNA Replication in Trypanosomatids: There Are No Easy Methods for Solving Difficult Problems. *Trends Parasitol.* **2017**, *33*, 858–874. [CrossRef]
46. Marques, C.A.; Dickens, N.J.; Paape, D.; Campbell, S.J.; McCulloch, R. Genome-wide Mapping Reveals Single-origin Chromosome Replication in Leishmania, a Eukaryotic Microbe. *Genome Biol.* **2015**, *16*, 230. [CrossRef] [PubMed]
47. da Silva, M.S.; Vitarelli, M.O.; Souza, B.F.; Elias, M.C. Comparative Analysis of The Minimum Number of Replication Origins in Trypanosomatids And Yeasts. *Genes* **2020**, *11*, 523. [CrossRef] [PubMed]
48. Harashima, H.; Dissmeyer, N.; Schnittger, A. Cell Cycle Control across the Eukaryotic Kingdom. *Trends Cell Biol.* **2013**, *23*, 345–356. [CrossRef] [PubMed]
49. Hu, H.; Liu, Y.; Zhou, Q.; Siegel, S.; Li, Z. The Centriole Cartwheel Protein SAS-6 in Trypanosoma brucei Is Required for Probasal Body Biogenesis and Flagellum Assembly. *Eukaryot. Cell* **2015**, *14*, 898–907. [CrossRef]
50. Wheeler, R.J.; Sunter, J.D.; Gull, K. Flagellar Pocket Restructuring through the Leishmania Life Cycle Involves a Discrete Flagellum Attachment zone. *J. Cell Sci.* **2016**, *129*, 854–867. [CrossRef]
51. Lodish, H.; Berk, A. Overview of the Cell Cycle, and Its Control. In *Molecular Cell Biology*, 4th ed.; W. H. Freeman: New York, NY, USA, 2000; Section 13.1.
52. Campbell, P.C.; De Graffenried, C.L. Alternate Histories of Cytokinesis: Lessons from the Trypanosomatids. *Mol. Biol. Cell* **2020**, *31*, 2631–2639. [CrossRef]
53. Hecker, H.; Gander, E.S. The Compaction Pattern of the Chromatin of Trypanosomes. *Biol. Cell* **1985**, *53*, 199–208. [CrossRef]
54. Hecker, H.; Betschart, B.; Bender, K.; Burri, M.; Schlimme, W. The Chromatin of Trypanosomes. *Int. J. Parasitol.* **1994**, *24*, 809–819. [CrossRef]
55. Wheeler, R.J.; Gluenz, E.; Gull, K. The Limits on Trypanosomatid Morphological Diversity. *PLoS ONE* **2013**, *8*, e79581. [CrossRef]
56. Shlomai, J. The Structure and Replication of Kinetoplast DNA. *Curr. Mol. Med.* **2005**, *4*, 623–647. [CrossRef]
57. Liu, B.; Liu, Y.; Motyka, S.A.; Agbo, E.E.C.; Englund, P.T. Fellowship of the Rings: The Replication of Kinetoplast DNA. *Trends Parasitol.* **2005**, *21*, 363–369. [CrossRef] [PubMed]
58. Valdivia, H.O.; Reis-Cunha, J.L.; Rodrigues-Luiz, G.F.; Baptista, R.P.; Baldeviano, G.C.; Gerbasi, R.V.; Dobson, D.E.; Pratlong, F.; Bastien, P.; Lescano, A.G.; et al. Comparative Genomic Analysis of *Leishmania* (Viannia) Peruviana and *Leishmania* (Viannia) braziliensis. *BMC Genom.* **2015**, *16*, 715. [CrossRef] [PubMed]

59. Chiurillo, M.A.; Beck, A.E.; Devos, T.; Myler, P.J.; Stuart, K.; Ramirez, J.L. Cloning and Characterization of Leishmania Donovani Telomeres. *Exp. Parasitol.* **2000**, *94*, 248–258. [CrossRef] [PubMed]
60. Fu, G.; Baker, D.C. Characterisation of Leishmania Telomeres Reveals Unusual Telomeric Repeats and Conserved Telomere-Associated Sequence. *Nucleic Acids Res.* **1998**, *26*, 2161–2167. [CrossRef] [PubMed]
61. Conte, F.F.; Cano, M.I.N. Genomic Organization of Telomeric and Subtelomeric Sequences of Leishmania (*Leishmania*) Amazonensis. *Int. J. Parasitol.* **2005**, *35*, 1435–1443. [CrossRef]
62. Genest, P.A.; Borst, P. Analysis of Telomere Length Variation in Leishmania over Time. *Mol. Biochem. Parasitol.* **2007**, *151*, 213–215. [CrossRef]
63. Pryde, F.E.; Gorham, H.C.; Louis, E.J. Chromosome ends: All the same under their caps. *Curr. Opin. Genet. Dev.* **1997**, *7*, 822–828. [CrossRef]
64. Ravel, C.; Wincker, P.; Bastien, P.; Blaineau, C.; Pagès, M. A polymorphic minisatellite sequence in the subtelomeric regions of chromosomes I and V in *Leishmania infantum*. *Mol. Biochem. Parasitol.* **1995**, *74*, 31–41. [CrossRef]
65. Dmitriev, P.V.; Petrov, A.V.; Dontsova, O.A. Yeast Telosome Complex: Components and Their Functions. *Biochem. Mosc.* **2003**, *68*, 718–734. [CrossRef] [PubMed]
66. da Silva, M.S.; Silveira, R.C.V.; Perez, A.M.; Monteiro, J.P.; Calderano, S.G.; Da Cunha, J.P.; Elias, M.C.; Cano, M.I.N. Nuclear DNA Replication in Trypanosomatid Protozoa. In *DNA Replication and Mutation*; Leitner, R., Ed.; Nova Science Publishers, Inc.: New York, NY, USA, 2012.
67. Lira, C.B.B.; de Siqueira Neto, J.L.; Khater, L.; Cagliari, T.C.; Peroni, L.A.; dos Reis, J.R.R.; Ramos, C.H.I.; Cano, M.I.N. LaTBP1: A Leishmania Amazonensis DNA-Binding Protein That Associates in vivo with Telomeres and GT-Rich DNA Using a Myb-like Domain. *Arch. Biochem. Biophys.* **2007**, *465*, 399–409. [CrossRef] [PubMed]
68. da Silva, M.S.; Perez, A.M.; Silveira, R.C.; Moraes, C.E.; Siqueira-Neto, J.L.; Freitas-Junior, L.H.; Cano, M.I.N. The Leishmania amazonensis TRF (TTAGGG Repeat Binding Factor) Homologue Binds and Co-localizes with Telomeres. *BMC Microbiol.* **2010**, *10*, 136. [CrossRef] [PubMed]
69. Fernandez, M.F.; Castellari, R.R.; Conte, F.F.; Gozzo, F.C.; Sabino, A.A.; Pinheiro, H.; Novello, J.C.; Eberlin, M.N.; Cano, M.I.N. Identification of Three Proteins That Associate in vitro with The Leishmania (*Leishmania*) amazonensis G-rich Telomeric Strand: G-telomeric Proteins in L. amazonensis. *Eur. J. Biochem.* **2004**, *271*, 3050–3063. [CrossRef]
70. Lira, C.B.B.; Siqueira Neto, J.L.; Giardini, M.A.; Winck, F.V.; Ramos, C.H.I.; Cano, M.I.N. LaRbp38: A Leishmania Amazonensis Protein That Binds Nuclear and Kinetoplast DNAs. *Biochem. Biophys. Res. Commun.* **2007**, *358*, 854–860. [CrossRef]
71. Siqueira-Neto, J.L.; Lira, C.B.B.; Giardini, M.A.; Khater, L.; Perez, A.M.; Peroni, L.A.; dos Reis, J.R.R.; Freitas-Junior, L.H.; Ramos, C.H.I.; Cano, M.I.N. Leishmania Replication Protein A-1 Binds in vivo Single-stranded Telomeric DNA. *Biochem. Biophys. Res. Commun.* **2007**, *358*, 417–423. [CrossRef]
72. Pavani, R.S.; Fernandes, C.; Perez, A.M.; Vasconcelos, E.J.R.; Siqueira-Neto, J.L.; Fontes, M.R.; Cano, M.I.N. RPA-1 from *Leishmania amazonensis* (LaRPA-1) Structurally Differs from Other Eukaryote RPA-1 and Interacts with Telomeric DNA via Its N-Terminal OB-Fold Domain. *FEBS Lett.* **2014**, *588*, 4740–4748. [CrossRef]
73. Cuesta, J.; Read, M.A.; Neidle, S. The design of G-quadruplex ligands as telomerase inhibitors. *Mini Rev. Med. Chem.* **2003**, *3*, 11–21. [CrossRef]
74. De Cian, A.; Grellier, P.; Mouray, E.; Depoix, D.; Bertrand, H.; Monchaud, D.; Teulade-Fichou, M.P.; Mergny, J.L.; Alberti, P. Plasmodium telomeric sequences: Structure, stability and quadruplex targeting by small compounds. *Chembiochem* **2008**, *9*, 2730–2739. [CrossRef]
75. Morea, E.G.O.; Viviescas, M.A.; Fernandes, C.A.H.; Matioli, F.F.; Lira, C.B.B.; Fernandez, M.F.; Moraes, B.S.; da Silva, M.S.; Storti, C.B.; Fontes, M.R.M.; et al. A Calmodulin-like Protein (LCALA) Is a New Leishmania Amazonensis Candidate for Telomere End-Binding Protein. *Biochim. Biophys. Acta BBA-Gen. Subj.* **2017**, *1861*, 2583–2597. [CrossRef]
76. Xu, Y. Chemistry in Human Telomere Biology: Structure, Function and Targeting of Telomere DNA/RNA. *Chem. Soc. Rev.* **2011**, *40*, 2719. [CrossRef] [PubMed]
77. Morea, E.G.O.; Vasconcelos, E.J.R.; Alves, C.S.; Giorgio, S.; Myler, P.J.; Langoni, H.; Azzalin, C.M.; Cano, M.I.N. Exploring TERRA During Leishmania major Developmental Cycle and Continuous in vitro Passages. *Int. J. Biol. Macromol.* **2021**, *174*, 573–586. [CrossRef]
78. van Leeuwen, F.; Taylor, M.C.; Mondragon, A.; Moreau, H.; Gibson, W.; Kieft, R.; Borst, P. β-D-Glucosyl-Hydroxymethyluracil Is a Conserved DNA Modification in Kinetoplastid Protozoans and Is Abundant in Their Telomeres. *Proc. Natl. Acad. Sci. USA* **1998**, *95*, 2366–2371. [CrossRef]
79. Pays, E.; Delauw, M.F.; Laurent, M.; Steinert, M. Possible DNA Modification in GC Dinucleotides of *Trypanosoma Brucei* Telomeric Sequences; Relationship with Antigen Gene Transcriptiond. *Nucleic Acids Res.* **1984**, *12*, 5235–5247. [CrossRef] [PubMed]
80. Bernards, A.; van Harten-Loosbroek, N.; Borst, P. Modification of Telomeric DNA in *Trypanosoma Brucei*; a Role in Antigenic Variation? *Nucleic Acids Res.* **1984**, *12*, 4153–4170. [CrossRef]
81. van Leeuwen, F.; Wijsman, E.R.; Kieft, R.; van der Marel, G.A.; van Boom, J.H.; Borst, P. Localization of the Modified Base J in Telomeric VSG Gene Expression Sites of Trypanosoma Brucei. *Genes Dev.* **1997**, *11*, 3232–3241. [CrossRef] [PubMed]
82. Borst, P.; Sabatini, R. Base J: Discovery, Biosynthesis, and Possible Functions. *Annu. Rev. Microbiol.* **2008**, *62*, 235–251. [CrossRef]
83. Shay, J.W.; Wright, W.E. Telomeres and Telomerase: Three Decades of Progress. *Nat. Rev. Genet.* **2019**, *20*, 299–309. [CrossRef]

84. Weinrich, S.L.; Pruzan, R.; Ma, L.; Ouellette, M.; Tesmer, V.M.; Holt, S.E.; Bodnar, A.G.; Lichtsteiner, S.; Kim, N.W.; Trager, J.B.; et al. Reconstitution of Human Telomerase with the Template RNA Component HTR and the Catalytic Protein Subunit HTRT. *Nat. Genet.* **1997**, *17*, 498–502. [CrossRef]

85. Beattie, T.L.; Zhou, W.; Robinson, M.O.; Harrington, L. Reconstitution of Human Telomerase Activity in Vitro. *Curr. Biol.* **1998**, *8*, 177–180. [CrossRef]

86. Harrington, L. Biochemical Aspects of Telomerase Function. *Cancer Lett.* **2003**, *194*, 139–154. [CrossRef]

87. Cano, M.I.N.; Dungan, J.M.; Agabian, N.; Blackburn, E.H. Telomerase in Kinetoplastid Parasitic Protozoa. *Proc. Natl. Acad. Sci. USA* **1999**, *96*, 3616–3621. [CrossRef]

88. Giardini, M.A.; Fernández, M.F.; Lira, C.B.B.; Cano, M.I.N. Leishmania amazonensis: Partial Purification and Study of The Biochemical Properties of The Telomerase Reverse Transcriptase Activity from Promastigote-stage. *Exp. Parasitol.* **2011**, *127*, 243–248. [CrossRef] [PubMed]

89. Giardini, M.A.; Lira, C.B.B.; Conte, F.F.; Camillo, L.R.; de Siqueira Neto, J.L.; Ramos, C.H.I.; Cano, M.I.N. The Putative Telomerase Reverse Transcriptase Component of Leishmania Amazonensis: Gene Cloning and Characterization. *Parasitol. Res.* **2006**, *98*, 447–454. [CrossRef]

90. Collins, K. Single-Stranded DNA Repeat Synthesis by Telomerase. *Curr. Opin. Chem. Biol.* **2011**, *15*, 643–648. [CrossRef]

91. Rouda, S.; Skordalakes, E. Structure of the RNA-Binding Domain of Telomerase: Implications for RNA Recognition and Binding. *Structure* **2007**, *15*, 1403–1412. [CrossRef] [PubMed]

92. Xia, J.; Peng, Y.; Mian, I.S.; Lue, N.F. Identification of Functionally Important Domains in the N-Terminal Region of Telomerase Reverse Transcriptase. *Mol. Cell. Biol.* **2000**, *20*, 5196–5207. [CrossRef]

93. Armbruster, B.N.; Banik, S.S.R.; Guo, C.; Smith, A.C.; Counter, C.M. N-Terminal Domains of the Human Telomerase Catalytic Subunit Required for Enzyme Activity in vivo. *Mol. Cell. Biol.* **2001**, *21*, 7775–7786. [CrossRef]

94. Huang, J.; Brown, A.F.; Wu, J.; Xue, J.; Bley, C.J.; Rand, D.P.; Wu, L.; Zhang, R.; Chen, J.J.-L.; Lei, M. Structural Basis for Protein-RNA Recognition in Telomerase. *Nat. Struct. Mol. Biol.* **2014**, *21*, 507–512. [CrossRef]

95. Jansson, L.I.; Akiyama, B.M.; Ooms, A.; Lu, C.; Rubin, S.M.; Stone, M.D. Structural Basis of Template-Boundary Definition in Tetrahymena Telomerase. *Nat. Struct. Mol. Biol.* **2015**, *22*, 883–888. [CrossRef] [PubMed]

96. Jiang, J.; Chan, H.; Cash, D.D.; Miracco, E.J.; Ogorzalek Loo, R.R.; Upton, H.E.; Cascio, D.; O'Brien Johnson, R.; Collins, K.; Loo, J.A.; et al. Structure of *Tetrahymena* Telomerase Reveals Previously Unknown Subunits, Functions, and Interactions. *Science* **2015**, *350*, aab4070. [CrossRef]

97. Chan, H.; Wang, Y.; Feigon, J. Progress in Human and *Tetrahymena* Telomerase Structure Determination. *Annu. Rev. Biophys.* **2017**, *46*, 199–225. [CrossRef]

98. Robart, A.R.; Collins, K. Human Telomerase Domain Interactions Capture DNA for TEN Domain-Dependent Processive Elongation. *Mol. Cell* **2011**, *42*, 308–318. [CrossRef]

99. Egan, E.D.; Collins, K. Biogenesis of Telomerase Ribonucleoproteins. *RNA* **2012**, *18*, 1747–1759. [CrossRef]

100. Sandhu, R.; Sanford, S.; Basu, S.; Park, M.; Pandya, U.M.; Li, B.; Chakrabarti, K. A Trans-Spliced Telomerase RNA Dictates Telomere Synthesis in Trypanosoma Brucei. *Cell Res.* **2013**, *23*, 537–551. [CrossRef]

101. Schmidt, J.C.; Cech, T.R. Human Telomerase: Biogenesis, Trafficking, Recruitment, and Activation. *Genes Dev.* **2015**, *29*, 1095–1105. [CrossRef]

102. Nguyen, K.T.T.; Wong, J.M.Y. Telomerase Biogenesis and Activities from the Perspective of Its Direct Interacting Partners. *Cancers* **2020**, *12*, 1679. [CrossRef]

103. Viviescas, M.A.; Cano, M.I.N.; Segatto, M. Chaperones and Their Role in Telomerase Ribonucleoprotein Biogenesis and Telomere Maintenance. *Curr. Proteom.* **2018**, *16*, 31–43. [CrossRef]

104. Weber, H.; Valbuena, J.R.; Barbhuiya, M.A.; Stein, S.; Kunkel, H.; García, P.; Bizama, C.; Riquelme, I.; Espinoza, J.A.; Kurtz, S.E.; et al. Small Molecule Inhibitor Screening Identified HSP90 Inhibitor 17-AAG as Potential Therapeutic Agent for Gallbladder Cancer. *Oncotarget* **2017**, *8*, 26169–26184. [CrossRef] [PubMed]

105. Nepomuceno-Mejía, T.; Florencio-Martínez, L.E.; Martinez-Calvillo, S. Nucleolar Division in the Promastigote Stage of Leishmania major Parasite: A Nop56 Point of View. *Biomed. Res. Int.* **2018**, *2018*, 1641839. [CrossRef] [PubMed]

106. Zamudio, J.R.; Mittra, B.; Chattopadhyay, A.; Wohlschlegel, J.A.; Sturm, N.R.; Campbell, D.A. *Trypanosoma brucei* Spliced Leader RNA Maturation by the Cap 1 2'-O-Ribose Methyltransferase and SLA1 H/ACA SnoRNA Pseudouridine Synthase Complex. *Mol. Cell. Biol.* **2009**, *29*, 1202–1211. [CrossRef]

107. Gupta, S.K.; Kolet, L.; Doniger, T.; Biswas, V.K.; Unger, R.; Tzfati, Y.; Michaeli, S. The *Trypanosoma brucei* Telomerase RNA (TER) Homologue Binds Core Proteins of the C/D SnoRNA Family. *FEBS Lett.* **2013**, *587*, 1399–1404. [CrossRef]

108. Greider, C. Telomerase Is Processive. *Mol. Cell. Biol.* **1991**, *11*, 4572–4580. [CrossRef]

109. Xin, H.; Liu, D.; Wan, M.; Safari, A.; Kim, H.; Sun, W.; O'Connor, M.; Songyang, Z. TPP1 Is a Homologue of Ciliate TEBP-beta And Interacts with POT1 to Recruit Telomerase. *Nature* **2007**, *445*, 559–562. [CrossRef]

110. Latrick, C.M.; Cech, T.R. POT1–TPP1 Enhances Telomerase Processivity by Slowing Primer Dissociation and Aiding Translocation. *EMBO J.* **2010**, *29*, 924–933. [CrossRef] [PubMed]

111. Aramburu, T.; Plucinsky, S.; Skordalakes, E. POT1-TPP1 Telomere Length Regulation and Disease. *Comput. Struct. Biotechnol. J.* **2020**, *18*, 1939–1946. [CrossRef]

112. Chiodi, I.; Mondello, C. Telomere-Independent Functions of Telomerase in Nuclei, Cytoplasm, and Mitochondria. *Front. Oncol.* **2012**, *2*, 133. [CrossRef] [PubMed]
113. Campelo, R.; Díaz Lozano, I.; Figarella, K.; Osuna, A.; Ramírez, J.L. Leishmania Major Telomerase TERT Protein Has a Nuclear/Mitochondrial Eclipsed Distribution That Is Affected by Oxidative Stress. *Infect. Immun.* **2015**, *83*, 57–66. [CrossRef] [PubMed]
114. Yagoubat, A.; Corrales, R.M.; Bastien, P.; Lévêque, M.F.; Sterkers, Y. Gene Editing in Trypanosomatids: Tips and Tricks in the CRISPR-Cas9 Era. *Trends Parasitol.* **2020**, *36*, 745–760. [CrossRef] [PubMed]
115. Yagoubat, A.; Crobu, L.; Berry, L.; Kuk, N.; Lefebvre, M.; Sarrazin, A.; Bastien, P.; Sterkers, Y. Universal Highly Efficient Conditional Knockout System in *Leishmania*, with a Focus on Untranscribed Region Preservation. *Cell. Microbiol.* **2020**, *22*, e13159. [CrossRef]
116. Beneke, T.; Gluenz, E. LeishGEdit: A Method for Rapid Gene Knockout and Tagging Using CRISPR-Cas9. In *Leishmania*; Clos, J., Ed.; Methods in Molecular Biology; Springer: New York, NY, USA, 2019; Volume 1971, pp. 189–210, ISBN 978-1-4939-9209-6.
117. Xi, L.; Schmidt, J.C.; Zaug, A.J.; Ascarrunz, D.R.; Cech, T.R. A Novel Two-step Genome Editing Strategy with CRISPR-Cas9 Provides New Insights into Telomerase Action and TERT Gene Expression. *Genome Biol.* **2015**, *16*, 231. [CrossRef]
118. Xia, B.; Amador, G.; Viswanatha, R.; Zirin, J.; Mohr, S.E.; Perrimon, N. CRISPR-Based Engineering of Gene Knockout Cells by Homology-Directed Insertion in Polyploid Drosophila S2R+ Cells. *Nat. Protoc.* **2020**, *15*, 3478–3498. [CrossRef] [PubMed]
119. Dreesen, O.; Li, B.; Cross, G. Telomere Structure and Shortening in Telomerase-deficient Trypanosoma brucei. *Nucleic Acids Res.* **2005**, *33*, 4536–4543. [CrossRef] [PubMed]
120. Haendeler, J.; Hoffmann, J.; Rahman, S.; Zeiher, A.M.; Dimmeler, S. Regulation of Telomerase Activity and Anti-Apoptotic Function by Protein-Protein Interaction and Phosphorylation. *FEBS Lett.* **2003**, *536*, 180–186. [CrossRef]
121. Lee, J.; Sung, Y.H.; Cheong, C.; Choi, Y.S.; Jeon, H.K.; Sun, W.; Hahn, W.C.; Ishikawa, F.; Lee, H.-W. TERT Promotes Cellular and Organismal Survival Independently of Telomerase Activity. *Oncogene* **2008**, *27*, 3754–3760. [CrossRef] [PubMed]
122. Zhang, A.; Zheng, C.; Hou, M.; Lindvall, C.; Li, K.-J.; Erlandsson, F.; Björkholm, M.; Gruber, A.; Blennow, E.; Xu, D. Deletion of the Telomerase Reverse Transcriptase Gene and Haploinsufficiency of Telomere Maintenance in Cri Du Chat Syndrome. *Am. J. Hum. Genet.* **2003**, *72*, 940–948. [CrossRef] [PubMed]
123. Zhou, J.; Ding, D.; Wang, M.; Cong, Y.-S. Telomerase Reverse Transcriptase in the Regulation of Gene Expression. *BMB Rep.* **2014**, *47*, 8–14. [CrossRef] [PubMed]
124. Zhu, J.; Liu, W.; Chen, C.; Zhang, H.; Yue, D.; Li, C.; Zhang, L.; Gao, L.; Huo, Y.; Liu, C.; et al. TPP1 OB-Fold Domain Protein Suppresses Cell Proliferation and Induces Cell Apoptosis by Inhibiting Telomerase Recruitment to Telomeres in Human Lung Cancer Cells. *J. Cancer Res. Clin. Oncol.* **2019**, *145*, 1509–1519. [CrossRef]

Article

M2 Muscarinic Receptor Activation Impairs Mitotic Progression and Bipolar Mitotic Spindle Formation in Human Glioblastoma Cell Lines

Maria Di Bari [1], Vanessa Tombolillo [1], Francesco Alessandrini [1], Claudia Guerriero [1], Mario Fiore [2], Italia Anna Asteriti [2], Emilia Castigli [3], Miriam Sciaccaluga [4], Giulia Guarguaglini [2], Francesca Degrassi [2] and Ada Maria Tata [1,5,*]

1 Department of Biology and Biotechnologies Charles Darwin, Sapienza University of Rome, 00185 Rome, Italy; Marina.Dibari@live.it (M.D.B.); vanessa.tombolillo@gmail.com (V.T.); francesco.alessandrini@northwestern.edu (F.A.); Claudia.Guerriero@uniroma1.it (C.G.)

2 Institute of Molecular Biology and Pathology, CNR, 00185 Rome, Italy; mario.fiore@uniroma1.it (M.F.); lia.asteriti@uniroma1.it (I.A.A.); giulia.guarguaglini@uniroma1.it (G.G.); francesca.degrassi@uniroma1.it (F.D.)

3 Department of Experimental Medicine, Section of Physiology and Biochemistry, University of Perugia, 06100 Perugia, Italy; emilia.castigli@gmail.com

4 Department of Medicine and Surgery, University of Perugia, 06100 Perugia, Italy; miriam.sciaccaluga@unipg.it

5 Research Centre of Neurobiology Daniel Bovet, 00185 Rome, Italy

* Correspondence: adamaria.tata@uniroma1.it

Citation: Di Bari, M.; Tombolillo, V.; Alessandrini, F.; Guerriero, C.; Fiore, M.; Asteriti, I.A.; Castigli, E.; Sciaccaluga, M.; Guarguaglini, G.; Degrassi, F.; et al. M2 Muscarinic Receptor Activation Impairs Mitotic Progression and Bipolar Mitotic Spindle Formation in Human Glioblastoma Cell Lines. *Cells* **2021**, *10*, 1727. https://doi.org/10.3390/cells10071727

Academic Editor: Zhixiang Wang

Received: 31 May 2021
Accepted: 6 July 2021
Published: 8 July 2021

Publisher's Note: MDPI stays neutral with regard to jurisdictional claims in published maps and institutional affiliations.

Abstract: Background: Glioblastoma multiforme (GBM) is characterized by several genetic abnormalities, leading to cell cycle deregulation and abnormal mitosis caused by a defective checkpoint. We previously demonstrated that arecaidine propargyl ester (APE), an orthosteric agonist of M2 muscarinic acetylcholine receptors (mAChRs), arrests the cell cycle of glioblastoma (GB) cells, reducing their survival. The aim of this work was to better characterize the molecular mechanisms responsible for this cell cycle arrest. Methods: The arrest of cell proliferation was evaluated by flow cytometry analysis. Using immunocytochemistry and time-lapse analysis, the percentage of abnormal mitosis and aberrant mitotic spindles were assessed in both cell lines. Western blot analysis was used to evaluate the modulation of Sirtuin2 and acetylated tubulin—factors involved in the control of cell cycle progression. Results: APE treatment caused arrest in the M phase, as indicated by the increase in p-HH3 (ser10)-positive cells. By immunocytochemistry, we found a significant increase in abnormal mitoses and multipolar mitotic spindle formation after APE treatment. Time-lapse analysis confirmed that the APE-treated GB cells were unable to correctly complete the mitosis. The modulated expression of SIRT2 and acetylated tubulin in APE-treated cells provides new insights into the mechanisms of altered mitotic progression in both GB cell lines. Conclusions: Our data show that the M2 agonist increases aberrant mitosis in GB cell lines. These results strengthen the idea of considering M2 acetylcholine receptors a novel promising therapeutic target for the glioblastoma treatment.

Keywords: M2 muscarinic receptor; glioblastoma; cell cycle; aberrant mitosis; mitotic spindle

1. Introduction

Glioblastoma multiforme (GBM) is the most malignant and frequent human brain tumor, representing more than 60% of all brain tumors in adults [1]. Glioblastoma (GB) cells have a highly infiltrative capacity, being able to spread into surrounding brain tissue by using the perivascular space [2]. Despite the great advances made in understanding the molecular alterations that occur in GB, there is not a definitive cure and mortality is still very high. Eighty-eight percent of patients affected by GB succumb to the disease within one to three years [3]. GB is normally treated with surgery, followed by radiotherapy and chemotherapy with temozolomide [4,5]. Therefore, research focused on the development of new therapeutic agents is clinically relevant today.

Muscarinic acetylcholine receptors (mAChRs) are G protein-coupled receptors widely distributed in the nervous system and in several mammalian organs [6]. mAChRs are expressed in several primary and metastatic tumors, such as breast [7], ovarian [8], and lung cancers [9] and in astrocytoma and glioblastoma cells [10–12]. In particular, M2 mAChRs appear to be involved in tumor behavior, negatively modulating the proliferation and the migration of cancer cells [13–15].

Recently, by in vitro studies, we showed that the selective stimulation of M2 mAChRs by the orthosteric agonist arecaidine propargyl ester (APE) inhibits cell cycle progression and decreases cell survival, both in GB cell lines [14,16] and in GB stem cells [17]. Furthermore, the activation of M2 mAChRs induces apoptosis and oxidative stress, predominantly in p53-mutated glioblastoma cell lines [18].

In recent years, a wide range of clinically employed and experimental anticancer agents arresting the cell cycle and triggering apoptosis have started emerging [19]. Like other solid tumors, GB cells show an aberrant cell cycle and an increased rate of proliferation [20], as well as marked aneuploidy [21], suggesting that one of the causes of chromosomal instability could be determined by deregulated checkpoints.

In the last decade, the phenomenon of mitotic catastrophe has achieved increasing prominence and, in recent years, this mechanism has gained importance as a possible therapeutic target in cancer treatment [22]. In 2012, mitotic catastrophe was defined by the International Nomenclature Committee on Cell Death as an oncosuppressive mechanism that prevents the genomic instability of cells through the induction of mitosis-related cell death or permanent cell cycle arrest [23]. At least three mechanisms of mitotic catastrophe have been described: (I) Activation of the death machinery while the cell is still in mitosis; (II) "mitotic checkpoint adaptation," i.e., arrested cells enter the interphase without chromosome segregation, and cell death is triggered at the following interphase; (III) development of the senescent phenotype after aberrant mitosis [24,25]. Depending on the status of cell cycle checkpoints, several cytotoxic agents could induce aberrant mitosis/mitotic catastrophe, activating different pathways. Moreover, mitotic catastrophe could also occur after abnormal re-entry of tumor cells into the cell cycle following prolonged growth arrest [26,27].

Previous results have shown that M2 mAChR activation by APE causes the proliferative arrest of U251MG and U87MG cells [16]. Given the ability of APE to increase ROS levels and to cause cytotoxic damage, including chromosomal aberrations [17], in the present work, we investigated in depth the cell cycle progression upon M2 receptor activation in these two cell lines. The data obtained show that M2 agonist treatment causes an arrest in cell cycle progression with an accumulation of cells during pro-metaphase/metaphase transition, resulting in a significant increase in abnormal mitosis and multipolar mitotic spindle formation.

2. Materials and Methods

2.1. Cell Cultures

U251MG and U87MG cell lines were maintained at 37 °C in an atmosphere with 10% CO_2. The cells were grown in DMEM (Sigma-Aldrich, St. Louis, MO, USA) plus 10% fetal bovine serum (Sigma-Aldrich, St. Louis, MO, USA), 50 µg/mL of streptomycin, 50 IU/mL of penicillin, 2 mM of glutamine (Sigma-Aldrich, St. Louis, MO, USA), and 1% no essential amino-acids (Sigma-Aldrich, St. Louis, MO, USA).

2.2. Pharmacological Treatment

The M2 agonist arecaidine propargyl ester hydrobromide (APE) was used as a preferential agonist of the M2 muscarinic receptor subtype. Pharmacological binding experiments and silencing of the receptor by short interference RNA demonstrated the selectivity of this agonist for the M2 receptor subtype [16,17].

2.3. Western Blot Analysis

Cells were homogenized in lysis buffer (Tris-EDTA 10 mM, 0.5% NP40, and NaCl 150 mM) containing protease inhibitor cocktail (Sigma-Aldrich, St. Louis, MO, USA). After protein extraction, the total amount of protein was determined by a Pierce BCA Protein Assay Kit (Thermo Fisher Scientific, Waltham, MA, USA) according to the manufacturer's protocol. The protein extracts were run on SDS-polyacrilamide gel (SDS-PAGE) and transferred to polyvinylidene difluoride (PVDF) sheets (Merck Millipore, Darmstadt, Germany). Membranes were blocked in 5% non-fat milk powder (Sigma-Aldrich, St. Louis, MO, USA) in PBS containing 0.1% Tween-20, and then incubated with the primary antibodies overnight at 4 °C. The primary antibodies used were SIRT-2 (Santa Cruz Biotechnologies, Dallas, TX, USA) and acetylated alpha-tubulin (Lys 40) (Sigma-Aldrich, St. Louis, MO, USA). The blots were washed three times with PBS + 0.1% Tween-20, then incubated with secondary antibodies conjugated to horseradish-peroxidase for 1 h at room temperature. The immunoreaction was revealed by ECL chemiluminescence reagent (Immunological Science, Rome, Italy). The bands were detected by exposition to Chemidoc (Molecular Imager ChemiDoc XRS + System with Image Lab Software; Biorad, CA, USA).

2.4. Flow Cytometry Analysis

The cells were plated at a density of 5×10^5 cells/dish. The day after, the cells, excluding the control samples, were treated either with 100 μM of APE (24 and 48 h of treatment) or with nocodazole (0.2 μg/mL) used as a positive control (24 h treatment). At the end of the treatments, cells were collected by trypsinization and fixed in PBS/ethanol (1:1; v/v). Samples were then incubated with monoclonal antibodies against phospho-histone H3 (p-HH3) (Ser10) (Merck Millipore, Darmstadt, Germany) for 60 min at room temperature, washed twice with 0.5% Tween-20 in PBS and incubated for 30 min with anti-mouse Alexa fluor 488-conjugated antibody (Invitrogen, Monza, Italy). Samples were washed with PBS and finally stained with 20 μg/mL propidium iodide for 15 min at room temperature. Flow cytometry analysis was performed using a flow cytometer Coulter Epics XL with 488 nm wavelength excitation. For each sample, at least 10^4 events were measured.

2.5. Immunocytochemistry

Cells were fixed with 4% paraformaldehyde for 20 min, permeabilized in PBS containing 0.1% Triton X-100 and incubated in 10% normal goat serum (NGS) for 1 h at room temperature. Coverslips were then incubated overnight at +4 °C with a human CREST antibody (Biolegend, San Diego, CA, USA), diluted 1:50 in PBS containing 0.1% Triton X-100 5% NGS, and then incubated in a Red-X anti-human antibody (1:1000, Jackson Immunoresearch Europe, Cambridge, U.K.) after extensive washing. After a 45 min incubation in a FITC conjugated anti-alpha tubulin antibody (1:100, Sigma-Aldrich, St. Louis, MO, USA), coverslips were washed three times with PBS + 1% BSA and twice with PBS, incubated with Hoechst 33,342 (1:15,000; $v:v$ in PBS) for 10 min and mounted with glycerol/PBS (3:1; v/v). In parallel experiments, an anti-acetylated tubulin antibody (1:500, Sigma-Aldrich, St. Louis, MO, USA) and a Red X anti-mouse antibody (1:1000, Jackson Immunoresearch Europe, Cambridge, UK) were used.

2.6. M2 Receptor Knock-Down

U251 cells were transfected with a pool of siRNAs selective for human M2 receptors (CHRM2), as previously described [14].

The siRNA duplexes were synthesized by Riboxx Life Sciences. The following sequences of siRNAs (Riboxx Life Sciences, Dresden, Germany) used were:

(siRNA1129-1) sense, 5'-AUUUACUACUAAAUCCUCCCCC-3'
antisense 5'-GGGGGAGGAUUUAGUAGUAAAU-3';

(siRNA1129-2) sense 5′-AUGUAGCCCAUUUCUUCCCCC-3′
antisense 5′-GGGGGAAGAAAUGGGCUACUA-3′;
(siRNA 1129-3) sense 5′-UCCUUUGAGUUUCAGGCUGCCCCC-3′
antisense 5′- GGGGGCAGCCUGAAACUCAAAGGA-3′;
(siRNA 1129-4) sense 5′-AGUUACACCUUGACCUAACCCCC-3′
antisense 5′-GGGGGUUAGGUCAAGGUGUAACU-3′

2.7. Time-Lapse Live Cell Imaging

Cells seeded in four-well micro-slides (ibiTreat, cod. 80426, Ibidi, Planegg, Germany) were observed under an Eclipse Ti inverted microscope (Nikon, Tokyo, Japan), using a 40× (Plan Fluor, 0.60 N.A., DIC) objective; during the whole observation, cells were kept in a microscope stage incubator (Basic WJ, Okolab, Pozzuoli, Italy), at 37 °C and 5% CO_2. DIC images were acquired every 7 min over 72 h using a DS-Qi1Mc camera and NIS-Elements AR 3.22 software (Nikon). Image and movie processing were performed with NIS-Elements HC 4.2.

2.8. Statistical Analysis

Data presented are the average ± SEM obtained from three independent experiments. Statistical analysis was performed by Student's *t*-tests, Mann–Whitney tests, and one way ANOVAs followed by Tukey's comparison post-hoc tests. For the evaluation of aberrant mitosis, the ratio between total abnormal metaphases/mitotic cells was calculated. Ten photographic fields for each sample were considered. Each sample was produced in triplicate. Data from time-lapse experiments were statistically analyzed using Mann–Whitney or χ^2 tests, as indicated. The results were considered statistically significant at $p < 0.05$ (*), $p < 0.01$ (**), $p < 0.001$ (***), and $p < 0.0001$ (****).

3. Results

3.1. M2 Receptor Activation Caused Accumulation of the GB Cell Lines in the M Phase

Our previous data indicated that APE induces G2/M arrest in the U251 glioblastoma cell line [16]. In order to verify whether APE was able to induce an arrest in the G2 or M phase, we evaluated—by FACS analyses—the expression of histone H3 phosphorylated at serine 10 (p-HH3), a specific mitotic marker (Figure 1A). Nocodazole, a drug that interferes with microtubule polymerization [28], was used as a positive control. As expected, nocodazole treatment (0.2 μg/mL) caused a significant increase in the percentage of cells positive for p-HH3. Interestingly, APE treatment (100 μM) increased the percentage of p-HH3-positive cells in a comparable manner to nocodazole, at least in terms of 24 h of treatment. As shown in Figure 1C, 30% of treated U251 cells appeared to have accumulated in the M phase. Forty-eight hours after treatment, the percentage of p-HH3-positive cells significantly decreased (18%), albeit remaining higher than untreated cells.

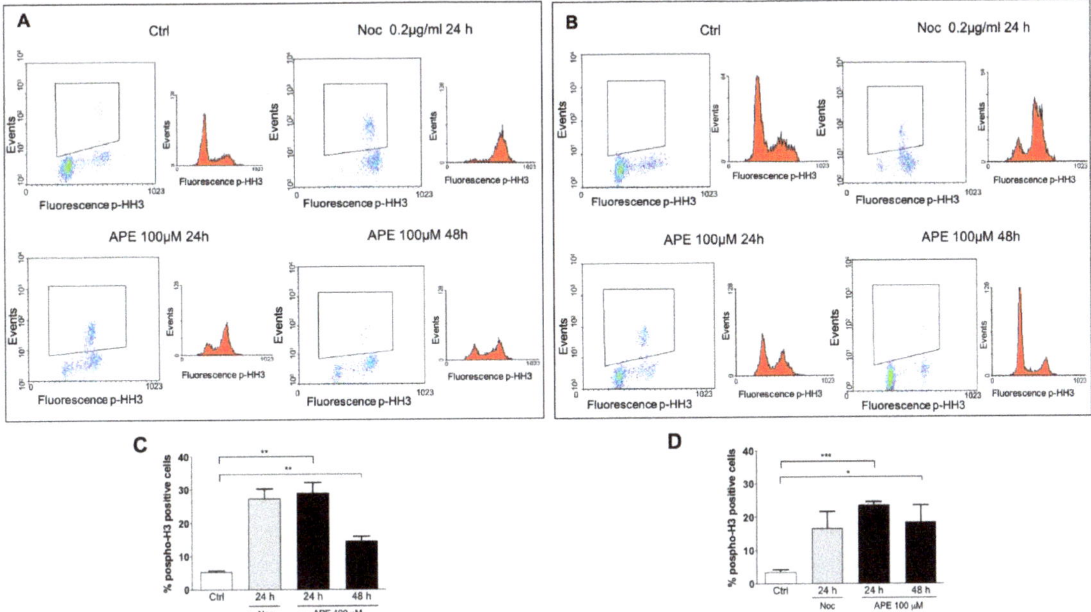

Figure 1. FACS analysis showing the expression of phospho-H3 (ser10) in the U251 (**A**) and U87 (**B**) cell lines. The graphs report the percentage of phospho-H3-positive cells upon treatment with 100 µM of APE or 0.2 µg/mL of nocodazole in the U251 (**C**) and U87 (**D**) cells. Data presented are the average ± SEM of three independent experiments conducted in triplicate. Student's *t*-test was used to statistically compare the different experimental conditions (*** $p < 0.001$, ** $p < 0.01$, and * $p < 0.05$).

As previously observed, U87 cells appeared mainly accumulated in the G1 phase after M2 receptor activation [16]. However, an increase in p-HH3-positive cells after APE treatment was observed in U87 cells. In this cell line, only 22% of cells were accumulated in the M phase, mainly 24 h after APE treatment (Figure 1B,D).

In order to confirm cell accumulation in the M phase, we performed a microscopic analysis on the GB cells after staining with the nuclear dye Hoechst 33342 to assess the frequency of cells in the mitotic stage. Figure 2A is a representative image of the U251 nuclei after APE treatment. Evaluating the nuclei organization and the chromosome distribution, we counted the number of the cells in the mitotic stage at both 24 and 48 h after M2 agonist treatment. As shown in Figure 2B,C, a progressive increase in cells in mitosis was evident after 24 h of APE treatment in both cell lines (U251 and U87). This increase appeared more consistent in U251 than in U87 cells. After longer treatment (48 h), we observed a significant decrease in the percentage of dividing cells.

To verify the direct involvement of the M2 receptor in this phenomenon, we evaluated the percentage of dividing cells in the U251 cell line, treated for 24 h with 100 µM of APE after M2 receptor silencing by siRNA transfection (Figure 2D). After M2 silencing, the percentage of dividing cells in the presence of APE was comparable to that observed in untreated cells, albeit the number of the cells resulted significantly reduced compared to non-transfected cells, most likely as a consequence of the toxicity caused by the transfection.

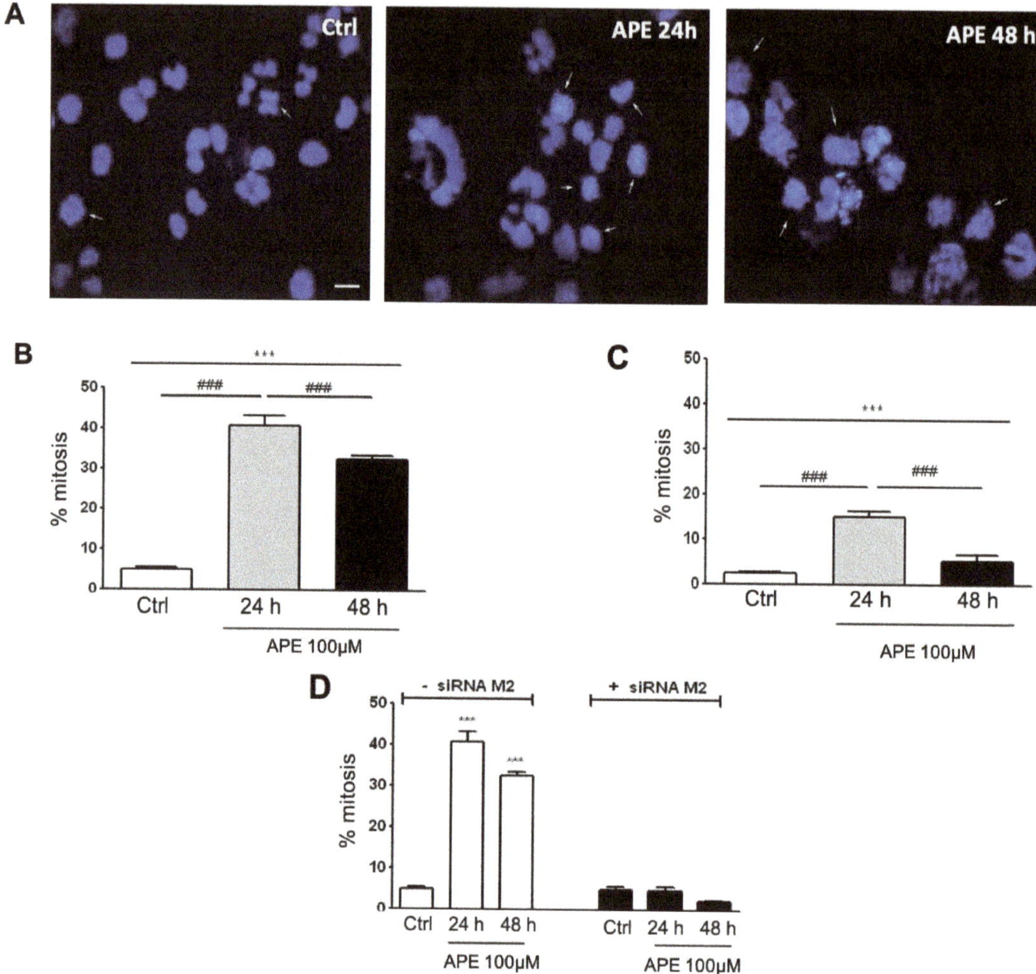

Figure 2. (**A**) Nuclei staining by Hoechst 33,342 in the U251 cells untreated or treated with APE (100 µM) for 24 and 48 h. The arrows indicate the cells in mitosis (bar = 10 µm). The graphs indicate the percentage of cells in mitosis 24 and 48 h after treatment with 100 µM of APE: (**B**) U251 cells and (**C**) U87 cells. (**D**) Percentage of cells in mitosis in the U251 cell line in untreated and APE-treated cells in the presence or absence of M2 receptor knock-down by siRNA transfection. An ANOVA test was used, followed by Tukey's test, to statistically compare all of the experimental conditions (*** $p < 0.001$; ### $p < 0.001$).

3.2. Analysis of Mitosis Progression

In order to directly follow the entrance to the mitotic stage, as well as the progression through mitosis, we videorecorded U251 cells for 72 h after treatment with 100 µM of APE. More than 80% of the cells entered mitosis during videorecording, both in the control and the APE-treated cultures, confirming that no G2 arrest was induced by the treatment. Nonetheless, mitotic entry may be delayed by APE, since we noticed that approximately 30% of the dividing cells entered mitosis between 24 and 72 h after treatment, while the majority (>95%) of control cells progressed to mitosis within the first 24 h (see video in Supplementary Materials). A strong delay/arrest in the pro-metaphase was observed (Figure 3A): While the average time from rounding-up to chromosome segregation onset was approximately 50 min in the control cells, APE-treated mitoses remained for approximately 22 h in a prometaphase-like state and were not able to divide normally. Only 15% of those cells were eventually able to originate two daughter cells through an abnormal division (second row of Figure 3B,C) (see video in Supplementary Materials). The majority of mitoses (>70%) were still arrested in a prometaphase-like state after 48 h and displayed membrane blebbing movements often associated with an elongated shape (third row of Figure 4B,C). At the end of the videorecording (72 h), more than 50% of the analyzed cells had exited mitosis without segregating chromosomes (fourth row of Figure 4B,C), suggesting a failure of cell division after the long blebbing phase (see video in Supplementary Materials).

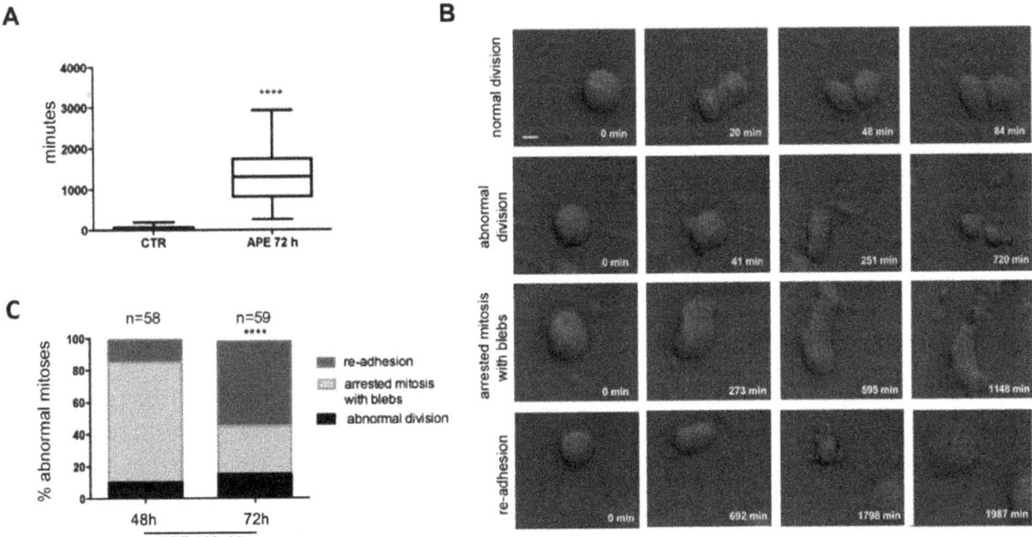

Figure 3. Time-lapse analysis of APE-treated cultures revealing abnormal mitotic phenotypes. The U251 cultures treated with 100 μM of APE were recorded by time-lapse for 72 h upon M2 agonist treatment, and 60–90 mitoses per condition were analyzed. (**A**) Time from mitotic round-up to chromosome segregation (**** $p < 0.0001$; Mann–Whitney test). (**B**) Representative single photograms are shown for a normal mitotic division (first row) and the different observed defects (second to fourth rows); minutes from round-up are indicated. Scale bar: 10 μm. (**C**) Histograms showing the percentage of abnormal mitoses in the APE-treated cultures, divided according to their fate at 48 and 72 h from the beginning of the treatment (**** $p < 0.0001$; Student's *t*-test comparing 48 vs. 72 h). All mitoses in the control cultures were normal.

3.3. APE-Arrested Glioblastoma Cells during Prometa-Metaphase Transition and Induced Multipolar Mitotic Spindles

To further investigate the effect of APE on mitosis progression, cells in the mitotic stage were classified as pro-metaphase/metaphase or anaphase, considering the position the chromosomes and their organization on the metaphase plate (Figure 4A,B). We chose to analyze the cells 30 h after treatment, since this is a time before the severe cell death observed after 48 h of APE treatment [16,18]. The data obtained indicate that the number of cells in the pro-metaphase/metaphase was higher in the APE-treated cells compared to the untreated cells (control) and was associated with a concomitant decrease in cells in the anaphase. This trend was observed in both cell lines, but it was statistically significant only in the U251 cell line (Figure 4B). Thus, our data indicate that APE consistently arrested glioblastoma cells prior to anaphase chromosome segregation.

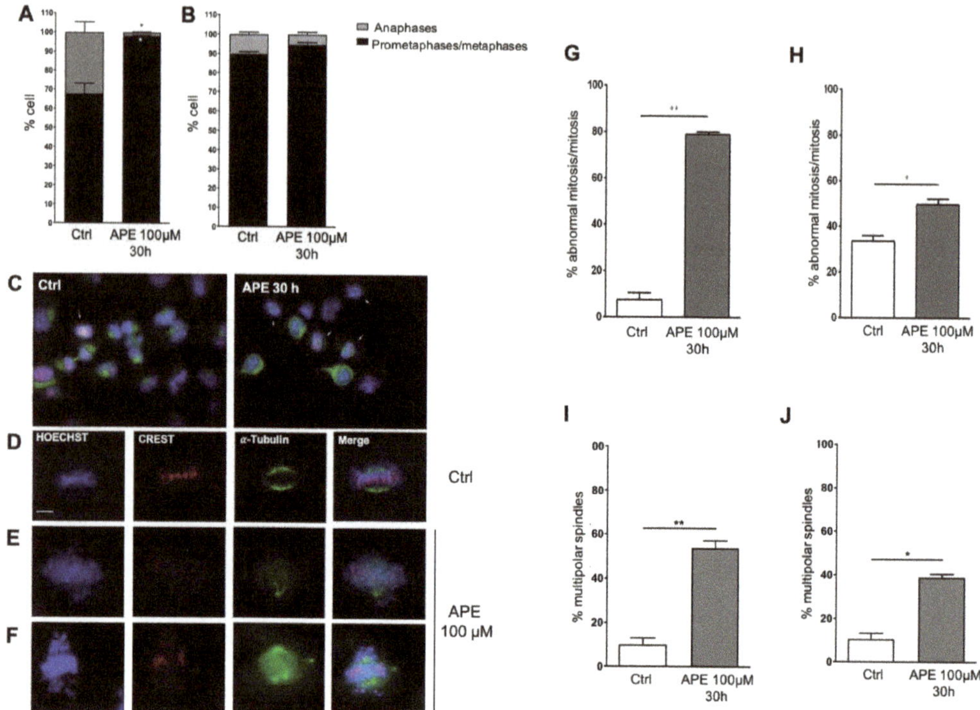

Figure 4. Percentage of pro-metaphases/metaphases and anaphases in the U251 (**A**) and in U87 (**B**) cells after treatment with 100 μM of APE for 30 h. Student's *t*-test was used to statistically compare the different experimental conditions (* $p < 0.05$). (**C–F**) Immunocytochemistry analysis of U251 cells untreated or treated with APE (100 μM) for 30 h. Cells were immunostained with anti-CREST (**red**) and anti-α-tubulin (**green**) antibodies and counterstained with Hoechst 33,258 (**blue**). Scale bars: 5 μm. (**G,H**) Percentage of the cells presenting abnormal mitosis in the control condition and after 30 h of treatment with 100 μM of APE, in the U251 and U87 cells, respectively. Student's *t*-test was used to statistically compare the different experimental conditions (** $p < 0.01$ and * $p < 0.05$). (**I,J**) Percentage of the cells with multipolar spindles in the control condition and 30 h after treatment with 100 μM of APE, in the U251 and U87 cells, respectively. Student's *t*-test was used to statistically compare the different experimental conditions (** $p < 0.01$ and * $p < 0.05$).

The effect of APE treatment on mitotic spindle structure was additionally examined in pro-metaphase/metaphase cells by immunofluorescence microscopy (Figure 4C–F). We used an antibody against alpha-tubulin to detect the microtubules forming the mitotic spindles and an antibody against kinetochore protein (CREST antibody) to visualize the centromere. The presence of bipolar, well-organized, overall fusiform spindles was evident in the untreated cells. Within these spindles, the highly condensed chromosomes appeared to be aligned along the equatorial plate (Figure 4D). On the contrary, strong abnormalities in mitosis were found in the cells treated with APE for 30 h. This treatment caused a significant increase in the percentage of abnormal mitoses in both the U251 and U87 cell lines compared to the control condition, as shown in Figure 4G,H. Thirty hours after APE treatment, some of the chromosomes appeared not aligned at the equatorial plate or with the spindles themselves, being asymmetric and often of multipolar origin (i.e., bipolar spindles) with clustered centrosomes (Figure 4E) or showing several centrosomes and a multipolar organization of chromosomes (Figure 4F). The graphs in Figure 4I,J show a significant increase in multipolar spindles in the APE-treated cells of both glioblastoma cell lines compared to the untreated cells.

3.4. Modulation of SIRT2 and Acetylated Tubulin Expression after APE Treatment

Several pieces of evidence have connected the activity of deacetylase SIRT2 to the control of mitotic progression. Indeed, SIRT2 appears to act in the mitotic checkpoint, blocking chromosome condensation in response to mitotic stress [29]. In light of these findings, we evaluated SIRT2 protein expression by Western blot analysis after APE treatment. As shown in Figure 5A–E, SIRT2 expression significantly increased in the U251 (Figure 5A,B) and U87 cells (Figure 5E,F), particularly 30 h after treatment with 100 µM of APE. Considering that SIRT2 deacetylates tubulin [30], we assessed the expression of acetylated tubulin by Western blot analysis, showing that its levels significantly decreased only in the U87 (Figure 5G,H) cells 24 and 30 h after APE (100 µM) treatment. Apparently, no significant modulation of acetylated tubulin was evident in the U251 cells (Figure 5C,D). Nevertheless, high-resolution immunodetection of acetylated microtubules on mitotic cells demonstrated that the spindle microtubules were extremely abnormal in the U251 treated cells. Meanwhile, the mitotic spindles of the untreated cells consisted of several arrays of acetylated microtubules, and the mitotic spindles in APE-treated cells were characterized by long fibers. These fibers were composed of bundled acetylated microtubules that passed over the chromosomes and did not attach the kinetochores (Figure 5I).

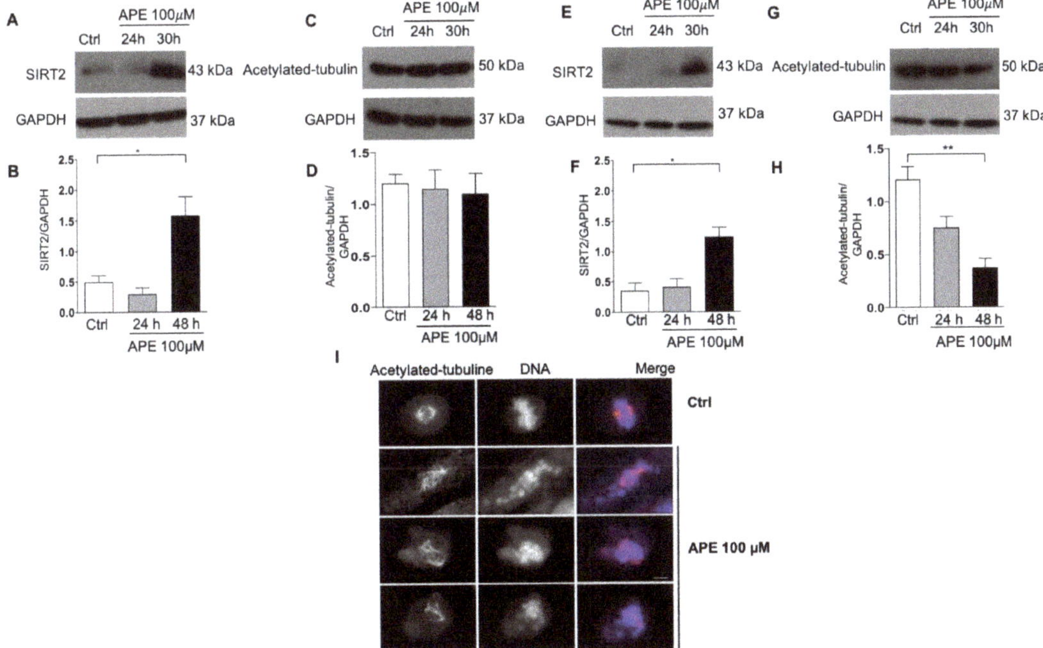

Figure 5. Western blot analysis of SIRT2 and acetylated tubulin expression in U251 and U87 cells treated with APE (100 µM) for 24 and 30 h. Representative blot for SIRT2 (**A,E**) and acetylated tubulin (**C,G**) in U251 and U87 cells, respectively. GAPDH was used as an internal reference protein. (**B,D,F,H**) Densitometric analysis obtained from three independent experiments in U251 (**B,D**) and U87 (**F,H**) cells, respectively. Student's *t*-test was used to statistically compare the different experimental conditions (** $p < 0.01$ and * $p < 0.05$). (**I**) Immunocytochemistry analysis of U251 cells untreated or treated with APE (100) µM for 30 h. Cells were immunostained with an anti-acetylated tubulin (**red**) antibody and counterstained with Hoechst 33,258 (**blue**). Scale bars: 5 µm.

4. Discussion

Glioblastoma cells are known to respond to cholinergic stimulation [6,31,32]. The effects of M2 receptor subtype activation on cell growth and survival have been largely characterized using two different glioblastoma cell lines (U251MG and U87MG) in primary cultures obtained from different human biopsies and in glioblastoma cancer stem cells [17]. In particular, we previously demonstrated that M2 receptor activation causes a cell cycle arrest in the G1/S and at G2/M phases in U87MG and U251MG cells, respectively [16]. In addition, cell viability analysis showed that the M2 orthosteric agonist APE causes cytotoxic effects (i.e., ROS production, double-strand DNA breaks, and chromosomes aberrations) and severe apoptosis, especially in p53-mutated cells such as U251MG [18].

Starting from these already described observations, in the present work we investigated in depth the modalities through which the M2 agonist APE triggers the cytostatic effects in glioblastoma cells.

Since M2 receptor activation arrests cell cycle progression, we analyzed the progression of the mitotic phase. In both glioblastoma cell lines (U251MG and U87MG), we observed an arrest at the mitotic stage, as shown by a significant increase in the percentage of p-HH3 (ser10)-positive cells after M2 agonist treatment. Mitotic arrest was much less evident in the U87MG cell line, in accordance with the higher levels of cells accumulated in the G1 phase than in the G2/M phase, as previously observed in this cell line [16]. Instead, the U251MG cells showed a larger increase in p-HH3-positive cells upon APE treatment, confirming an accumulation of cells in the M phase rather than in the G2 phase, as previously suggested by FACS analysis results [16]. This different behavior may be dependent on the genetic background of two the different glioblastoma cell lines; in fact the presence of wildtype or mutated p53 in U87 and in U251 cells, respectively, may be responsible for the cell accumulation in the different phased of the cell cycle.

Nuclei staining and immunocytochemistry analysis also confirmed that APE caused an accumulation of cells in the M phase. Moreover, a significant increase in abnormal mitosis and an accumulation of cells during pro-metaphase/metaphase transition were observed, accompanied by a reduction in the number of cells in the anaphase. Additionally, the untreated cells formed normal bipolar spindles and the chromosomes appeared aligned along the equatorial plate during the metaphase. Conversely, in the APE-treated cells, asymmetric spindle formation was observed, together with incorrect chromosome alignment along the equatorial plate. In addition, M2 agonist treatment increased the formation of pseudobipolar, tripolar, and multipolar spindles.

To demonstrate the direct involvement of the M2 receptor in abnormal mitosis induction, we silenced *CHRM2* expression by siRNA transfection in the U251 cells, with the cell line showing a higher percentage of abnormal mitosis after M2 agonist treatment. In this condition, the number of abnormal mitoses in the untreated and APE-treated cells was comparable. These results, in accordance with previous studies [16,17], confirm the selectivity of the M2 agonist APE for M2 receptors and show that APE is unable to cause a mitotic arrest in the absence of M2 receptors.

Several studies have shown that sirtuins, particularly Sirtuin 2 (SIRT2), are among the factors involved in mitotic progression in the normal cell cycle [30]. SIRT2 is an NAD+-dependent histone deacetylase [33]. SIRT2 overexpression has been shown to prolong the M phase and delay mitotic exit [34]. It has been reported that SIRT2 mRNA expression is reduced in approximately 70% of human gliomas [35]. Like SIRT1, SIRT2 confers protection against genotoxic and oxidative stress by reducing cellular ROS levels through the induction of the mitochondrial anti-oxidant enzymes manganese superoxide dismutase (MnSOD) [36] and by blocking chromosome condensation, acting at the level of the mitotic checkpoint [29].

According to the literature data [35], SIRT2 appears to be strongly downregulated, both in U87 and U251 cell lines. Interestingly, APE treatment induced a significant upregulation of SIRT2 expression 30 h after treatment, in line with the increased accumulation of the cells in the M phase. Moreover, SIRT2 activity may contribute to the formation of deacetylated unstable microtubules [37]. In our experiments, we observed a decrease in acetylated tubulin expression, particularly in the U87 cells. This phenomenon may lead to spindle fragmentation, which contributes to the formation of spindle extra poles and abnormal chromosome segregation. Although U251 did not show a significant modulation of acetylated tubulin after M2 agonist treatment, by immunocytochemistry analysis for acetylated tubulin, we observed that in U251 cells treated with APE, the acetylated microtubules passed over the chromosomes and did not attach the kinetochores. Previous data have demonstrated that APE treatment induces oxidative stress, DNA damage, and chromosomal instability, particularly in the U251 cell line [18]. These results suggest that the inability of microtubules to bind chromosomes may also be dependent on the possible alterations that may occur on DNA or other proteins associated with the kinetochores that prevent the interaction between chromosomes and microtubules. Altogether, these alterations may negatively influence chromosomes' organization and their attachment to microtubules of the mitotic spindle and their inability of divide cells. The dramatically extended mitotic arrest and the highly abnormal phenotypes observed by time-lapse microscopy showed that glioblastoma cells are unable to further proliferate after APE treatment. This suggests that mitotic catastrophic and apoptotic events may occur in subsequent mitosis.

5. Conclusions

Glioblastoma cells are characterized by an aberrant cell cycle [20] and marked aneuploidy [21]. Therefore, mitotic catastrophe induction may be an interesting oncosuppressive mechanism with the aim of inducing mitosis-related cell death or permanent cell cycle arrest [23]. The data obtained in the present work, together with previous data [14,16–18,38,39], may allow to hypothesize that the activation of the M2 receptor through selective agonists causes two significant effects in glioblastoma cells: (1) A cytostatic effect, impairing cell proliferation and inducing cell cycle arrest; (2) a cytotoxic effect, increasing apoptosis and cell death. The two effects should be strictly correlated and both contribute to impairment of glioblastoma cell proliferation and survival.

These results, together with the data obtained in other tumor types (i.e., neuroblastoma, breast cancer, and urothelial cancer) [13,15,40,41], highlight the M2 muscarinic receptor as a new strategic therapeutic target in cancer therapy. Therefore, a special effort to identify new selective ligands able to bind this receptor with more efficacy and to reduce the possible side effects appears clinically relevant and may open new therapeutic perspectives for glioblastoma treatment, as well as for the treatment of other tumors.

Supplementary Materials: The following are available online at https://www.mdpi.com/article/10.3390/cells10071727/s1, Video S1: Abnormal cell division in U251 cells.

Author Contributions: Conception and design: A.M.T. and E.C.; acquisition of data: M.D.B., F.A., V.T., M.F., C.G., M.S. and F.D.; time-lapse experiments: I.A.A. and G.G.; analysis and interpretation of data: A.M.T., M.D.B., F.D., I.A.A. and G.G.; drafting of the article: M.D.B., C.G. and A.M.T.; critical revision of the article: A.M.T., F.D., G.G., E.C. and F.A. All authors have read and agreed to the published version of the manuscript.

Funding: No external funds were received.

Informed Consent Statement: Not applicable.

Acknowledgments: This work was supported by Ateneo Sapienza and Network CIB 2018 funds to A.M.T. Time-lapse experiments were performed at the IBPM Nikon Reference Centre.

Conflicts of Interest: The authors declare no conflict of interest.

References

1. Rock, K.; McArdle, O.; Forde, P.; Dunne, M.; Fitzpatrick, D.; O'Neill, B.; Faul, C. A clinical review of treatment outcomes in glioblastoma multiforme—The validation in a non-trial population of the results of a randomised Phase III clinical trial: Has a more radical approach improved survival? *Br. J. Radiol.* **2012**, *85*, e729–e733. [CrossRef]
2. Calabrese, C.; Poppleton, H.; Kocak, M.; Hogg, T.L.; Fuller, C.; Hamner, B.; Oh, E.Y.; Gaber, M.; Finklestein, D.; Allen, M.; et al. A Perivascular Niche for Brain Tumor Stem Cells. *Cancer Cell* **2007**, *11*, 69–82. [CrossRef]

3. Ostrom, Q.T.; Cioffi, G.; Gittleman, H.; Patil, N.; Waite, K.; Kruchko, C.; Barnholtz-Sloan, J.S. CBTRUS Statistical Report: Primary Brain and Other Central Nervous System Tumors Diagnosed in the United States in 2012–2016. *Neuro Oncol.* **2019**, *21*, v1–v100. [CrossRef]
4. Stupp, R.; Mason, W.P.; van den Bent, M.J.; Weller, M.; Fisher, B.; Taphoorn, M.J.; Belanger, K.; Brandes, A.A.; Marosi, C.; Bogdahn, U.; et al. Radiotherapy plus Concomitant and Adjuvant Temozolomide for Glioblastoma. *N. Engl. J. Med.* **2005**, *352*, 987–996. [CrossRef]
5. Ohka, F.; Natsume, A.; Wakabayashi, T. Current Trends in Targeted Therapies for Glioblastoma Multiforme. *Neurol. Res. Int.* **2012**, *2012*, 878425. [CrossRef] [PubMed]
6. Tata, A.M. Muscarinic Acetylcholine Receptors: New Potential Therapeutic Targets in Antinociception and in Cancer Therapy. *Recent Pat. CNS Drug Discov.* **2008**, *3*, 94–103. [CrossRef]
7. Jiménez, E.; Montiel, M. Activation of MAP kinase by muscarinic cholinergic receptors induces cell proliferation and protein synthesis in human breast cancer cells. *J. Cell. Physiol.* **2005**, *204*, 678–686. [CrossRef] [PubMed]
8. Oppitz, M.; Möbus, V.; Brock, S.; Drews, U. Muscarinic Receptors in Cell Lines from Ovarian Carcinoma: Negative Correlation with Survival of Patients. *Gynecol. Oncol.* **2002**, *85*, 159–164. [CrossRef] [PubMed]
9. Song, P.; Sekhon, H.S.; Lu, A.; Arredondo, J.; Sauer, D.; Gravett, C.; Mark, G.P.; Grando, S.A.; Spindel, E.R. M3 Muscarinic receptor antagonists inhibit small cell lung carcinoma growth and mitogen-activated protein kinase phosphorylation induced by acetylcholine secretion. *Cancer Res.* **2007**, *67*, 3936–3944. [CrossRef]
10. Guizzetti, M.; Costa, P.; Peters, J.; Costa, L.G. Acetylcholine as a mitogen: Muscarinic receptor-mediated proliferation of rat astrocytes and human astrocytoma cells. *Eur. J. Pharmacol.* **1996**, *297*, 265–273. [CrossRef]
11. Matute, C.; Arellano, R.O.; Conde-Guerri, B.; Miledi, R. mRNA coding for neurotransmitter receptors in a human astrocytoma. *Proc. Natl. Acad. Sci. USA* **1992**, *89*, 3399–3403. [CrossRef] [PubMed]
12. Gurwitz, D.; Razon, N.; Sokolovsky, M.; Soreq, H. Expression of muscarinic binding sites in primary human brain tumors. *Dev. Brain Res.* **1984**, *14*, 61–70. [CrossRef]
13. Pacini, L.; De Falco, E.; Di Bari, M.; Coccia, A.; Siciliano, C.; Ponti, D.; Pastore, A.L.; Petrozza, V.; Carbone, A.; Tata, A.M.; et al. M2muscarinic receptors inhibit cell proliferation and migration in urothelial bladder cancer cells. *Cancer Biol. Ther.* **2014**, *15*, 1489–1498. [CrossRef] [PubMed]
14. Ferretti, M.; Fabbiano, C.; Di Bari, M.; Ponti, D.; Calogero, A.; Tata, A.M. M2 muscarinic receptors inhibit cell proliferation in human glioblastoma cell lines. *Life Sci.* **2012**, *91*, 1134–1137. [CrossRef]
15. Lucianò, A.M.; Perciballi, E.; Fiore, M.; Del Bufalo, D.; Tata, A.M. The Combination of the M2 Muscarinic Receptor Agonist and Chemotherapy Affects Drug Resistance in Neuroblastoma Cells. *Int. J. Mol. Sci.* **2020**, *21*, 8433. [CrossRef]
16. Ferretti, M.; Fabbiano, C.; Di Bari, M.; Conte, C.; Castigli, E.; Sciaccaluga, M.; Ponti, D.; Ruggieri, P.; Raco, A.; Ricordy, R.; et al. M2 receptor activation inhibits cell cycle progression and survival in human glioblastoma cells. *J. Cell. Mol. Med.* **2013**, *17*, 552–566. [CrossRef]
17. Alessandrini, F.; Cristofaro, I.; Di Bari, M.; Zasso, J.; Conti, L.; Tata, A.M. The activation of M2 muscarinic receptor inhibits cell growth and survival in human glioblastoma cancer stem cells. *Int. Immunopharmacol.* **2015**, *29*, 105–109. [CrossRef]
18. Di Bari, M.; Tombolillo, V.; Conte, C.; Castigli, E.; Sciaccaluga, M.; Iorio, E.; Carpinelli, G.; Ricordy, R.; Fiore, M.; Degrassi, F.; et al. Cytotoxic and genotoxic effects mediated by M2 muscarinic receptor activation in human glioblastoma cells. *Neurochem. Int.* **2015**, *90*, 261–270. [CrossRef]
19. Galluzzi, L.; Vitale, I.; Vacchelli, E.; Kroemer, G. Cell Death Signaling and Anticancer Therapy. *Front. Oncol.* **2011**, *1*. [CrossRef]
20. Naqvi, A.Z.; Mahjabeen, I.; Ameen, S.; Ahmad, M.W.; Khan, A.U.; Akram, Z.; Kayani, M.A. Genetic and expression variations of cell cycle pathway genes in brain tumor patients. *Biosci. Rep.* **2020**, *40*, BSR20190629. [CrossRef]
21. Bigner, S.H.; Vogelstein, B. Cytogenetics and Molecular Genetics of Malignant Gliomas and Medulloblastoma. *Brain Pathol.* **1990**, *1*, 12–18. [CrossRef]
22. Mc Gee, M.M. Targeting the Mitotic Catastrophe Signaling Pathway in Cancer. *Mediat. Inflamm.* **2015**, *2015*, 146282. [CrossRef]
23. Galluzzi, L.; Vitale, I.; Abrams, J.M.; Alnemri, E.S.; Baehrecke, E.H.; Blagosklonny, M.V.; Dawson, T.M.; Dawson, V.L.; El-Deiry, W.S.; Fulda, S.; et al. Molecular definitions of cell death subroutines: Recommendations of the Nomenclature Committee on Cell Death 2012. *Cell Death Differ.* **2012**, *19*, 107–120. [CrossRef]
24. Weaver, B.A.; Cleveland, D.W. Decoding the links between mitosis, cancer, and chemotherapy: The mitotic checkpoint, adaptation, and cell death. *Cancer Cell* **2005**, *8*, 7–12. [CrossRef]
25. Vitovcova, B.; Skarkova, V.; Rudolf, K.; Rudolf, E. Biology of Glioblastoma Multiforme—Exploration of Mitotic Catastrophe as a Potential Treatment Modality. *Int. J. Mol. Sci.* **2020**, *21*, 5324. [CrossRef]
26. Bataller, M.; Portugal, J. Mitotic Catastrophe as a Consequence of Chemotherapy. *Anti-Cancer Agents Med. Chem.* **2006**, *6*, 589–602. [CrossRef]
27. Vitale, I.; Galluzzi, L.; Castedo, M.; Kroemer, G. Mitotic catastrophe: A mechanism for avoiding genomic instability. *Nat. Rev. Mol. Cell Biol.* **2011**, *12*, 385–392. [CrossRef] [PubMed]
28. De Brabander, M.J.; Van de Veire, R.M.L.; Aerts, F.E.M.; Borgers, M.; Janssen, P.A.J. The effects of methyl (5-(2-thienylcarbonyl)-1H-benzimidazol-2-yl) carbamate, (R 17934; NSC 238159), a new synthetic antitumoral drug interfering with microtubules, on mammalian cells cultured in vitro. *Cancer Res.* **1976**, *36*, 905–916.

29. Inoue, T.; Hiratsuka, M.; Osaki, M.H.; Yamada, H.; Kishimoto, I.; Yamaguchi, S.; Nakano, S.; Katoh, M.; Ito, H.; Oshimura, M. SIRT2, a tubulin deacetylase, acts to block the entry to chromosome condensation in response to mitotic stress. *Oncogene* **2007**, *26*, 945–957. [CrossRef] [PubMed]

30. Inoue, T.; Hiratsuka, M.; Osaki, M.; Oshimura, M. The Molecular Biology of Mammalian SIRT Proteins: SIRT2 Functions on Cell Cycle Regulation. *Cell Cycle* **2007**, *6*, 1011–1018. [CrossRef]

31. So, E.C.; Huang, Y.-M.; Hsing, C.-H.; Liao, Y.-K.; Wu, S.-N. Arecoline inhibits intermediate-conductance calcium-activated potassium channels in human glioblastoma cell lines. *Eur. J. Pharmacol.* **2015**, *758*, 177–187. [CrossRef]

32. Thompson, E.G.; Sontheimer, H. Acetylcholine Receptor Activation as a Modulator of Glioblastoma Invasion. *Cells* **2019**, *8*, 1203. [CrossRef]

33. North, B.J.; Marshall, B.L.; Borra, M.T.; Denu, J.M.; Verdin, E. The Human Sir2 Ortholog, SIRT2, Is an NAD^+-Dependent Tubulin Deacetylase. *Mol. Cell* **2003**, *11*, 437–444. [CrossRef]

34. Dryden, S.C.; Nahhas, F.A.; Nowak, J.E.; Goustin, A.-S.; Tainsky, M.A. Role for Human SIRT2 NAD-Dependent Deacetylase Activity in Control of Mitotic Exit in the Cell Cycle. *Mol. Cell. Biol.* **2003**, *23*, 3173–3185. [CrossRef]

35. Hiratsuka, M.; Inoue, T.; Toda, T.; Kimura, N.; Shirayoshi, Y.; Kamitani, H.; Watanabe, T.; Ohama, E.; Tahimic, C.G.; Kurimasa, A.; et al. Proteomics-based identification of differentially expressed genes in human gliomas: Down-regulation of SIRT2 gene. *Biochem. Biophys. Res. Commun.* **2003**, *309*, 558–566. [CrossRef]

36. Wang, F.; Nguyen, M.; Qin, F.X.-F.; Tong, Q. SIRT2 deacetylates FOXO3a in response to oxidative stress and caloric restriction. *Aging Cell* **2007**, *6*, 505–514. [CrossRef]

37. Szabó, A.; Oláh, J.; Szunyogh, S.; Lehotzky, A.; Szenasi, T.; Csaplár, M.; Schiedel, M.; Lőw, P.; Jung, M.; Ovádi, J. Modulation of Microtubule Acetylation by the Interplay of TPPP/p25, SIRT2 and New Anticancer Agents with Anti-SIRT2 Potency. *Sci. Rep.* **2017**, *7*, 17070. [CrossRef]

38. Cristofaro, I.; Alessandrini, F.; Spinello, Z.; Guerriero, C.; Fiore, M.; Caffarelli, E.; Laneve, P.; Dini, L.; Conti, L.; Tata, A.M. Cross Interaction between M2 Muscarinic Receptor and Notch1/EGFR Pathway in Human Glioblastoma Cancer Stem Cells: Effects on Cell Cycle Progression and Survival. *Cells* **2020**, *9*, 657. [CrossRef]

39. Cristofaro, I.; Spinello, Z.; Matera, C.; Fiore, M.; Conti, L.; De Amici, M.; Dallanoce, C.; Tata, A.M. Activation of M2 muscarinic acetylcholine receptors by a hybrid agonist enhances cytotoxic effects in GB7 glioblastoma cancer stem cells. *Neurochem. Int.* **2018**, *118*, 52–60. [CrossRef]

40. Lucianò, A.M.; Mattei, F.; Damo, E.; Panzarini, E.; Dini, L.; Tata, A.M. Effects mediated by M2 muscarinic orthosteric agonist on cell growth in human neuroblastoma cell lines. *Pure Appl. Chem.* **2019**, *91*, 1641–1650. [CrossRef]

41. Español, A.J.; Salem, A.; Di Bari, M.; Cristofaro, I.; Sanchez, Y.; Tata, A.M.; Sales, M.E. The metronomic combination of paclitaxel with cholinergic agonists inhibits triple negative breast tumor progression. Participation of M2 receptor subtype. *PLoS ONE* **2020**, *15*, e0226450. [CrossRef]

Review

Incorporation of 5-Bromo-2′-deoxyuridine into DNA and Proliferative Behavior of Cerebellar Neuroblasts: All That Glitters Is Not Gold

Joaquín Martí-Clúa 

Unidad de Citología e Histología, Departament de Biologia Cellular, de Fisiologia i d'Immunologia, Facultad de Biociencias, Institut de Neurociències, Universidad Autónoma de Barcelona, Bellaterra, 08193 Barcelona, Spain; Joaquim.marti.clua@uab.es; Tel.: +34-93-581-1666

Abstract: The synthetic halogenated pyrimidine analog, 5-bromo-2′-deoxyuridine (BrdU), is a marker of DNA synthesis. This exogenous nucleoside has generated important insights into the cellular mechanisms of the central nervous system development in a variety of animals including insects, birds, and mammals. Despite this, the detrimental effects of the incorporation of BrdU into DNA on proliferation and viability of different types of cells has been frequently neglected. This review will summarize and present the effects of a pulse of BrdU, at doses ranging from 25 to 300 µg/g, or repeated injections. The latter, following the method of the progressively delayed labeling comprehensive procedure. The prenatal and perinatal development of the cerebellum are studied. These current data have implications for the interpretation of the results obtained by this marker as an index of the generation, migration, and settled pattern of neurons in the developing central nervous system. Caution should be exercised when interpreting the results obtained using BrdU. This is particularly important when high or repeated doses of this agent are injected. I hope that this review sheds light on the effects of this toxic maker. It may be used as a reference for toxicologists and neurobiologists given the broad use of 5-bromo-2′-deoxyuridine to label dividing cells.

Keywords: prenatal life; perinatal life; 5-bromo-2′-deoxyuridine; cell cycle; cerebellar neuroepithelium; external granular layer; neurogenetic timetables; neurogenetic gradients; apoptosis

Citation: Martí-Clúa, J. Incorporation of 5-Bromo-2′-deoxyuridine into DNA and Proliferative Behavior of Cerebellar Neuroblasts: All That Glitters Is Not Gold. *Cells* **2021**, *10*, 1453. https://doi.org/10.3390/cells10061453

Academic Editor: Zhixiang Wang

Received: 12 May 2021
Accepted: 7 June 2021
Published: 10 June 2021

Publisher's Note: MDPI stays neutral with regard to jurisdictional claims in published maps and institutional affiliations.

1. Introduction

The thymidine analogue 5-bromo-2′-deoxyuridine (BrdU) is a pyrimidine 2′-deoxyribonucleoside compound having 5-bromouracil as the nucleobase. This agent is permanently incorporated into the DNA during the synthetic phase of the cell cycle. It has been argued that gene duplication, DNA repair or apoptotic cellular events might contribute to BrdU labeling in vivo. Therefore, this marker is considered as an indicator of DNA synthesis and not the capacity to divide [1–3].

BrdU has provided new opportunities for analyzing the proliferative behavior of neuroblasts and to infer the generation of neurons during the development of the central nervous system at the baseline level and under several experimental conditions that disturb the basal status [4–6]. Currently, BrdU labeling is the most widely used procedure for studying cell-cycle phases and durations as well as to identify dividing neuroblasts and follow their fate. The random incorporation of this agent into the DNA disturbs its composition and sequence and, therefore, BrdU presents in a myriad of negative effects because its toxicity affects the number and distribution of tagged cells [3,7–10]. The detrimental effects of BrdU administration are present from prenatal life to adulthood. In this context, when the adult neurogenesis is considered, the possibility of false labeling or the incorrect interpretation of such labeling have indicated that the existence of functional neurogenesis in the adult human brain is questionable. For example, it has been reported that in postmortem brain samples from cancer patients injected with BrdU, some BrdU-stained cells were co-labeled with NeuN in the ventricular–subventricular zone and the dentate gyrus of the hippocampal formation. It was reported in the same study that the amount of BrdU-reactive cells decreased with the longest interval between BrdU administration and histological evaluation. From these results, it was proposed that, in humans, continued neurogenesis exists. However, these data can be interpreted in a different way, e.g., a process not associated with neuroblasts proliferation. Damaged neurons, attempting to repair themselves or due to the activation of apoptotic events, incorporate BrdU and continue to die so that eventually the longer the waiting period before histological analysis, less BrdU-positive cells there are [11]. On the other hand, a way to study, in humans, the process of generating neurons from adult neural precursors is to detect, in the same tissue section, specific antigens for cell proliferation and doublecortin-reactive cells. The controversy is generated when BrdU-immunoreactive cells can be explained by processes not associated with cell division (DNA repair or apoptotic events) and doublecortin-stained cells do not present substantial proliferative activity. These results can be interpreted as the existence of young neurons, which maturation, in the human brain, might take years. Thus, a slow and delayed development of young neurons may replace neurogenetic processes in the adulthood [11,12]. Despite these data, BrdU has been considered that it is relatively benign and the consequences on the progression of the cell-cycle, migration, and fate are usually negligible or understated.

In light of the above, I began a set of experiments to determine the effects of BrdU exposure during the prenatal and perinatal development of the cerebellum. I show here the results of my research indicating that the effect of the incorporation of BrdU into newly synthesized DNA may lead to inaccurate results. Three aspects were addressed. In the first of these, it was studied whether a single injection of BrdU at doses ranging from 25 to 300 µg/g, modifies the development of the cerebellar neuroepithelium. In the second, it was compared the effects of a single-dose of BrdU, at doses ranging from 50 to 300 µg/g, on cell cycle parameters and phase durations in the cerebellar external granular layer neuroblasts. In the third procedure, to know whether there are any differences between BrdU or tritiated thymidine ([^{3}H]TdR) labeling, it was determined whether the administration of several doses of both markers, by a progressively delayed cumulative labeling method in utero, modify the developmental timetables and neurogenetic gradients of PCs and DCN neurons.

2. The Thymidine Analogue 5-Bromo-2′-Deoxyuridine: An Overview

5-iodo-2′deoxyuridine, 5-chloro-2′deoxyuridine, 5-ethynyl-2′-deoxyuridine, and BrdU are halogenated thymidine analogs [7]. The latter is a chemically synthesized bromine-labeled base analogue that competes with thymidine for incorporation into DNA. Once integrated into the new DNA, BrdU will remain in place and be passed down to daughter cells following mitosis [8,13]. Since the introduction of monoclonal antibodies against BrdU [14], an increasing number of immunostaining assays have been described for the detection of BrdU-labeled cells [10,15]. This nucleoside has provided new advances for investigating several issues during the development of the central nervous system, including developmental timetables, neurogenetic gradients, cell cycle parameters, and cell lineage in a variety of animals such as insects [16], birds [17], and mammals [18–20].

BrdU is generally thought to be a relatively benign substitute for the endogenous thymidine. The effects of the incorporation this thymidine analogue on cell proliferation, migration, and differentiation are frequently neglected. BrdU can cause unforeseen problems [21]. In this context, it has been revealed that this agent is an anticancer drug, and in combination with secondary stressors, such as ionizing radiation, BrdU presents adverse consequences for cancer cells [22]. In line with this scenario, previous reports have denoted negative effects of BrdU incorporation on the proliferative dynamics of mouse and human fibroblasts, and adult neural progenitor cells in vitro [8,23–25].

Several studies have indicated that BrdU produces sister-chromatid exchanges and double-strand breaks [26]. Other authors have shown that the exposure of rodent neural stem cells to BrdU gives rise to a decrease in the methylation of the DNA [27]. It has also been demonstrated that this synthetic halogenated pyrimidine induces alterations during the growth of the chick dorsal telencephalon [28]. In mammalian embryos, BrdU is toxic to cultured embryos [29], and has a harmful effect on the embryonic development of the neocortex, the striatum [30,31], the cerebral cortex [3], and the cerebellum [9,10]. Deleterious effects on neuron fate [3,7], as well as senescence in several cell types [32,33] including neuronal progenitors [24] have also been reported. These results suggest that BrdU may not be as safe as we thought. Despite that, the effects of the incorporation of this genotoxic agent into DNA on cell proliferation, migration and differentiation are not taken into account. This is particularly important when high doses of BrdU are injected.

3. The Cerebellum: A Model to Assess the Effects of Bromodeoxyuridine Administration

The cerebellum is an ancient and a prominent part of the vertebrate central nervous system. It presents a great variability in size and complexity from cyclostomes to humans [34–36]. In mammalians, the cerebellum presents a central vermis and hemispheres laterally situated. The first of these has a characteristic pattern of folia separated by fissures running perpendicular to the antero–posterior axis that is distinct from the hemispheres [37,38]. The cerebellum shows a large cortical component, the cerebellar cortex, and a set of nuclei located close to the roof of the fourth ventricle deep within the white matter or cerebellar medulla. They are called deep cerebellar nuclei or deep nuclei [39,40].

In adults, the cerebellar cortex shows a regular trilaminate architecture comprising a ganglionar layer sandwiched between an outer molecular layer and an inner granular layer. The neuronal component of these layers is integrated in a nucleocortical network [41,42], and presents distinctive dendritic arborizations, axons, and perikaryal as well as molecular markers [43–46]. This regular cytoarchitecture has allowed us to ascertain the successive phases in the ontogeny of the cerebellum, and to analyze the cellular and molecular mechanisms involved in its development [39,47].

Previous neuroembryological studies have shown that the cerebellum arises from the rhombomere 1, which comprises the most anterior part of the hindbrain [37,48]. Expression of the repressive homeobox genes *orthodenticle homolog 2* (Otx2) and *gastrulation brain homeobox 2* (Gbx2) establish the midbrain–hindbrain boundary and form the isthmic organizer [49–51]. It has been observed that the isthmus, the neural tissue located at the midbrain–hindbrain junction, plays an important role in the establishment of the cerebellar territory [52]. The organizing activity of the isthmic tissue is mediated by the expression of the *fibroblast growth factor 8* gene, which encode the fibroblast growth factor 8, a protein required for orchestrating multiples stages of cerebellar development [51]. The expression of the *fibroblast growth factor 8* gene is regulated by several genes, which are involved in the control of the cerebellum development. Among these genes are the engrailed 1 and 2 [53], and the Pax 2 and Pax 5 [37,54].

The establishment of the cerebellar territory is followed by the formation of two primary germinative areas: the cerebellar neuroepithelium, defined by Ptf1a-expression progenitors, and the rhombic lip, defined by progenitor cells expressing Atoh1 [55]. Many lines of evidence have indicated that GABAergic neurons such as Purkinje cells (PCs), inhibitory interneurons and some deep cerebellar nuclei (DCN) neurons originate from the cerebellar neuroepithelium. By contrast, glutamatergic neurons including DCN neurons, unipolar brush cells and the projection neurons of the precerebellar nuclei arise from progenitors located in the rhombic lip [37,47]. The granule cells are also glutamatergic. Their neuroblasts precursors also arise from the rhombic lip, but they migrate over the surface of the cerebellar anlage to form a temporary matrix, the external granule layer (EGL) [45,56]. These data have revealed that cerebellar neurogenesis is a process strictly compartmentalized.

The cerebellum has been selected as a model to study and interpret the effects of BrdU because this encephalic region is highly vulnerable to intoxication and poisoning. It is the main target of several environmental toxins including mercury, lead, manganese, and toluene/benzene derivates, and drugs such as anticonvulsants, lithium salts, and antineoplastics [57]. The analysis of the cerebellar development can serve as a model to know the effects of BrdU on the development of other areas of the central nervous system.

4. Effect of Bromodeoxyuridine Exposure on the Cerebellar Neuroepithelium

During the early development of the cerebellum, neuroblasts located in a specialized germinal matrix, the neuroepithelium, gives rise to several types of neurons including interneurons, PCs and some DCN neurons [47]. Previous research has revealed that that the dose of BrdU usually used in many laboratories (50 µg/g) is well tolerated, when administered in the embryonic period, and produces no cytotoxic effects [58,59]. These observations deserve attention because other data from the literature have been reported that the administration of BrdU affects the cell cycle progression of neural stem cells [60]. In addition to that, when injected in the prenatal life, low doses of BrdU compromise the number and distribution of spatial distribution of cells in the cerebral cortex of monkeys [3]. Moreover, there are also proofs revealing that the administration of this marker in utero alters neuroblast proliferation, migration, and patterning of the cerebellum [9,10]. These results suggest that BrdU may not be as safe as we thought. In this section, evidences are presented indicating that high doses of BrdU interferes with cell proliferation and promotes apoptotic cellular events in the neuroepithelium.

Pregnant rats were injected at embryonic day (E) 13 with a single dose of saline or BrdU at doses ranging from 25 to 300 µg/g. Embryos were removed by caesarian at regular intervals from 5 to 30 h after agent exposure. To determine the effect of BrdU administration on cerebellar neuroepithelial cells several parameters were quantified: (I) density of mitotic figures, (II) BrdU, and (III) proliferating cell nuclear antigen (PCNA)-positive cells. Results indicate that, irrespective of the analyzed survival times, the density of the afore-mentioned parameters was close in animals exposed to doses of BrdU ranging from 25 to 75 µg/g (Figure 1). Interestingly, no statistical differences were observed in comparison with rats administered with saline. However, when doses of 100 to 300 µg/g of BrdU were studied, it was observed the analyzed parameters were decreased in comparison to the saline group. The dose of 300 µg/g of BrdU produced the most detrimental effects (Figure 1). These results have indicated that a single low dose of BrdU (25 to 75 µg/g) does not alter the proliferative behavior of the neuroepitelial neuroblasts. Despite this, a more protracted effect on neuroepithelial neuroblasts cannot be excluded, i.e., cell differentiation and final fate.

On the other hand, to discover whether BrdU-injection leads to apoptotic degeneration, the density of terminal deoxynucleotidyl transferase dUTP nick-end labeling (TUNEL) and active caspase-3 were quantified. Transmission electron microscopy was carried out to confirm apoptotic cell death. Results showed that, in comparison to saline injected rats and irrespective of the analyzed survival times, the density of TUNEL-positive and caspase-3 reactive neuroblasts were similar in animals exposed to doses of BrdU ranging from 25 to 75 µg/g of BrdU (Figure 2). No statistical differences were seen in comparison to rats administered with saline. When doses of 100 to 300 µg/g of BrdU were considered, the density of these parameters increased, indicating that high doses of this brominated thymidine led to the activation of apoptotic cellular events (Figure 2).

In order to verify the apoptotic state of neuroepithelial cells, neuroblasts treated with saline of BrdU (100 to 300 µg/g) were examined with the transmission electron microscopy (TEM). The earliest ultrastructural alterations were the presence of nuclear chromatin clumps of high electron density in close contact with the inner nuclear envelope. Another important feature was the presence of clusters of approximately round electron-dense apoptotic bodies containing very condensed dark chromatin masses. In some cases, the apoptotic bodies were broken, and their content extruded into the cytoplasm (Figure 3).

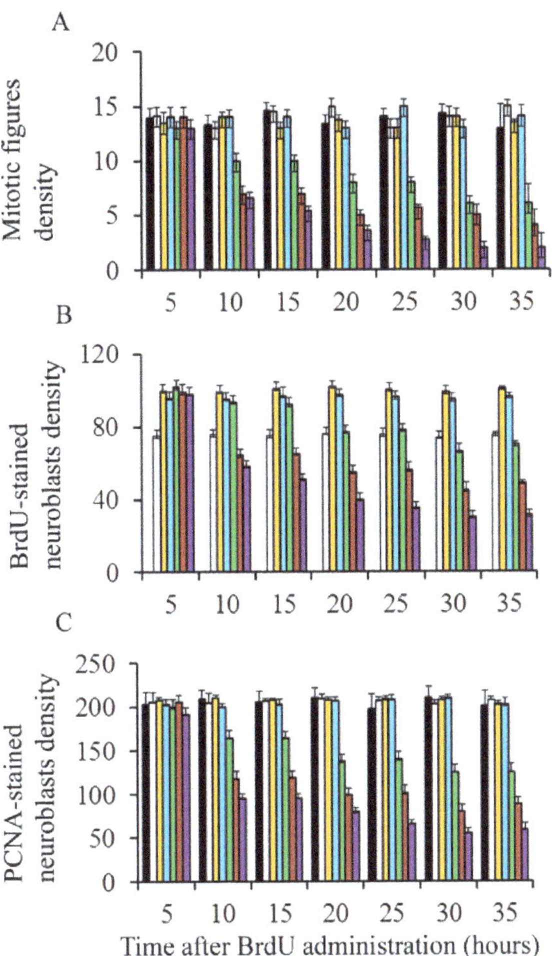

Figure 1. Mean values for mitotic figure (**A**), BrdU-reactive neuroblasts (**B**), and PCNA-stained cells density (**C**) in the cerebellar neuroepithelium of rodents injected with saline (black columns), 25 µg/g (white columns), 50 µg/g (yellow columns), 75 µg/g (blue columns), 100 µg/g (green columns), 200 µg/g (red columns), and 300 µg/g of BrdU (purple columns).

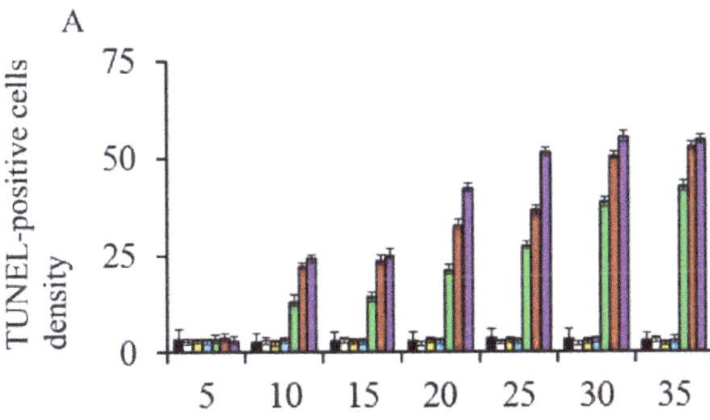

Figure 2. *Cont.*

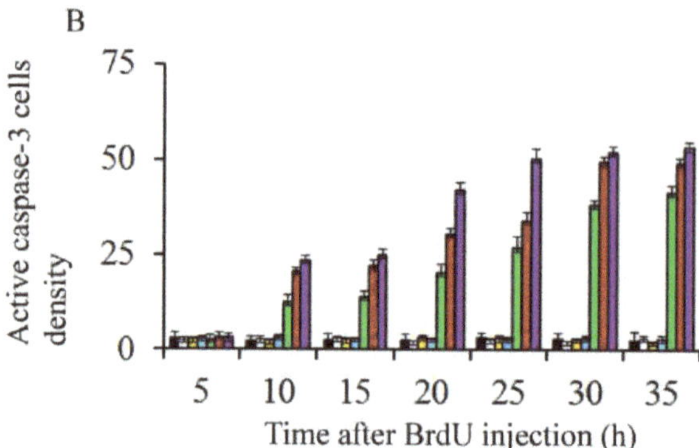

Figure 2. Mean values for TUNEL-stained (**A**) and active caspase 3-reactive neuroblasts (**B**) in the cerebellar neuroepithelium of rodents injected with saline (black columns), 25 µg/g (white columns), 50 µg/g (yellow columns), 75 µg/g (blue columns), 100 µg/g (green columns), 200 µg/g (red columns), and 300 µg/g of BrdU (purple columns).

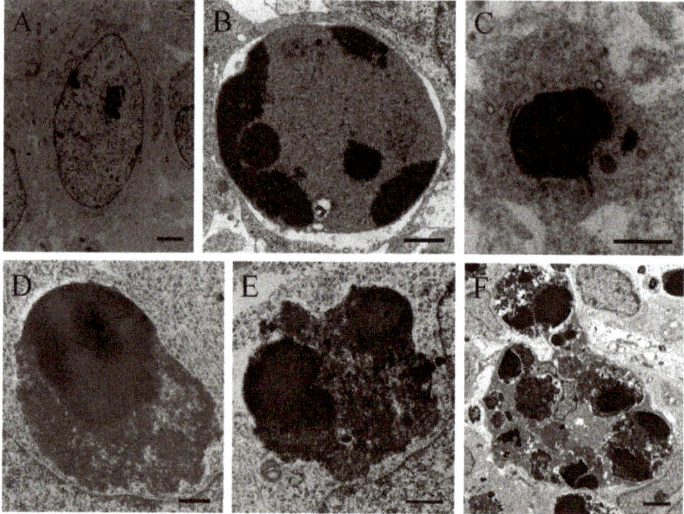

Figure 3. Electron micrograph of a healthy cell (**A**) and apoptotic cellular profile (**B–F**) from the neuroepithelium 10 h following BrdU treatment (100 µg/g). (**C**) Ultrastructural morphology of an apoptotic cell. The masses of compact chromatin display a high electron density and a homogeneous texture. They are associated with an intact nuclear envelope. (**D–F**) Typical electron-dense apoptotic bodies. (**D–E**) Examples of apoptotic bodies releasing their contents into the cytoplasm. Copyright © 2021, Wiley. Adapted with permission from Rodríguez-Vázquez, L.; Martí, J. (2000). Administration of 5-bromo-2′deoxyuridine interferes with neuroblast proliferation and promotes apoptotic cell death in the rat cerebellar neuroepithelium. *J. Comp. Neurol.* **2021**, *529*, 1081–1096. Scale bar: 2 µm (**A**), 1 µm (**B**), 2.5 µm (**C**), 1 µm (**D**), 0.5 µm (**E**), 2 µm (**F**).

At least, two observations follow from these data. (I) A single low dose of BrdU (25 to 75 µg/g) alters neither the cell-cycle progression nor promotes apoptosis, and (II) high doses of BrdU (100 to 300 µg/g) during the early prenatal life activates apoptotic cellular events, leading to an important depletion of neuroblasts.

5. Effects of Bromodeoxyuridine Exposure on the Cerebellar External Granular Layer Neuroblasts

The EGL is a transient proliferative matrix of growing cerebellar folia located beneath the piamater. In most of the vertebrates, this structure increases in thickness during an initial period of time as a result of proliferative activity of its neuroblasts [45,56]. Previous reports have revealed that BrdU is a radiosensitizer [22]. However, its therapeutic potential as an antitumoral agent is independent of secondary stressors. This is because BrdU inhibits cancer cell proliferation both in vitro and in vivo [22,61,62]. It has been reported that this agent alters cell cycle dynamics in neurosphere cultures derived from adult rat brain [8]. In this section, evidences are provided revealing that, at postnatal day (P) 9, a single high dose of BrdU interferes with the cell cycle progression of the EGL neuroblasts.

Rats were injected with BrdU at doses ranging from 50 to 300 µg/g. This thymidine analogue was supplied according to two schedules: in the first of these, rats were allowed to survive for 0.5, 1, 1.5, and 2 h after BrdU administration. In the second, animals were sacrificed 1 h later, and at regular intervals from 2 h to 26 h (spaced at two-hour intervals) following BrdU administration. The labeling index (LI) was determined as a percentage of BrdU-stained interphase nuclei per total number of scored EGL neuroblasts. Mitotic figures were examined for the presence or absence of 3,3'-diaminobenzidine-H_2O_2 reaction product in BrdU immunostained sections. The duration and phases of the cell cycle were inferred from the graphic representation of the percentage of BrdU-reactive mitoses plotted as a function of survival time after agent injection [15,20,63].

Figure 4 summarizes, in rats injected with BrdU at doses ranging from 50 to 300 µg/g, the variation in the frequency of labeled neuroblasts after different survival times (from 0.5 to 2 h). Data analysis indicated that, from 0.5 to 2 h after marker injection, the LI was highest using 50 µg/g of BrdU. It was followed by 100 and then by 200 µg/g. Does of 300 µg/g provided the lowest values.

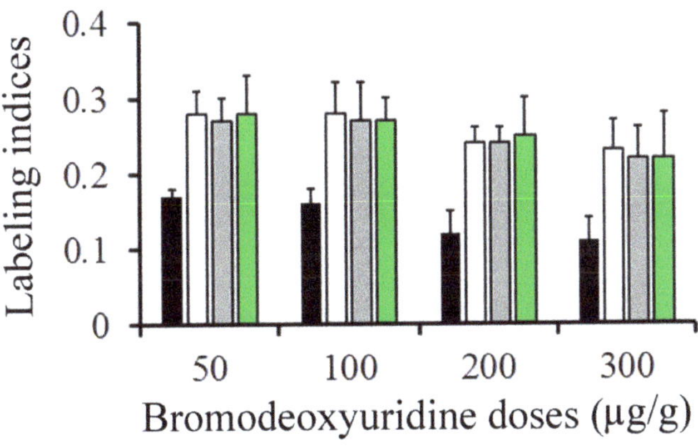

Figure 4. Mean values of labeling indices in the external granular layer. Animals were sacrificed at 0.5 (black columns), 1 (white columns), 1.5 (grey columns), and 2 h (green columns) after a single injection of bromodeoxyuridine at doses ranging from 50 to 300 µg/g.

Incorporation of BrdU into proliferating neuroblasts labels a cohort of asynchronous cycling cells in S-phase. Phase durations and cell-cycle length of the EGL neuroblasts can be quantitatively examined from the rhythmic appearance and disappearance of labeled mitotic cells. The variation that took place in the fraction of labeled mitoses from 1 to 26 h after BrdU exposure is indicated in Figure 5A. Cell kinetic parameters were inferred from Figures 4 and 5A. The time required for completion of each phase of the cell cycle is displayed in Figure 5B. Current results denoted that different doses of BrdU affect the parameters used for estimated the time required to complete the entire cell cycle.

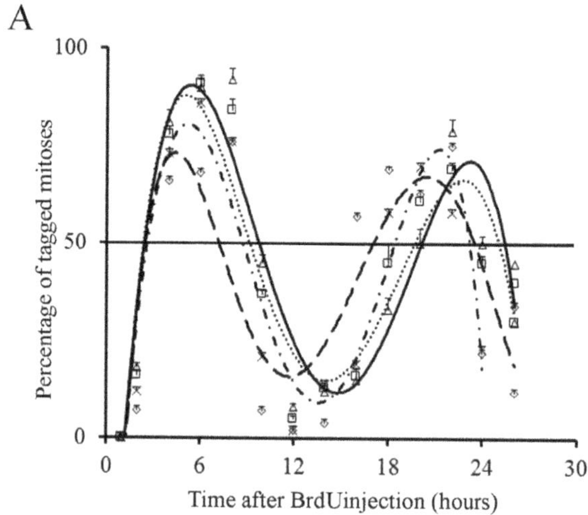

Parameters (hours)	5-Bromo-2'-deoxyuridine doses (μg/g)			
	50	100	200	300
T_C	18.4	18.1	16.6	15.6
T_S	8.0	7.3	6.2	5.3
T_{G2}	2.2	2.0	1.8	1.7
T_{G1}	7	7.7	7.9	9.1
T_M	1.2	1.1	0.7	0.5
GF	0.64	0.67	0.67	0.70

Figure 5. (**A**) Frequency of labeled mitoses in the external granular layer at successive times following a single injection of 50 μg/g (open triangles and solid black lines), 100 μg/g (open squares and dotted line), 200 μg/g (crosses and short broken lines), and 300 μg/g of BrdU (diamonds and large broken lines). Data are expressed as mean ± SEM. (**B**) Values for phase durations and cell-cycle length of the EGL cells. T_C: duration of the whole cell cycle. T_S: duration of the S-phase. T_{G2}: duration of the G2 phase. T_{G1}: duration of the G1 phase. T_M: duration of the mitotic phase. GF: growth fraction.

One fact emerges from these experiments. A single dose of 50 μg/g of BrdU has no apparent harmful effects on the cell cycle progression, which suggest that this dose is appropriate, and it provides accurate results. Higher doses (100 to 300 μg/g) altered the detection of BrdU-immunoreactive cells. As the duration and phases of the cell cycle are inferred in accordance with BrdU detection, an effect on this detection can render the measurement of the cell cycle inaccurate.

6. Inferring Purkinje Cells and Deep Cerebellar Neurons Developmental Timetables with Bromodeoxyuridine or Tritiated Thymidine. What Is the Most Suitable Marker?

BrdU and tritiated thymidine ([³H]TdR) are two markers of DNA synthesis. They have produced important insights into the neural mechanisms of the central nervous system development including cell kinetics, neuron production and migration [15,18,19,39,64,65]. Both nucleosides have advantages and disadvantages. For example, both BrdU and [³H]TdR produce cytotoxic and teratologic effects when high doses of these are supplied [9,13,30]. In addition, the intensity of [^{3}H]TdR labeling is stoichiometric while BrdU is not [66–68]. BrdU can rapidly be revealed by immunohistochemical procedures, whilst [^{3}H]TdR autoradiography is more expensive, requires a special laboratory and it takes a lot of days to develop a picture.

Previous reports have indicated that, in the cerebral cortex of macaque monkeys exposed to either BrdU or [^{3}H]TdR as embryos, quantitative differences in the number and placement of tagged neurons exist [3]. In this section, proofs are presented indicating that repeated injections of BrdU produce, in comparation with several [^{3}H]TdR administrations, systematic differences in the pattern of PCs and deep neurons neurogenesis as well as in spatial location of these macroneurons.

To study the generation and spatial location of rodents PCs and DCN neurons, both markers were administered following a progressively delayed cumulative labeling method [9,64]. This method consists of injecting pregnant dams with BrdU or [^{3}H]TdR in an overlapping series in accordance with the following time-windows: embryonic day (E)11–12, E12–13, E13–14, and E14–15. Six intraperitoneal injections of 6 mg BrdU (10 mg/mL in sterile saline with 0.007 N sodium hydroxide) 8 h apart were supplied [9]. Whereas, [^{3}H]TdR (5 μCi/g of body weight) was subcutaneously supplied. Two doses were administered on consecutive days [4,69]. PCs were analyzed in four compartments of the cerebellar cortex (vermis, paravermis, and medial and lateral hemispheres). DCN neurons were studied in the fastigial, interpositus, and dentate nuclei. Animals were sacrificed in adulthood.

Variations in the percentage of tagged neurons after different labeling times with BrdU or [^{3}H]TdR are shown in Figures 6 and 7 for PCs and DCN neurons, respectively. It is observed that, in each time-window and in each cerebellar compartment, the proportion of [^{3}H]TdR labeled PCs is higher that the percentage of BrdU tagged PCs. Similar results were obtained when DCN neurons were taken into account.

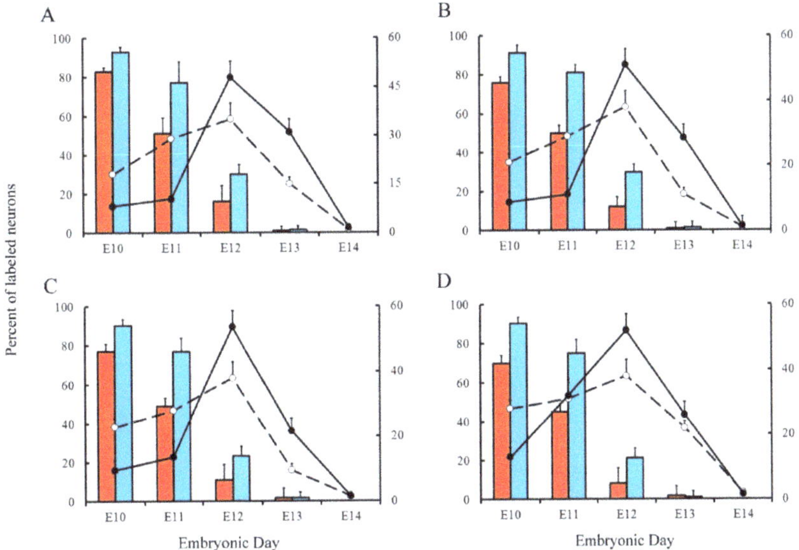

Figure 6. Comparison of Purkinje cells neurogenetic patterns in animals injected with bromodeoxyuridine (red columns) or [^{3}H]TdR (blue columns) at the level of the vermis (**A**), paravermis (**B**), medial (**C**), and lateral hemispheres (**D**). Developmental timetables of Purkinje cells were inferred using bromodeoxyuridine (open circles and dashed lines) or [^{3}H]TdR (closed circles and solid black lines). E: embryonic day. Values are expressed as mean ± SEM.

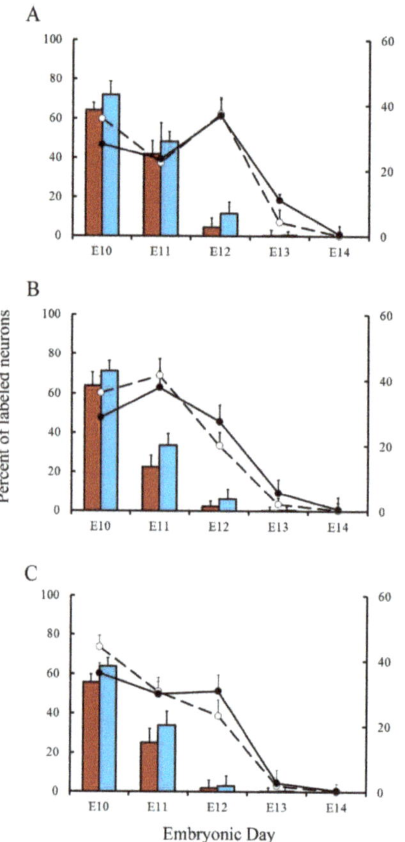

Figure 7. Comparison of deep nuclei neurons neurogenetic patterns in animals injected with bromodeoxyuridine (red columns) or [³H]TdR (blue columns) at the level of the fastigial (**A**), interpositus (**B**), and dentate (**C**) nuclei. Developmental timetables of deep nuclei neurons were inferred using bromodeoxyuridine (open circles and dashed lines) or [³H]TdR (closed circles and solid black lines). E: embryonic day. Values are expressed as mean ± SEM.

In a subsequent study, the neurogenetic timetables of PCs and DCN neurons were built. In Figure 6 (for PCs) and Figure 7 (for DCN neurons), the percentages of produced neurons are plotted against the time. My results have revealed that, with the exception of medial hemispheres at E11, more BrdU than [³H]TdR labeled PCs are produced on embryonic day 10 and 11 in the remain studied areas. The same takes place in each deep cerebellar nucleus studied at E10. However, when frequencies of newborn macroneurons were inferred from E12 to E13, more PCs were produced when the radioactive precursor was used. Similar results were seen on E12 when all the deep nuclei were examined.

The cerebellum develops following a well-known spatiotemporal sequence of events. There is evidence indicating that PCs and DCN neurons are distributed throughout the cerebellum according to precise neurogenetic gradients [37]. By examining the above-inferred times of neuron origin, it is possible to classify PCs and DCN neurons among those produced from E10 to E11 (early-born) and those generated from E12 to E14 (late-generated). In Figure 8 (for PCs) and Figure 9 (for DCN neurons), the proportions of each population fractions are compared for BrdU and [³H]TdR. My results have shown that, in the analyzed cerebellar compartments as well as in each deep nucleus, proportions of early-produced BrdU-reactive cells were always higher than percentages of neurons tagged with [³H]TdR. The opposite pattern was found when late-produced neurons were studied, revealing that significant differences in the settling pattern of PC and DCN neurons exit depending on whether BrdU and [³H]TdR are used.

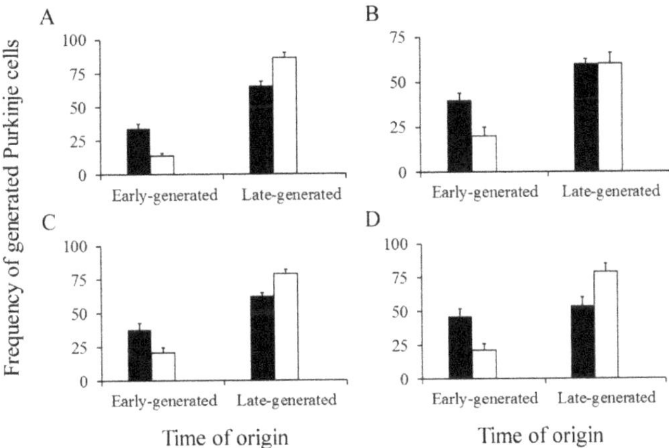

Figure 8. Percentage of early and late-produced Purkinje cells scored at the level of the vermis (**A**), paravermis (**B**), medial (**C**), and lateral hemispheres (**D**). Black columns indicate animals injected with bromodeoxyuridine and white columns animals administered with [³H]TdR. Values are expressed as mean ± SEM.

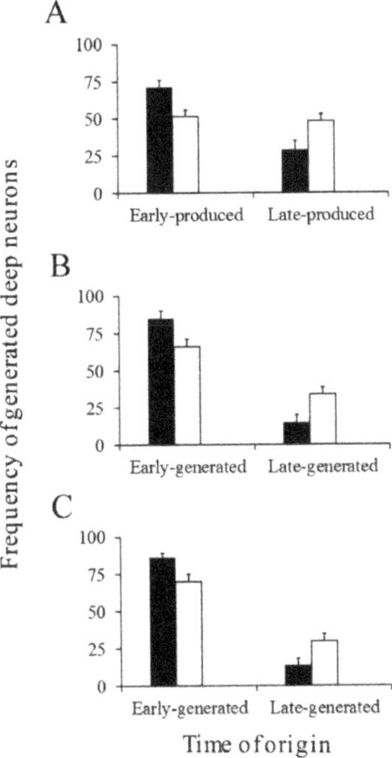

Figure 9. Percentage of early and late-generated deep nuclei neurons scored at the level of the fastigial (**A**), interpositus (**B**), and dentate (**C**) nuclei. Black columns indicate animals injected with bromodeoxyuridine and white columns animals administered with [³H]TdR. Values are expressed as mean ± SEM.

It can be drawn as a general conclusion that, in comparison to several doses of [^{3}H]TdR, to infer neurogenetic timetables and neurogenetic gradients using several doses of BrdU is less accurate. This is presumably due to the BrdU toxicity. The incorporation of this marker into DNA produces base pairing of the bromouracil with guanine instead of adenine. This substitution is involved in mutations and breaks in double-stranded DNA leading to detrimental effects on the differentiation and survival of neuroblasts [8] and on neuron fate and function [3,26,28]. Interestingly, I have found no damage to cell-cycle progression after [^{3}H]TdR exposure. This was expected because the difference between [^{3}H]TdR and the thymine is only an extra neutron in a hydrogen atom. Therefore, as suggested by Duque and Rakic [3,7], DNA with [^{3}H]TdR possibly reflects closer DNA in the non-injected animal.

7. Conclusions

The synthetic halogenated pyrimidine analogue BrdU is a marker of DNA synthesis. BrdU has provided important clues about the development of the central nervous system under different experimental contexts. Despite the fact that BrdU toxicity has been demonstrated, it is generally thought that this agent is a benign substitute for the [^{3}H]TdR. The effect of the incorporation of BrdU on proliferating neuroblasts is frequently neglected by investigators. This current review has implications for the interpretation of the results obtained by this marker to investigate the cell cycle of neuron precursors as well as the genesis of neurons. This is particularly important when a single high dose of BrdU is used because effects on proliferative behavior and the activation of apoptotic cellular events may lead to false results in the identification of proliferative neuroblasts. The same occurs when repeated injections of BrdU are supplied following a progressively delayed cumulative labeling method. Thus, data obtained with this thymidine analogue should be interpreted with caution. It is proposed that to label proliferating neural progenitors, a single pulse of BrdU (until 50 µg/g) has no apparent harmful effects. I hope that this current review may be also useful when studying and interpreting cell birth with BrdU labeling in the rest of the central nervous system.

Funding: This research received no external funding.

Institutional Review Board Statement: Not applicable.

Informed Consent Statement: Not applicable.

Data Availability Statement: There are no special databases associated with this manuscript.

Conflicts of Interest: The author declares no conflict of interest.

References

1. Rakic, P. Pre and post-development neurogenesis in primates. *Clinic. Neurosci. Res.* **2002**, *2*, 29–39. [CrossRef]
2. Breunig, J.J.; Arellano, J.I.; Macklis, J.D.; Rakic, P. Everything that glitters isn't gold: A critical review of postnatal neural precursor analyses. *Cell Stem Cell* **2007**, *1*, 612–627. [CrossRef] [PubMed]
3. Duque, A.; Rakic, P. Different effects of BrdU and ^{3}H-Thymidine incorporation into DNA on cell proliferation, position and fate. *J. Neurosci.* **2011**, *31*, 15205–15217. [CrossRef]
4. Martí, J.; Santa-Cruz, M.C.; Serra, R.; Hervás, J.P. Systematic differences in time of cerebellar-neuron origin derived from bromodeoxyuridine immunoperoxidase staining protocols and tritiated thymidine autoradiographic: A comparative study. *Int. J. Dev. Neurosci.* **2015**, *47*, 216–228. [CrossRef] [PubMed]
5. Martí, J.; Molina, V.; Santa-Cruz, M.C.; Hervás, J.P. Developmental injury to the cerebellar cortex following hydroxyurea treatment in early postnatal life: An immunohistochemical and electron microscopic study. *Neurotox. Res.* **2017**, *31*, 187–203. [CrossRef]
6. Rodríguez-Vázquez, L.; Vons, O.; Valero, O.; Martí, J. Hydroxyurea exposure and developmental of the cerebellar external granular layer: Effects on granule cell precursors, Bergmann glial and microglial cells. *Neurotox. Res.* **2019**, *35*, 387–400. [CrossRef] [PubMed]
7. Duque, A.; Rakic, P. Identification of proliferating and migration cell by BrdU and other thymidine analogs: Benefits and limitations. In *Immunocytochemistry and Related Techniques*; Merighi, A., Lossi, L., Eds.; Humana Press: New York, NY, USA, 2015; Volume 101, Chapter 7, pp. 123–139.

8. Lehner, B.; Sandner, B.; Marschallinger, J.; Lehner, C.; Furtner, T.; Couillard-Despres, S.; Rivera, F.J.; Brockhoff, G.; Bauer, H.C.; Weidner, N.; et al. The dark side of BrdU in neural stem cell biology: Detrimental effects on cell cycle, differentiation and survival. *Cell Tissue Res.* **2011**, *345*, 313–328. [CrossRef] [PubMed]

9. Sekerkova, G.; Ilijic, E.; Mugnaini, E. Bromodeoxyuridine administered during neurogenesis of the projection neurons causes cerebellar defects in rat. *J. Comp. Neurol.* **2004**, *470*, 221–239. [CrossRef] [PubMed]

10. Rodríguez-Vázquez, L.; Martí, J. Administration of 5-bromo-2′deoxyuridine interferes with neuroblast proliferation and promotes apoptotic cell death in the rat cerebellar neuroepithelium. *J. Comp. Neurol.* **2021**, *529*, 1081–1096. [CrossRef]

11. Duque, A.; Spector, R. A balanced evaluation of the evidence for adult neurogenesis in humans: Implications for the Neuropsychiatry disorders. *Brain Struct. Funct.* **2019**, *224*, 2281–2295. [CrossRef] [PubMed]

12. La Rosa, C.; Parolisi, R.; Bonfanti, L. Brain structural plasticity: From adult neurogenesis to immature neurons. *Front. Neurosci.* **2020**, *14*, 75. [CrossRef] [PubMed]

13. Nowakowski, R.S.; Lewin, S.B.; Miller, M.W. Bromodeoxyuridine immunohistochemical determination of the lengths of the cell cycle and the DNA-synthetic phase for an anatomically defined population. *J. Neurocytol.* **1989**, *18*, 311–318. [CrossRef] [PubMed]

14. Gratzner, H.G. Monoclonal antibody to 5-bromo- and 5-iododeoxyuridine. A new reagent for detection of DNA replication. *Science* **1982**, *218*, 474–475. [CrossRef] [PubMed]

15. Molina, V.; Rodríguez-Vázquez, L.; Owen, D.; Valero, O.; Martí, J. Cell cycle analysis in the rat external granular layer evaluated by several bromodeoxyuridine immunoperoxidase staining protocols. *Histochem. Cell Biol.* **2017**, *148*, 477–488. [CrossRef] [PubMed]

16. Li, G.; Hidalgo, A. Adult neurogenesis in the *Drosophila* brain: The evidence and the void. *Int. J. Mol. Sci.* **2020**, *21*, 6653. [CrossRef]

17. Larson, T.A.; Thatra, N.M.; Hou, D.; Hu, R.A.; Brenowitz, E.A. Seasonal change in neuronal turnover in a forebrain nucleus in adult songbirds. *J. Comp. Neurol.* **2019**, *527*, 767–779. [CrossRef]

18. Lanctot, A.A.; Guo, Y.; Le, Y.; Edens, B.M.; Nowakowski, R.S.; Feng, Y. Loss of brap results in premature G1/S phase transition and impeded neural progenitor differentiation. *Cell Rep.* **2017**, *20*, 1148–1160. [CrossRef]

19. Rash, B.G.; Duque, A.; Morozov, Y.M.; Arellano, J.I.; Micali, N.; Rakic, P. Gliogenesis in the outer subventricular zone promotes enlargement and gyrification of the primate cerebrum. *Proc. Natl. Acad. Sci. USA* **2019**, *116*, 7089–7094. [CrossRef]

20. Martí, J.; Rodríguez-Vázquez, L. An immunocytochemical approach to the analysis of the cell division cycle in the cerebellar neuroepithelium. *Cell Cycle* **2020**, *19*, 2451–2459. [CrossRef]

21. Costandi, M. Warning on neural technique. *Nature* **2011**. [CrossRef]

22. Levkoff, L.H.; Marshall, G.P., 2nd; Ross, H.H.; Caldeira, M.; Reynolds, B.A.; Cakiroglu, M.; Mariani, C.L.; Streit, W.J.; Laywell, E.D. Bromodeoxyuridine inhibits cancer cell proliferation in vitro and in vivo. *Neoplasia* **2008**, *10*, 804–816. [CrossRef] [PubMed]

23. Michishita, E.; Nakabayashi, K.; Suzuki, T.; Kaul, S.C.; Ogino, H.; Fujii, M.; Mitsui, Y.; Ayusawa, D. 5-Bromodeoxyuridine induces senescence-like phenomena in mammalian cells regardless of cell type or species. *J. Biochem.* **1999**, *126*, 1052–1059. [PubMed]

24. Ross, H.H.; Levkoff, L.H.; Marshall, G.P.; Caldeira, M.; Steindler, D.A.; Reynolds, B.A.; Laywell, E.D. Bromodeoxyuridine induces senescence in neural stem and progenitor cells. *Stem Cells* **2008**, *26*, 3218–3227. [CrossRef]

25. Caldwell, M.A.; He, X.; Svendsen, C.N. 5-Bromo-2′-deoxyuridine is selectively toxic to neural precursors in vitro. *Eur. J. Neurosci.* **2005**, *22*, 2965–2970. [CrossRef]

26. Taupin, P. BrdU immunohistochemistry for studying adult neurogenesis: Paradigms, pitfalls, limitations, and validation. *Brain Res. Rev.* **2007**, *53*, 198–214. [CrossRef]

27. Schneider, L.; di Fagagna, F.A. Neural stem cells exposed to BrdU lose their global DNA methylation and undergo astrocytic differentiation. *Nucleic Acids Res.* **2012**, *40*, 5332–5342. [CrossRef] [PubMed]

28. Rowell, J.J.; Ragsdale, C.W. BrdU birth dating can produce errors in cell fate specification in chick brain development. *J. Histochem. Cytochem.* **2012**, *60*, 801–810. [CrossRef] [PubMed]

29. Biggers, W.J.; Barnea, E.R.; Sanyal, M.K. Anomalous neural differentiation induced by 5-bromo-2′-deoxyuridine during organogenesis in the rat. *Teratology* **1987**, *35*, 63–75. [CrossRef] [PubMed]

30. Kolb, B.; Pedersen, B.; Ballermann, M.; Gibb, R.; Whishaw, I.Q. Embryonic and postnatal injections of bromodeoxyuridine produce age-dependent morphological and behavioral abnormalities. *J. Neurosci.* **1999**, *19*, 2337–2346. [CrossRef]

31. Kuwagata, M.; Ogawa, T.; Nagata, T.; Shioda, S. The evaluation of early embryonic neurogenesis after exposure to the genotoxic agent 5-bromo-2′-deoxyuridine in mice. *Neurotoxicology* **2007**, *28*, 780–789. [CrossRef] [PubMed]

32. Minagawa, S.; Nakabayashi, K.; Fujii, M.; Scherer, S.W.; Ayusawa, D. Early BrdU-responsive genes constitute a novel class of senescence-associated genes in human cells. *Exp. Cell Res.* **2005**, *304*, 552–558. [CrossRef] [PubMed]

33. Suzuki, T.; Minagawa, S.; Michishita, E.; Ogino, H.; Fujii, M.; Mitsui, Y.; Ayusawa, D. Induction of senescence-associated genes by 5-bromodeoxyuridine in HeLa cells. *Exp. Gerontol.* **2001**, *36*, 465–474. [CrossRef]

34. Jacobs, B.; Johnson, N.L.; Wahl, D.; Schall, M.; Maseko, B.C.; Lewandowski, A.; Raghanti, M.A.; Wicinski, B.; Butti, C.; Hopkins, W.D.; et al. Comparative neuronal morphology of the cerebellar cortex in afrotherians, carnivores, cetartiodactyls, and primates. *Front. Neuroanat.* **2014**, *8*, 24.

35. Sugahara, F.; Murakami, Y.; Pascual-Anaya, J.; Kuratani, S. Reconstructing the ancestral vertebrate braim. *Dev. Growth Differ.* **2017**, *59*, 163–174. [CrossRef]

36. Macrì, S.; Di-Poï, N. Heterochronic developmental shifts underlying squamate cerebellar diversity unveil the key features of amniote cerebellogenesis. *Front. Cell. Dev. Biol.* **2020**, *8*, 593377. [CrossRef]

37. Sillitoe, R.V.; Joyner, A.L. Morphology, molecular codes, and circuitry produce the three-dimensional complexity of the cerebellum. *Annu. Rev. Cell Dev. Biol.* **2007**, *23*, 549–577. [CrossRef] [PubMed]

38. Sudarov, A.; Joyner, A.L. Cerebellum morphogenesis: The foliation pattern is orchestrated by multi-cellular anchoring centers. *Neural Dev.* **2007**, *2*, 26. [CrossRef]

39. Altman, J.; Bayer, S.A. *Development of the Cerebellar System: In Relation to Its Evolution, Structure and Functions*; CRC Press, Inc.: Boca Raton, FL, USA, 1997.

40. Green, M.J.; Wingate, R.J.T. Developmental origins of diversity in cerebellar output nuclei. *Neural Dev.* **2014**, *9*, 1. [CrossRef]

41. Houck, B.D.; Person, A.L. Cerebellar loops: A review of the nucleocortical pathway. *Cerebellum* **2014**, *13*, 378–385. [CrossRef]

42. Cerminara, N.L.; Lang, E.J.; Sillitoe, R.V.; Apps, R. Redefining the cerebellar cortex as an assembly of non-uniform Purkinje cell microcircuits. *Nat. Rev. Neurosci.* **2015**, *16*, 79–93. [CrossRef]

43. Schilling, K.; Oberdick, J.; Rossi, F.; Baader, S.L. Besides Purkinje cells and granule neurons: An appraisal of the cell biology of the interneurons of the cerebellar cortex. *Histochem. Cell Biol.* **2008**, *130*, 601–615. [CrossRef] [PubMed]

44. Marzban, H.; Del Bigio, M.R.; Alizadeh, J.; Ghavami, S.; Zachariah, R.M.; Rastegar, M. Cellular commitment in the developing cerebellum. *Front. Cell. Dev.* **2015**, *8*, 450. [CrossRef] [PubMed]

45. Consalez, G.G.; Goldowitz, D.; Casoni, F.; Hawkes, R. Origins, development, and compartmentation of the granule cell of the cerebellum. *Front. Neural Circuits* **2021**, *14*, 611841. [CrossRef] [PubMed]

46. Tam, W.Y.; Wan, G.X.; Cheng, A.S.K.; Cheung, K.K. In search of molecular markers for cerebellar neurons. *Int. J. Mol. Sci.* **2021**, *22*, 1850. [CrossRef] [PubMed]

47. Leto, K.; Arancillo, M.; Becker, E.B.; Buffo, A.; Chiang, C.; Ding, B.; Dobyns, W.B.; Dusart, I.; Haldipur, P.; Hatten, M.E.; et al. Consensus paper: Cerebellar development. *Cerebellum* **2016**, *15*, 789–828. [CrossRef]

48. Nakamura, H.; Sato, T.; Suzuki-Hirano, A. Isthmus organizer for mesencephalon and metencephalon. *Dev. Growth Differ.* **2008**, *50*, S113–S118. [CrossRef]

49. Joyner, A.L.; Liu, A.; Millet, S. Otx2, Gbx2 and Fgf8 interact to position and maintain a mid-hindbrain organizer. *Curr. Opin. Cell Biol.* **2000**, *12*, 736–741. [CrossRef]

50. Carletti, B.; Rossi, F. Neurogenesis in the cerebellum. *Neuroscientist* **2008**, *14*, 91–100. [CrossRef]

51. Beckinghausen, J.; Sillitoe, R.V. Insights into cerebellar development and connectivity. *Neurosci. Lett.* **2019**, *688*, 2–13. [CrossRef] [PubMed]

52. Sotelo, C. Cellular and genetic regulation of the development of the cerebellar system. *Prog. Neurobiol.* **2004**, *72*, 295–339. [CrossRef]

53. Orvis, G.D.; Hartzell, A.L.; Smith, J.B.; Barraza, L.H.; Wilson, S.L.; Szulc, K.U.; Turnbull, D.H.; Joyner, A.L. The engrailed homeobox genes are required in multiple cell lineages to coordinate sequential formation of fissures and growth of the cerebellum. *Dev. Biol.* **2012**, *367*, 25–39. [CrossRef] [PubMed]

54. Wassef, M.; Joyner, A.L. Early mesencephalon/metencephalon patterning and development of the cerebellum. *Perspect Dev. Neurobiol.* **1997**, *5*, 3–16.

55. Carter, R.A.; Bihannic, L.; Rosencrane, C.; Hadley, J.L.; Tong, Y.; Phoenix, T.N.; Natarajan, S.; Easton, J.; Northcott, P.A.; Gawad, C. A single-cell transcriptional atlas of the developing murine cerebellum. *Curr. Biol.* **2018**, *28*, 2910–2920. [CrossRef]

56. Chédotal, A. Should I stay or should I go? Becoming a granule cell. *Trends Neurosci.* **2010**, *33*, 163–172. [CrossRef] [PubMed]

57. Manto, M. Toxic agents causing cerebellar ataxias. *Handb. Clin. Neurol.* **2012**, *103*, 201–213.

58. Takahashi, T.; Nowakowski, R.S.; Caviness, V.S. BrdU as an S-phase marker for quantitative studies of cytokinetic behavior in the murine cerebral ventricular zone. *J. Neurocytol.* **1992**, *21*, 185–197. [CrossRef]

59. Takahashi, T.; Goto, T.; Miyama, S.; Nowakowski, R.S.; Caviness, V.S. Sequence of neuron origin and neocortical laminar fate: Relation to cell cycle of origin in the development murine cerebral wall. *J. Neurocytol.* **1999**, *19*, 10357–10371. [CrossRef]

60. Hancock, A.; Priester, C.; Kidder, E.; Keith, J.R. Does 5-bromo-2′-deoxyuridine (BrdU) disrupt cell proliferation and Neuronal maturation in the adult rat hippocampus in vivo? *Behav. Brain Res.* **2009**, *199*, 218–221. [CrossRef]

61. Yoshihara, T.; Hibi, S.; Tsunamoto, K.; Todo, S.; Imashuku, S. Increased levels of alfa V-associated integrins in association with growth inhibition of cultured tumor cells by bromodeoxyuridine. *Anticancer Res.* **1997**, *17*, 833–838.

62. Masterson, J.C.; O'Dea, S. 5-bromo-2-deoxyuridine activates DNA damage signaling responses and induces a senescence-like phenotype in p16-null lung cancer cells. *Anticancer Drugs* **2007**, *18*, 1053–1068. [CrossRef] [PubMed]

63. Lauder, J.M. The effects of early hypo- and hyperthyroidism on the development of rat cerebellar cortex. III. Kinetics of cell proliferation in the external granular layer. *Brain Res.* **1977**, *126*, 31–51. [CrossRef]

64. Bayer, S.A.; Altman, J. Directions in neurogenetic gradients and patterns of anatomical connections in the telencephalon. *Prog. Neurobiol.* **1987**, *29*, 57–106. [CrossRef]

65. Torii, M.; Sasaki, M.; Chang, Y.-W.; Ishii, S.; Waxman, S.G.; Kocsis, J.D.; Rakic, P.; Hashimoto-Torii, K. Detection of vulnerable neurons damaged by environmental insults in utero. *Proc. Natl. Acad. Sci. USA* **2017**, *114*, 2367–2372. [CrossRef] [PubMed]

66. Nowakowski, R.S.; Hayes, N.L. New neurons: Extraordinary evidence or extraordinary conclusion? *Science* **2000**, *288*, 771. [CrossRef]

67. Rakic, P. Neurogenesis in adult primate neocortex: An evaluation of the evidence. *Nat. Rev. Neurosci.* **2002**, *3*, 65–71. [CrossRef] [PubMed]

68. Rakic, P. Adult neurogenesis in mammals: An identity crisis. *J. Neurosci.* **2002**, *22*, 614–618. [CrossRef] [PubMed]
69. Martí-Clúa, J. Developmental timetables and gradients of neurogenesis in cerebellar Purkinje cells and deep glutaminergic neurons: A comparative study between the mouse and the rat. *Anat. Rec.* **2021**. Online ahead of print. [CrossRef] [PubMed]

Article

Comparing Biochemical and Raman Microscopy Analyses of Starch, Lipids, Polyphosphate, and Guanine Pools during the Cell Cycle of *Desmodesmus quadricauda*

Šárka Moudříková [1,2], Ivan Nedyalkov Ivanov [3,4], Milada Vítová [3], Ladislav Nedbal [2], Vilém Zachleder [3], Peter Mojzeš [1,2] and Kateřina Bišová [3,*]

1 Institute of Physics, Faculty of Mathematics and Physics, Charles University, Ke Karlovu 5, CZ-12116 Prague 2, Czech Republic; sarka.moudrikova@gmail.com (Š.M.); mojzes@karlov.mff.cuni.cz (P.M.)
2 Institute of Bio- and Geosciences/Plant Sciences (IBG-2), Forschungszentrum Jülich, Wilhelm-Johnen-Straße, D-52428 Jülich, Germany; l.nedbal@fz-juelich.de
3 Laboratory of Cell Cycles of Algae, Centre Algatech, Institute of Microbiology of the Czech Academy of Sciences, Novohradská 237, CZ-37981 Třeboň, Czech Republic; ivanov@alga.cz (I.N.I.); vitova@alga.cz (M.V.); zachleder@alga.cz (V.Z.)
4 Faculty of Science, University of South Bohemia, Branišovská 1760, CZ-37005 České Budějovice, Czech Republic
* Correspondence: bisova@alga.cz; Tel.: +420-384-340-485

Citation: Moudříková, Š.; Ivanov, I.N.; Vítová, M.; Nedbal, L.; Zachleder, V.; Mojzeš, P.; Bišová, K. Comparing Biochemical and Raman Microscopy Analyses of Starch, Lipids, Polyphosphate, and Guanine Pools during the Cell Cycle of *Desmodesmus quadricauda*. *Cells* **2021**, *10*, 62. https://doi.org/10.3390/cells10010062

Received: 28 November 2020
Accepted: 29 December 2020
Published: 3 January 2021

Publisher's Note: MDPI stays neutral with regard to jurisdictional claims in published maps and institutional affiliations.

Abstract: Photosynthetic energy conversion and the resulting photoautotrophic growth of green algae can only occur in daylight, but DNA replication, nuclear and cellular divisions occur often during the night. With such a light/dark regime, an algal culture becomes synchronized. In this study, using synchronized cultures of the green alga *Desmodesmus quadricauda*, the dynamics of starch, lipid, polyphosphate, and guanine pools were investigated during the cell cycle by two independent methodologies; conventional biochemical analyzes of cell suspensions and confocal Raman microscopy of single algal cells. Raman microscopy reports not only on mean concentrations, but also on the distribution of pools within cells. This is more sensitive in detecting lipids than biochemical analysis, but both methods—as well as conventional fluorescence microscopy—were comparable in detecting polyphosphates. Discrepancies in the detection of starch by Raman microscopy are discussed. The power of Raman microscopy was proven to be particularly valuable in the detection of guanine, which was traceable by its unique vibrational signature. Guanine microcrystals occurred specifically at around the time of DNA replication and prior to nuclear division. Interestingly, guanine crystals co-localized with polyphosphates in the vicinity of nuclei around the time of nuclear division.

Keywords: microalgae; *Desmodesmus quadricauda*; cell cycle; starch; lipids; polyphosphate; guanine; confocal Raman microscopy

1. Introduction

Green algae dividing by multiple fission such as *Chlorella sp.*, *Chlamydomonas sp.* and the chlorococcal alga *Desmodesmus quadricauda*, which was chosen here as an experimental organism, grow and multiply rapidly [1], making them ideal model organisms for algal biotechnology as well as cell biology [1]. During the day, their cells grow rapidly without interruption from cell division, which occurs at night. This feature leads to natural synchronization of such algae by diurnally alternating day light. The same light/dark regime can be used in the laboratory, yielding highly synchronized cultures composed of uniformly aged and sized cells at the same phase of life and cell cycle. Thus, culture synchronization facilitates studies of suspensions of algal cells that are all in well-defined phases of their life and cell cycles.

The cell cycle, i.e., the period between two cell divisions, consists of two parts: growth and reproduction. Each reproductive sequence is composed of DNA replication, nuclear division and cell division. Cells of most organisms divide into two daughter cells and their cell cycle consists of a single growth and reproductive sequence. The cell cycle of *D. quadricauda*, and of some other chlorococcal and volvocal algae, may consist of one or several pairs of growth and reproductive sequences, thus leading to so called multiple fission. In *D. quadricauda*, DNA replication and nuclear division of the same reproductive sequence are temporally separated by several hours. In this organism, the cells are routinely multinuclear, as cell division is represented by one to several rounds of protoplast fission, closely following each other after nuclear division of the last reproductive sequence (Figure 1). Depending on the number of growth and reproductive sequences, a single mother cell of *D. quadricauda* can produce two, four, or eight daughter cells. The daughter cells originating from one mother cell remain connected by a shared cell wall in a structure called a coenobium. The obligate coenobial growth of this species provides a convenient means of monitoring growth, by simply counting cells in a particular coenobia and noting its two-fold, four-fold, or higher multiple fission pattern. Furthermore, the coenobia of *D. quadricauda* are particularly suitable for long-term microscopic studies, including confocal Raman microscopy, since they can be easily immobilized between a microscope slide and coverslip with their two largest dimensions running parallel to the slide surface, and, consequently, cellular structures of all cells of each coenobium can be found in the same focal plane.

In green algae dividing by multiple fission, the photosynthetic assimilation of carbon dioxide supports growth of most cell structures during the light phase as well as the accumulation of energy and other reserves that are required for or distributed by cell division in the dark phase. Thus, increasing cell volume, accumulation of bulk RNA and protein content or increasing energy and carbon stores can all serve to quantify growth. Furthermore, growth is accompanied by an increase in non-carbon energy rich compounds such as polyphosphates (polyP) that can serve both as components of nucleic acids (DNA, RNA) and as an energy store. Recently, guanine crystals were identified in the algal cells by Raman microscopy [2]. Although they remain rather enigmatic, they seem to be ubiquitous among microalgae [3] and might serve either as a non-specific nitrogen depot [3] or directly as a precursor for nucleic acid synthesis. In multiple fission, cell growth is composed of several growth sequences. Each of them is completed by reaching a threshold critical size, in which the cells approximately double their original mass, and double their original RNA, protein, starch, and lipid contents. The cells then enter into their reproductive sequence by attaining commitment point [4,5], thus tightly coupling the completion of the growth sequence to the reproductive events [6]. Once committed, the reproductive sequence can be completed in darkness without external supplies of energy and carbon. At the same time, depending on growth conditions, the cells can enter another growth sequence that, once completed, would allow another reproductive sequence to occur before a new coenobium is formed. In this way, individual growth sequences consecutively follow each other whilst reproductive sequences run concurrently and overlap within one cell cycle in individual cell. Thus, DNA replication of the second reproductive sequence follows the nuclear division of the first reproductive sequence and precedes the nuclear division of the same (second) sequence but also cell division of the first (preceding) reproductive sequence (Figure 1). The number of reproductive sequence/s as well as the extent of their overlap within a single cell cycle is determined by growth rate, which is governed by environmental conditions so that at higher growth rates, more reproductive sequences are started [4,5,7–11]. Under favorable conditions, *D. quadricauda* cells can consecutively replicate DNA in as many as three separate rounds and divide their nuclei, becoming bi-, tetra-, and octonuclear (Figure 1) [4,7,12–14].

Coordination between growth and reproductive sequences takes place within a short time period, commitment point, when, by reaching a critical threshold, cell size is translated into entry into the reproductive sequence. To better understand how entry of the reproductive sequence and its progression are dependent on the reserves of energy, carbon, other major biogenic elements, and essential biomolecules, it is necessary to quantify those reserves in different phases of the cell cycle, preferably their mutual relationships, spatial contexts, and temporal evolution. At the macroscopic level of the cell suspension, conventional analytical methods to study the algal biomass can be used [1,8]. To analyze spatial and temporal distribution of different macromolecules at the level of individual cells, the biomolecules can be stained or fluorescently labeled and visualized by means of optical and fluorescence microscopy [15–17]. However, when applying standard approaches, each biomolecular compound is usually labeled and monitored individually. Consequently, information about spatial and temporal relationships between the different substances at the level of individual cells is averaged and lost.

Raman microscopy is a label-free method that combines the advantages of molecular specificity of Raman spectroscopy with a high spatial resolution of optical confocal microscopy [18]. Raman microscopy has been widely used in biology in recent years [19], because it is relatively simple to use, with no need to introduce artificial staining or labelling to the sample, and it is able to detect several chemical components simultaneously, largely non-destructively and with a sub-cellular resolution. In microalgae, extensive use of Raman microscopy has long been hampered by strong autofluorescence of the photosynthetic apparatus, which reaches intensities several orders of magnitude higher than the desired Raman signal. This narrowed the applicability of Raman microscopy to studies of carotenoids that exhibit a resonantly enhanced Raman signal [20–22], or to algal cells with high levels of neutral lipids [23–25].

To observe a conventional Stokes Raman signal of other biomolecules such as starch, proteins, and polyP, high power in-focus photobleaching has been commonly used to remove the fluorescent background [26,27]. However, in-focus photobleaching greatly slows the acquisition of Raman maps and often results in thermal decomposition and burning of the inspected cells. We have recently demonstrated that wide-area, low-intensity photobleaching of a whole algal cell (or even multiple cells at once) by a defocused laser beam shortens the time needed for photobleaching to several seconds per cell, and enables one to reliably observe and identify starch, lipid, and polyP granules within individual microalgal cells, with sub-cellular spatial resolution [28]. The analytical strength of Raman microscopy has been further demonstrated by identifying hitherto unknown guanine microcrystals in two species of microalgae [2].

In this paper, we analyzed the dynamics of starch, lipid, polyP, and guanine pools during the cell cycle of *D. quadricauda* by simultaneously using biochemical analysis of a synchronized algal suspension, and by conventional fluorescent and Raman microscopy of individual algal cells from a synchronized algal culture. The power of Raman imaging was demonstrated in *D. quadricauda* cells by identifying guanine microcrystals occurring close to dividing nuclei in synchrony with polyP grains.

2. Materials and Methods

2.1. Experimental Organism

The chlorococcal alga *Desmodesmus quadricauda* was obtained from the Culture Collection of Autotrophic Organisms, Institute of Botany (CCALA, Czech Academy of Sciences, Třeboň, Czech Republic).

2.2. Algae Cultivation and Synchronization of Cultures

The algal cultures were routinely (every 3–4 weeks) sub-cultured on agar plates cultivated at 25 °C in continuous light with an incident light intensity at the surface of the Petri dishes of 150 μmol (photons)·m^{-2} s^{-1} of photosynthetically active radiation from a panel of fluorescent tubes. Such stock cultures were grown for 5–7 days and then stored at 15 °C with dim light until used or sub-cultured. To prepare synchronized cultures, the microalgae were inoculated either from an agar plate or from a liquid culture and cultivated in a glass tube (3 cm diameter) placed in a temperature-controlled water bath at

30 °C. Cultivation medium according to Zachleder and Šetlík [29] was used. The culture was aerated with air enriched to 2% CO_2 (flow rate 350 mL·min^{-1}). The glass tubes were exposed to an incident light intensity of 500 μmol (photons)·m^{-2}·s^{-1} of photosynthetically active radiation from a panel of fluorescent tubes for biochemical experiments and to an incident light intensity of 150 μmol (photons)·m^{-2}·s^{-1} from a panel of warm-white light-emitting diodes for Raman experiments.

For synchronization, the cultures were observed by light microscopy for two or three cycles to set the correct length of both the light and dark periods under given conditions so that the cells divided mostly into eight daughter cells (connected in eight-celled coenobia). The time for darkening the cells was when about 10% of cells started their first protoplast fission. The length of the dark period was chosen to allow all cells of the population to release their daughter cells and then the duration of the light and dark periods was kept constant. In the cultivation conditions described above, the synchronization conditions were 15/9 h light/dark cycle, which were maintained for at least two to three cycles before the experiments were initiated. The cell density at the beginning of the light period was maintained by dilution at below 1.0–1.5 × 10^6 cells·mL^{-1} to avoid light limitation. For details on the synchronization procedure see Hlavová, et al. [30]. On the day of the experiments, the synchronized daughter cells were diluted to 1.0–1.5 × 10^6 cells·mL^{-1} to avoid light-limitation and used as inocula for experimental cultures. For biochemical experiments, the cultures were cultivated in planar glass parallel cultivation cuvettes with a light path of 2.5 cm and inner volume 2 L, illuminated with an incident light intensity of 500 μmol (photons)·m^{-2}·s^{-1} of photosynthetically active radiation provided by a panel of fluorescent tubes. For Raman experiments, the cultures were grown in 3 cm diameter glass tubes, illuminated with an incident light intensity of 150 μmol (photons)·m^{-2}·s^{-1} of photosynthetically active radiation from a panel of warm-white light-emitting diodes. Due to differences in cultivations, the experimental cultures for biochemical experiments contained exclusively eight-celled coenobia while cultures for Raman experiments contained both eight- and four-celled coenobia. Only the two innermost cells of eight-celled coenobia were used for Raman mapping. The cultures were sampled hourly for biochemical experiments and as permitted for analysis in the Raman experiments (for details see below).

2.3. 4′,6-diamidine-2′-phenylindole Dihydrochloride (DAPI) Staining

Nuclei were stained with the fluorochrome 4′,6-diamidine-2′-phenylindole dihydrochloride (DAPI) and observed through a fluorescent microscope using the method described by Zachleder and Cepák [31] and Hlavová, Vítová, and Bišová [30]. Twenty microliters of DAPI solution (5 μg/mL in 0.25% (*w/v*) sucrose, 1 mM EDTA, 0.6 mM spermidine, 0.05% (*v/v*) mercaptoethanol, 10 mM Tris–HCl, pH 7.6) were added to the frozen cell pellet, vortexed and kept for 20–30 min in the dark at room temperature. The stained cells were observed using an Olympus microscope with 360–370 nm excitation and 420–460 emission filters.

2.4. Cell Size and Number

The cell suspension for Coulter counter measurements was stored frozen (−20 °C). Prior to measurement, it was allowed to thaw at room temperature and vortexed thoroughly for several minutes. According to the stage of the cell cycle, 0.4 or 0.8 mL of cell culture were diluted appropriately with ISOTON II solution to a total volume of 10 mL. Two mL of the sample (approximately 10^4–10^5 cells per mL in the final solution, in accordance to the Coulter counter instructions) were measured using a Multisizer 3 (Beckman Coulter, Brea, CA, USA), aperture size 100 μm. Cell number was also determined in the Bürker counting chamber (Meopta, Přerov, Czech Republic).

2.5. Assessment of Commitment Points and Cell Division Curves

To determine whether and how many commitment points were passed, the cells were sampled at appropriate time intervals and incubated at 30 °C in darkness. At the end of the cell cycle, the percentages of binuclear daughter cells, four- and eight-celled daughter coenobia, and undivided mother cells were estimated by light and/or fluorescence microscopy [30]. The values obtained by the assay of samples were plotted against the times of sampling. The curves are termed commitment curves. The proportion of mother cells, sporangia and daughter coenobia were determined by light microscopy in cells fixed in Lugol solution (1 g I, 5 g KI, 100 mL H_2O) at a final concentration of 10 µL of Lugol solution per 1 mL of cell suspension. Cell division and daughter cell release curves were obtained by plotting the cumulative percentages as a function of sampling time.

2.6. Polyphosphate Visualization

For polyP visualization, the samples were taken at designated time points during the cell cycle. In each sample, 1 mL of cell suspension was centrifuged for 2 min at $5000\times g$, the supernatant was removed and the sample was stored at -20 °C until staining. For staining, the method of Ota, et al. [32] was used.

2.7. Biochemical Analyses

2.7.1. Estimation of Bulk RNA, DNA, and Protein

The procedure of Wanka [33], as modified by Lukavský, et al. [34], was used for the extraction of total nucleic acids. The samples were centrifuged in 10 mL centrifuge tubes (4 min at $2800\times g$), which also served for storage of the samples. The pellet of algal cells was stored under 1 mL of ethanol at 20 °C. The algae were extracted five times with 0.2 M perchloric acid in 50% ethanol for 50 min at 20 °C and three times with an ethanol–ether mixture (3:1) at 70 °C for 10 min. Such pre-extracted samples were stored in ethanol. Total nucleic acids were extracted and hydrolyzed in 0.5 M perchloric acid at 60 °C for 5 h. After hydrolysis, concentrated perchloric acid was added to achieve a final concentration of 1 M perchloric acid in the sample. The absorbance of total nucleic acids in the supernatant was read at 260 nm (A_{260}).

The light activated reaction of diphenylamine with hydrolyzed DNA, as described by Decallonne and Weyns [35] was used with the following modification [36]. The diphenylamine reagent (4% w/v diphenylamine in glacial acetic acid) was mixed with the samples of total nucleic acid extracts in a ratio of 1:1 and the mixture in the test tubes was illuminated from two sides with fluorescent lamps (Tesla Z, 40 W). The incident radiation from each side was 20 $W{\cdot}m^{-2}$. After 6 h of illumination at 40 °C, the difference between the A_{600} and A_{700} was calculated. The RNA content was calculated as the difference between total nucleic acid and DNA content.

The sediment remaining after nucleic acid extraction was used for protein determination. It was hydrolyzed in 1 N NaOH for 1 h at 70 °C. The protein concentration in the supernatant after centrifugation of the hydrolysate (15 min, $5300\times g$, room temperature) was estimated by BCA assay (Thermo Fisher) according to manufacturer's instructions. The same procedure was carried out for the calibration curve set using different concentrations of bovine serum albumin.

2.7.2. Starch Assay

Cell pellets containing approximately 2×10^6 cells${\cdot}mL^{-1}$ were harvested during the cell cycle, washed with SCE buffer (100 mM sodium citrate, 2.7 mM EDTA-Na_2, pH 7 (citric acid)), snap frozen in liquid nitrogen and stored at -20 °C. After thawing the pellets, cells were disintegrated by adding 200 µL of distilled water and 300 µL of zirconium beads (0.7 mm in diameter) followed by vigorous vortexing for 15 min at room temperature. Depigmentation of the samples was done by adding 1 mL of 80% (v/v) ethanol to the pellet and incubating in a water bath at 68 °C for 15 min after which the samples were centrifuged for 2 min at $14,000\times g$ and the supernatant was removed. The depigmentation procedure was repeated 3 or 4 times (or until the pellet was completely discolored). After that, 1 mL of porcine pancreas α-amylase (Sigma-Aldrich, Prague, Czech Republic) solution (0.5 mg${\cdot}mL^{-1}$ w/v in 0.1 M sodium phosphate buffer (pH 6.9)) was added to each sample and they were incubated for 1 h at 37 °C. The samples were centrifuged for 2 min at 14,000× g and the supernatant was used for the quantification of reducing sugars through the DNSA color reaction as described by Miller [37]. In short, 500 µL of supernatant were mixed with 500 µL dinitrosalicylic acid (DNSA) solution (1% (w/v) 3,5-DNSA, 30% (w/v) potassium sodium tartrate tetrahydrate, 20% (v/v) 2 M sodium hydroxide) and incubated for 5 min at 105 °C on a heat block. Following a cooling period of 10 min at room temperature, the mixture was diluted five-fold with distilled water, after which the A_{570} values of the samples were measured. The concentration of starch was estimated through a calibration curve of potato starch (Lach-Ner, Czech Republic) digested with α-amylase.

2.7.3. Total Phosphate Assay

Cell pellets containing at least 4×10^7 cells, were harvested at designated time points during the cell cycle, washed with distilled water, and stored at −20 °C. After thawing, 1 mL of distilled water was added to the pellets in addition to 300 µL of zirconium beads (0.7 mm in diameter). The cells were disrupted by vigorous vortexing for 15 min at 4 °C. The samples were centrifuged for 2 min at 14,000× g, after which the supernatant was discarded. The pellets were washed three times with a 5% (w/v) solution of sodium hypochlorite. After the last wash, 200 µL of 0.67% (w/v) potassium persulfate were added to the pellet. The samples were finally autoclaved at 121 °C for 20 min. After cooling, 200 µL of sample were pipetted on a 96 micro well plate together with 8 µL of an ammonium molybdate tetrahydrate solution (1.2% (w/v) ammonium molybdate tetrahydrate, 4.8% (w/v) potassium antimonyl tartarate sesquihydrate, and 16% (v/v) sulfuric acid) and 2 µL of 7.2% (w/v) L-ascorbic acid. The plate was incubated for 20 min in darkness at room temperature, after which the A_{880} value for each sample was measured in a plate reader (Infinite F200, Tecan Trading AG, Männedorf, Switzerland). For quantification of total phosphate and polyP, a calibration curve was constructed using a phosphate standard for ion chromatography (Sigma-Aldrich, Prague, Czech Republic).

2.7.4. Polyphosphate Assays

Samples for the estimation of polyP were collected, stored, disintegrated and depigmented similarly to the samples for total phosphates. After depigmentation with a 5% (w/v) solution of sodium hypochlorite, 100 µL of distilled water were added to the pellet. Following an incubation time of 5 min, the samples were centrifuged for 2 min at 14,000× g and the resulting supernatant was collected. The elution with distilled water was repeated twice in order to ensure the optimal extraction of polyP from the pellet. Precipitation of polyP was by addition of 1.8 mL of 100% ethanol to the collected supernatant. The samples were centrifuged for 10 min. at 14,000× g after which the supernatant was removed. The resulting pellet was then re-suspended in 500 µL of distilled water. The hydrolysis of polyP to orthophosphates was achieved by adding 100 µL of 4% (w/v) potassium persulfate after which the samples were autoclaved for 20 min at 121 °C. After cooling, the samples were treated and analyzed as described in the quantification of total phosphates [32].

2.8. Raman Analyses

2.8.1. Sample Preparation

To study the content and intracellular distribution of storage biomolecules during the cell cycle using Raman microscopy, fresh living cells were needed because of problematic photobleaching of autofluorescence in chemically fixed or frozen cells. Samples from a synchronized culture of *D. quadricauda* were taken every 30–60 min, as permitted by the duration of the Raman scanning, from the start of the light phase (T = 0:00 h), considered as the beginning of the cell cycle, up to the end of the cycle (T = 24:00 h); the first sample was taken 15 min after the start of the light phase.

For Raman measurements, 0.5 mL of the cell suspension was centrifuged (2000× g for ca. 15 s), the supernatant was discarded and a part of the pellet was mixed with ca. 20 µL of low-gelling agarose (Sigma Aldrich, 2% solution, T = 39 °C). A few µL of the agar mixture were placed between a quartz slide and a quartz coverslip, and these were sealed with Covergrip sealant (Biotium, Hayward, CA, USA). Sample preparation took about 13 min from taking the sample to the start of the Raman measurement.

2.8.2. Raman Measurement

Raman maps were acquired with a WITec alpha300 RSA confocal Raman microscope (WITec, Ulm, Germany) equipped with an oil-immersion objective UPlanFLN 100×, NA 1.30 (Olympus, Tokyo, Japan). The spectra were excited with a 532 nm laser (excitation power of ca. 20 mW at the sample). The lateral and axial resolutions of the Raman microscope (according to standard test of Raman confocality by means of silicon wafer [28,38]) were ca. 250 and 900 nm, respectively. A scanning step of 125 nm in both x and y directions (thus below the Rayleigh diffraction limit of the experimental setup) was used, with 0.1 s acquisition time per pixel. The spectra were measured immediately in the range of 220–3850 cm^{-1}. This range covers characteristic (<1800 cm^{-1}) as well as stretching (>2800 cm^{-1}) vibrations of biomolecules and water. Prior to Raman mapping, a wide-area photobleaching of the entire algal coenobium by a defocused 532-nm laser beam was applied, as described previously, to eliminate the strong autofluorescence of chlorophyll [28]. At each sampling point, a two-dimensional Raman map of the two innermost cells of a randomly selected eight-celled coenobium was acquired.

2.8.3. Data Treatment

Raman maps were treated with Project Four Plus (WITec, Germany) and our own scripts for GNU Octave [39] and MATLAB (MathWorks, Natick, MA, USA) [40]. Firstly, signals of cosmic rays were removed using automated and manual functions of Project Four Plus. Next, to compensate eventual variations in laser power and efficiency of signal collection, Raman maps of different cells were normalized to the common intensity scale as follows: the pixels of the map obviously belonging to the surrounding medium were identified automatically according to the missing band of carbon–hydrogen stretching vibrations. For each Raman map, the average spectrum of the surrounding medium was calculated and the integral intensity of the oxygen-hydrogen stretching band of water (background corrected) was used as the standard intensity for uniform normalization of Raman maps. Spectral regions with characteristic Raman bands of respective biomolecules were then selected for starch (457–507 cm^{-1}, band maximum at 479 cm^{-1}), lipids (2836–2869 cm^{-1}, band maximum at 2854 cm^{-1}), polyphosphate (1143–1190 cm^{-1}, band maximum at 1159 cm^{-1}), guanine (613–684 cm^{-1}, band maximum at 651 cm^{-1}) and the region of carbon–hydrogen stretching vibrations of all biomolecules containing C-H bonds (2795–3060 cm^{-1}, band maximum at around 2935 cm^{-1}). The non-Raman background was subtracted as a bisector connecting mean intensities at the last 4–10 spectral points delimiting the upper and lower ends of the regions. Spectra with high levels of the respective biomolecules and the appropriate wavenumber regions are shown in Figure S1.

For visualization of distributions of biomolecules in *D. quadricauda* cells, the intensity of each band (after background correction) was calculated by integrating the band's area. The integral intensities in Raman maps are expressed in a color scale and, for a given biomolecule, the same scale was used in all maps. The darkest color shade corresponds to the mean intensity found in the surrounding medium and the lightest one to the maximum intensity. The only exception is the intensity of carbon–hydrogen stretching vibrations, where the darkest shade corresponds to zero.

To estimate the cellular content of respective biomolecules, pixels corresponding to both cells on each Raman map were identified using manual functions of Project Four Plus. The signals from every 2 × 2 pixels were merged to reduce the noise, and the background-corrected bands of biomolecules were integrated. For each Raman map i, the mean value m_i, and the standard error s_i of the integral intensity in the merged pixels of the surrounding medium were calculated. For each pair of cells, the integral intensities were summed for pixels where the signal was greater than $m_i + 7s_i$. Remaining pixels displaying a signal below this threshold were excluded from the summation.

3. Results

3.1. Cell Cycle Characteristics

Cultures of *D. quadricauda* were synchronized by a diurnal, alternating light/dark regime. During cultivation, cell cycle progression was analyzed hourly, as well as changes in bulk RNA, protein, DNA, starch, polyP, total phosphates, and lipid contents. These standard physiological and biochemical experiments set up a baseline, with which the results from Raman microscopy were correlated. At a given light intensity, half of the population attained the first commitment point shortly after being put into the light. By the 6th and 10th hour, more than 50% of the population attained the second and third commitment points, respectively. Attainment of each of the commitment points was followed by DNA replication, nuclear division, and cell division. Within the population, attainment of the second and third commitment points were closely followed by completing nuclear division into two and four nuclei, respectively. The third nuclear division together with the first protoplast division occurred shortly before darkening the cells at the 15th hour of the cell cycle. The cell cycle was concluded after about 22 h when all mother coenobia released daughter coenobia (Figure 1).

3.2. Bulk Analysis of Cell Composition

Biochemical analysis showed that accumulation of RNA (Figure 2A) progressed rapidly in the early light period, followed by a lower rate of RNA synthesis in the late light period and in the dark. The total protein content increased with a small delay after RNA accumulation in the light and, in the dark period, protein accumulation stopped. Later in the night, the level of protein even decreased slightly (Figure 2B). A significant increase in DNA content was observed only after the 7th hour of the cell cycle, which coincides with the midpoint of the first round of nuclear division. The increase in DNA synthesis then continued until the end of the cell cycle, when the DNA content had increased about 8-fold over the initial values (Figure 2C). Starch started to be synthesized immediately after the cultures were exposed to light and continued steadily throughout the light period (Figure 2D). The rate of starch accumulation was faster until the 10th hour when the cells attained the third commitment point. It slowed from then on as the cells underwent the second and third nuclear divisions. In the dark, the starch reserves were catabolized to support nuclear and cellular divisions that were taking place in the absence of photosynthetic energy conversion.

Total phosphate and polyP content increased soon after transfer into the light (Figure 2E,F). The increase continued until about the 7th hour when DNA replication started and the first round of nuclear division was completed in half of the population. Thereafter, both total and polyP levels plateaued.

PolyP granules can be also detected using conventional fluorescence microscopy after staining with high concentrations of DAPI [32]. PolyP granules were visible as groups of yellow spots on both sides of the bluish nuclei in the longitudinal direction (Figure 3). They were already present in the first hour of the cell cycle (Figure 3A). They seemed to multiply both in number and in extent of fluorescence in the next two hours and their localization was, conspicuously, at the outer side of two newly formed nuclei (Figure 3B). With time, both their fluorescence and number decreased (Figure 3C). By the seventh hour of the cell cycle (Figure 3D) and one hour later (Figure 3E), they were represented by groups of tiny yellow spots close to the outer cell edge in the vicinity of nuclei. From the ninth hour, the spots became even smaller (Figure 3F) and in most cells, they disappeared by the 10th hour (Figure 3G). In about 10 percent of cells, the polyP granules persisted for one more hour and were localized both to the perinuclear area and to the outer cell edge (Figure 3H). Even in these cells, the granules disappeared by the 11th hour of the cell cycle.

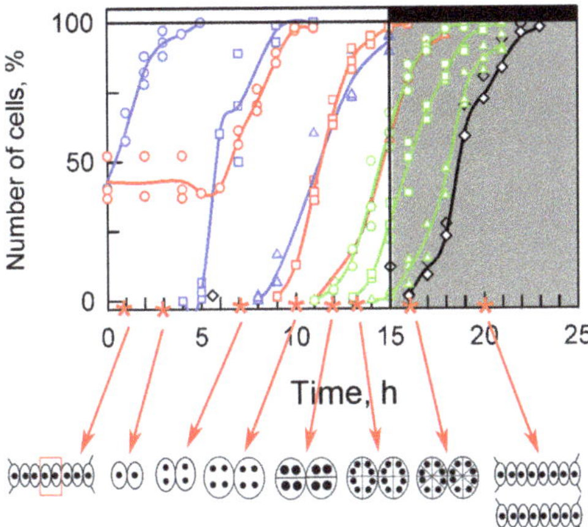

Figure 1. Dynamics of growth and the cell cycle in synchronized cultures of *D. quadricauda*. Upper part: Time courses of individual commitment points, nuclear division, protoplast fission and daughter cell release. Blue lines: cumulative percentage of cells that attained the commitment point for the first (circles), second (squares) and third (triangles) reproductive sequences, respectively; red lines: cumulative percentage of cells, in which the first (circles), second (squares), and third (triangles) nuclear divisions were terminated; green lines: cumulative percentage of cells, in which the first (circles), second (squares) and third (triangles) cell divisions were terminated, respectively; black line, empty diamonds: percentage of cells that released daughter coenobia. Light (15 h) and dark periods (9 h) are marked by stripes above panels and separated by vertical lines. The lines represent the means of at least three independent experiments. The raw values are plotted as dots and the line connects the mean values of the experiments. All values were calculated per parental cell, even after their division (17:00 to 22:00 h). Lower part: Schematic representation of the cells at the time-points (denoted by red asterisks) corresponding to the set of samples analyzed by Raman microscopy in Figure 4 (full set of Raman measurements is available in Figure S5). Schematic pictures of the cells indicate changes in their sizes. The full circles inside illustrate the ploidy and number of nuclei during the cell cycle. Larger circles indicate a doubling of DNA. For more details see text. Modified after Zachleder, et al. [41], Bišová and Zachleder [42], and Zachleder, Bišová, and Vítová [1].

3.3. Raman Estimates of Cell Volume and Chemical Composition

Raman spectra acquired from individual pixels of the specimen consist of spectral contributions from all chemical compounds present in the respective voxels, which complicates spectral analyses. The number of contributing chemical compounds, especially in the case of dense and well-separated subcellular structures—such as lipid bodies, starch grains, polyP bodies, or crystalline inclusions—can be decreased by increasing the confocality and spatial resolution of Raman microscope. In our experiments, due to high confocality and spatial resolution reaching diffraction limit, objects separated by ca 250 nm can be resolved and Raman spectra of voxels smaller than 0.4 μm^3 were effectively collected. This suppresses the spectral contribution of the neighboring structures. The Raman spectra within individual voxels thus contained spectral contributions of limited number of compounds and were easily interpretable by comparing them with the spectra of chemically-pure reference compounds (Figures S2 and S3). Furthermore, using advanced multivariate methods, even Raman spectra of the voxels containing a greater number of the contributing compounds can be decomposed into linearly independent spectral components corresponding to different chemically-pure species or cell structures (Figure S4).

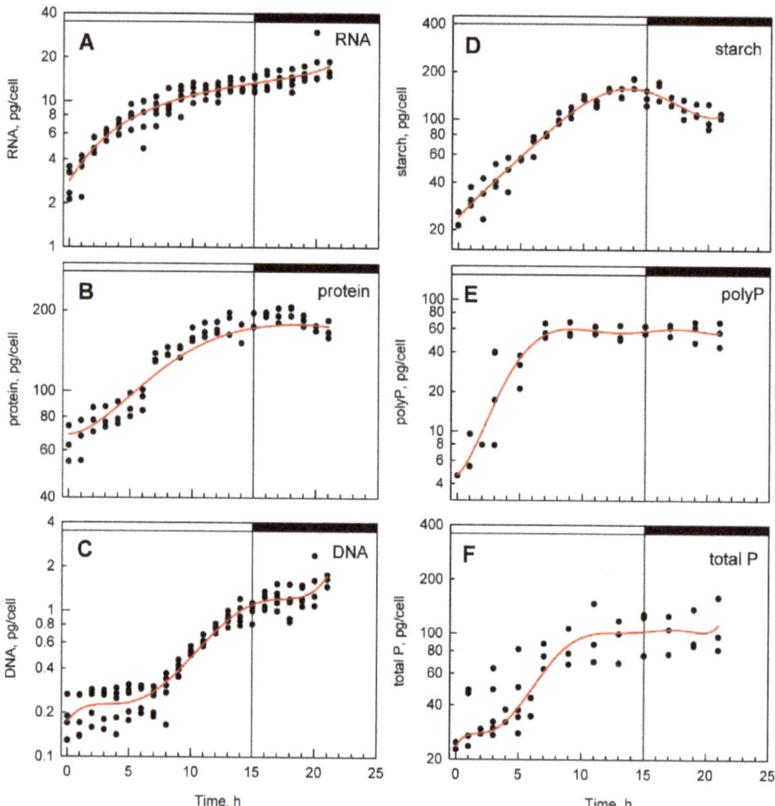

Figure 2. Time courses of accumulation of RNA (**A**), protein (**B**), DNA (**C**), starch (**D**), polyP (**E**), and total phosphates (**F**) in synchronized cultures of *D. quadricauda*. Light (15 h) and dark periods (7 h) are marked by stripes above panels and separated by a vertical line. Values from five different experiments are shown as dots, the smoothed red line representing the mean of five experiments in each panel. Values were calculated per parental cell even if divided (17:00 to 22:00 h).

The biochemical analyses and Raman experiments were not carried out on samples from the same cultivations because of equipment location at two remote workplaces, time and methodological constraints and different requirements on the live status of the cells. Raman microscopy was carried out immediately, using living algal cells from aliquots taken during the cell cycle, whereas biochemical analyses were carried out on samples that were accumulated and stored during the cycle.

The cultivation set-ups were the same for the two experimental methods, with the exception of light intensity, which was about three-fold lower for the Raman measurements. This led to changes in the proportion of four-celled and eight-celled coenobia within the population. The cultures used for biochemical experiments were composed exclusively of eight-celled coenobia, while the cultures for Raman measurements contained both four-celled and eight-celled coenobia. To overcome this limitation, the innermost two cells of randomly selected eight-celled coenobia were chosen for all measurements. These cells are usually the largest ones within the coenobium and the most advanced in the cell cycle. In this way, we ensured that the two analyses were carried out on as similar cells as possible. Raman microscopy mapping was performed 35-times during the 24-h cell cycle. All acquired Raman maps are presented in Figure S5. A subset of the maps is shown in Figure 4C; the age of the cells is schematically depicted in Figure 1 as well as in Figure 4A,B.

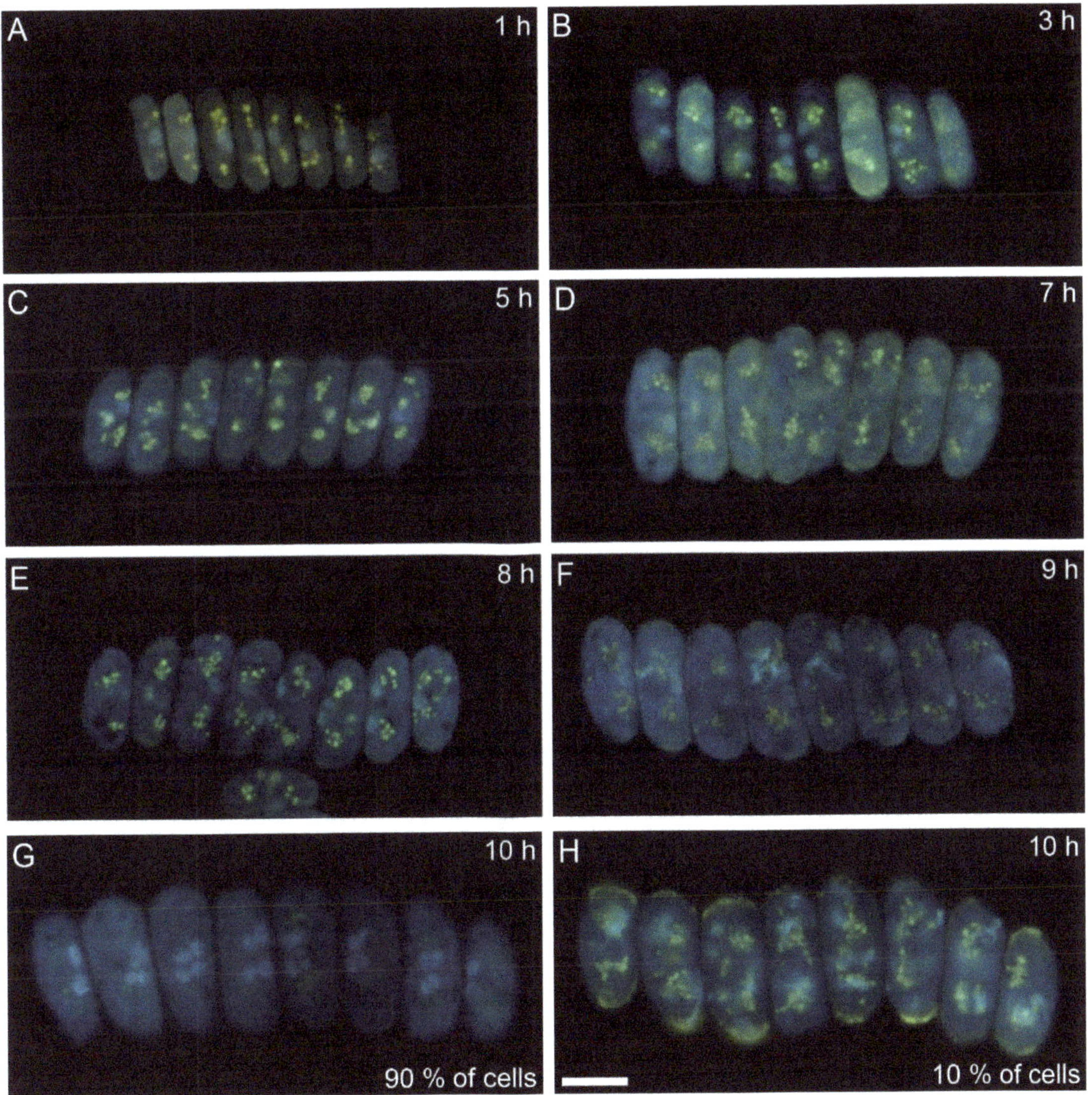

Figure 3. Fluorescence micrographs of coenobia stained by a high concentration of DAPI at different time points during the cell cycle. (**A**) 1st h, (**B**) 3rd h, (**C**) 5th h, (**D**) 7th h, (**E**) 8th h, (**F**) 9th h, (**G**) and (**H**) 10th h. G corresponds to situation in 90% of the cells in the synchronized population while H corresponds to situation in 10% of the cells in the synchronized population. Polyphosphates are visible as yellow spots; the bluish spots are nuclei. Bar is 10 μm.

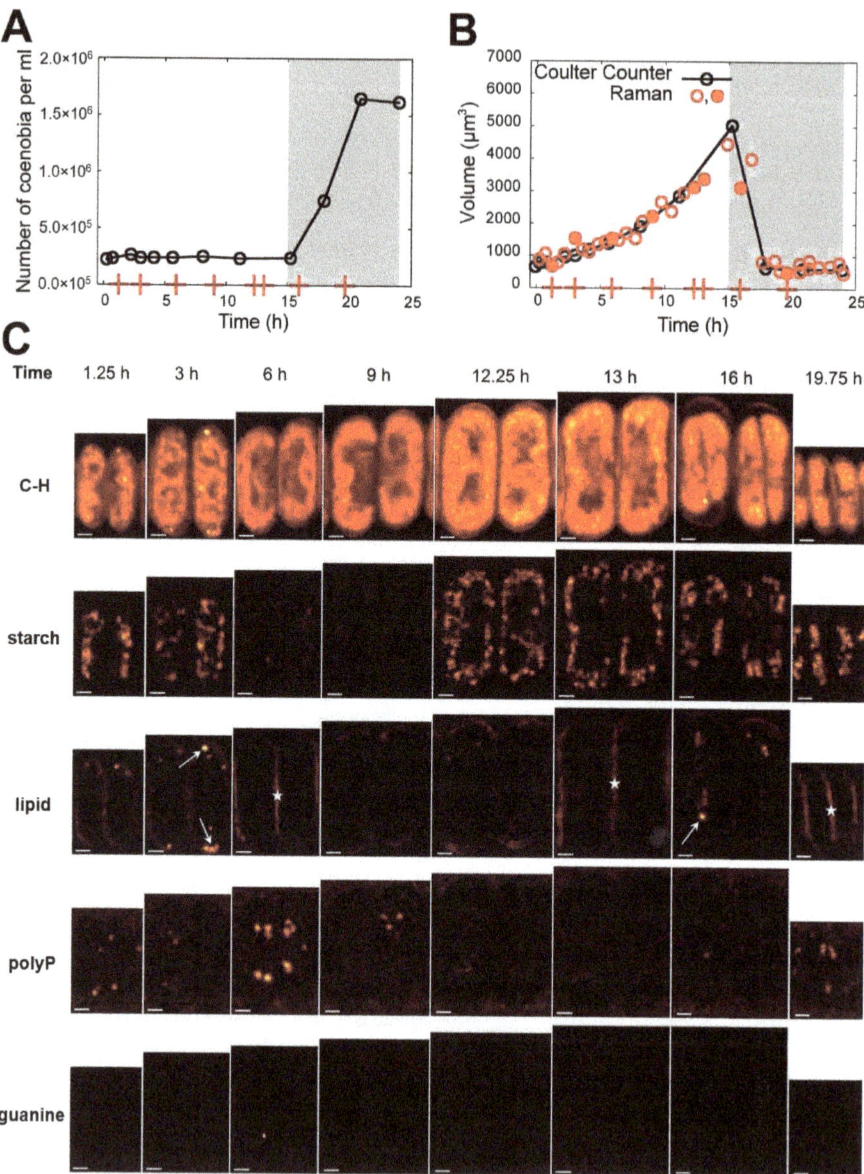

Figure 4. Time course of the number of coenobia in the cell culture of *D. quadricauda* as determined by Coulter counter (**A**). Time course of volumes of the coenobia determined by Coulter counter (black, median volume) and Raman microscopy (red) (**B**). For the time-point T = 18:05 h coinciding with the cell division of a majority of the cells, the Coulter counter volume of daughter cells is shown. The cells that are shown in panel (**C**) are indicated by filled circles. Raman maps showing the distribution of carbon–hydrogen (C-H) groups, starch, lipids, and polyP in the two innermost cells of eight-celled coenobia during the cell cycle (**C**). Time from start of the cycle is indicated in the top row. For a given biomolecule, the color scale is the same for all Raman maps. The white bars correspond to 2 μm. The red crosses on the time axis in panels A, B indicate times at which the cells shown in panel C were taken from the culture. The grey area in panels A, B indicates the dark period of the cycle. In panel C, lipids, the arrows depict lipid bodies, the asterisks depict membrane lipids. Spectra of both structures contain the same spectral proxy, Raman band located at ca 2854 cm^{-1}.

To verify the synchrony of the population used for Raman microscopy, the number of coenobia per mL of the cell suspension (Figure 4A) and median volumes of the cells (Figure 4B) were obtained by Coulter counter. To relate the bulk population analysis with the Raman microscopy results, cell volumes corresponding to the eight-celled coenobia provided by the Coulter counter measurement were compared with Raman estimates of the volume of the two innermost cells of the mapped coenobium. These estimates were calculated as follows: the number of pixels belonging to the cells, identified according to the carbon–hydrogen stretching band as described in Material and Methods, was raised to the power of 1.5 and renormalized to fit the Coulter counter values in a least squared sense. The number of pixels is directly proportional to the cross sectional area of each cell. As can be confirmed by Figure 4B, Raman estimates of the volume, although of only the two innermost cells of a randomly selected coenobium, fit well to the median volume determined for hundreds of thousands of cells by Coulter counter, indicating a good synchrony of the culture and justifying the simplifications.

For all 35 samples taken during the 24 h cycle for the Raman measurements, the distribution of starch, lipids and polyphosphate in the Raman maps were obtained by integrating the intensity of a Raman band characteristic for a given biomolecular species in each pixel spectrum, as described in Material and Methods (see also Figure S1). Additionally, the band of carbon–hydrogen (C-H) stretching vibrations was used to visualize the concentration of total organic matter (proteins, carbohydrates, lipids, RNA, DNA) in the cells.

The distributions of carbon–hydrogen stretching vibrations, starch, lipids, and polyP are shown in Figure 4C for selected time-points where the cells exhibited some characteristic features. Those particular time-points are indicated in Figure 4A,B by red crosses on the horizontal axes and their cell volumes are indicated by filled circles in Figure 4B. For all 35 samples measured here, the biomolecular distributions are shown in Figure S5.

As can be seen from Figures 4C and S5, the cells already contained some starch reserves and a few lipid and polyP bodies just after the dark phase of the previous cycle. The starch content was increasing from the beginning of the light phase, reaching its first maximum at around T = 2:15 h after the light onset (Figure S5), and then declining again. The Raman band located at ca 2854 cm^{-1}, which was used as spectral proxy for quantification of lipids, detects both lipid bodies and membrane lipids. The two structures are in Figure 4C, panel lipids discriminated by arrows (lipids bodies) and asterisks (membrane lipids). Lipid bodies, localized to the periphery of the cells (Figure 4C, panel lipids, arrows), increased in number and lipid content with about a 45 min delay behind the starch content. Finally, the polyP content peaked at T = 6:00 h, when the cells contained multiple polyP bodies but exhibited nearly no starch or lipid bodies identifiable by Raman microscopy.

After T = 6:00 h, the polyP bodies slowly dissolved and almost no starch or lipid bodies were visible in the cells. The carbon–hydrogen vibrations visible in interiors of the cells thus are derived predominantly from proteins (Figures 4 and S5). At around T = 10:30 h, the cells started to accumulate starch again. The starch content and cell volume grew until the end of the light period at T = 15:00 h, when some of the cells were already dividing. Some lipid bodies can be seen in the cells even during cell division (Figure S5). The released daughter cells contained some starch and a few polyP bodies until the end of the experiment at T = 24:00 h. In some daughter cells, a few lipid bodies can be identified (Figure S5).

A more detailed sampling at the early phase of the cell cycle, prior to the onset of DNA replication (between T = 4:45 h and T = 6:30 h) showed guanine bodies coinciding and possibly co-localizing with the polyP granules (Figure 5), and the cells exhibited guanine bodies between T = 4:45 h and T = 6:00 h. Small guanine bodies of a lower signal intensity were observed at T = 7:45 h, 9:45 h, and 16:00 h (Figure S5) but not in the daughter cells (Figure S5).

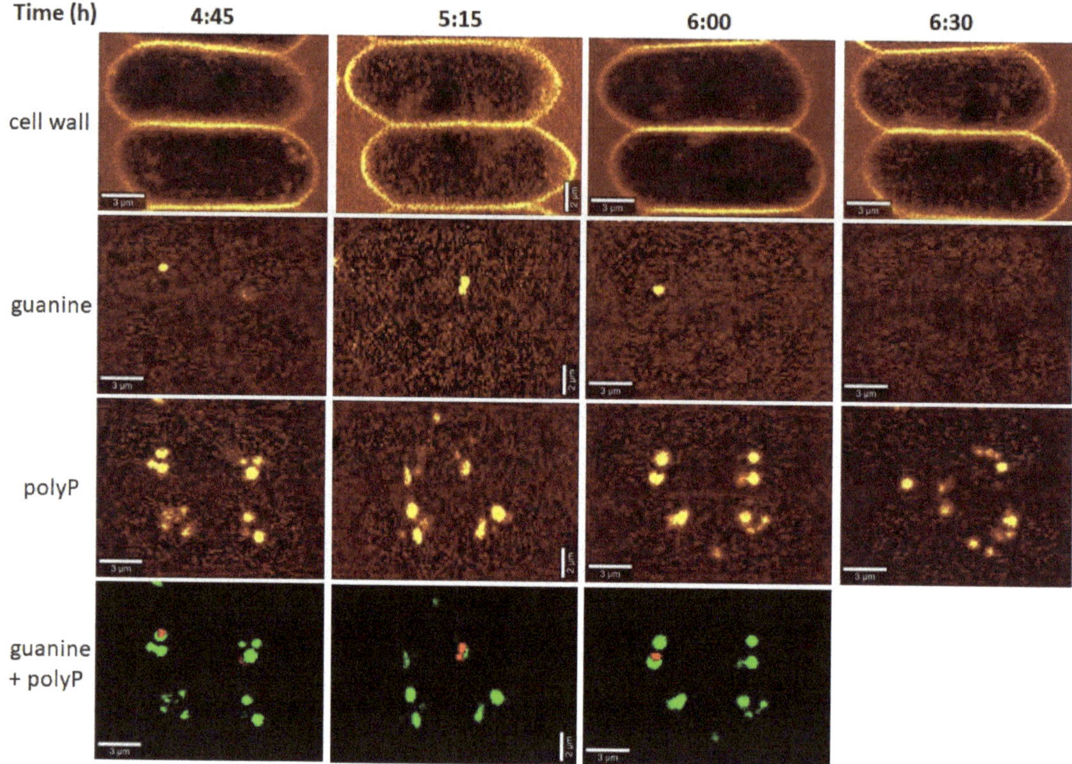

Figure 5. Raman maps showing the distribution of the cell walls, guanine and polyP bodies in two innermost cells of eight-celled coenobia during the early phase of the cell cycle. Time from the start of the cycle is indicated in the top row. For a given biomolecule, the color scale is the same for all Raman maps. The cell wall structure serves to visualize cell edges. It contains spectral signature of cellulose and membrane lipids, for details see Figure S4. The white bars correspond to 2 μm for Time = 5:15 h or 3 μm for other time-points. The bottom panel shows an overlay of localization of guanine (red) and polyphosphates (green).

To show the dynamics of the storage compounds (described qualitatively above) in a semi-quantitative way, Raman estimates of the starch, lipids and polyP content per single cell of the coenobium have been calculated by averaging contributions of both the measured cells (for details see Material and Methods) and are shown in Figure 6. The neutral lipid content was determined only from the Raman signal inside the cells. The cell's edges (Figure 4) corresponding to membrane lipids, with the same Raman band located at ca 2854 cm^{-1} as spectral proxy, were ignored. Outlying values have been excluded for the purpose of this figure and the data points were connected with splines to guide the eye.

3.4. Comparison of Raman and Bulk Biochemical Analyses

We have presented the results of cell content estimates by bulk biochemical methods and by Raman mapping of individual cells in synchronous cultures. The two methods differed in the biomolecules they are were able to detect and quantify. The bulk biochemical methods could easily separate RNA, DNA, and proteins (Figure 2) due to their different chemical properties. In contrast, Raman estimates of these three macromolecules are limited by a lack of intense and specific spectral signatures as well as by a relatively low density in the mapping voxels due to a high degree of hydration. On the other hand, Raman microscopy is surprisingly effective in detecting lipid droplets (Figures 4 and 7).

The same size of lipid droplets is hardly detectable by conventional Nile Red staining and fluorescence microscopy. Furthermore, we have tried to measure lipid content using a fluorescent plate reader in cells stained with Nile Red. In this way, more cells are measured at the same time, thus multiplying the signal. Even then, we were unable to reliably detect any changes in lipid content. The changes detected were within a few percent of the total lipid content, and not statistically significant. Furthermore, Raman spectroscopy was able to detect microcrystalline guanine, for which there are no alternatives for reliable in situ detection.

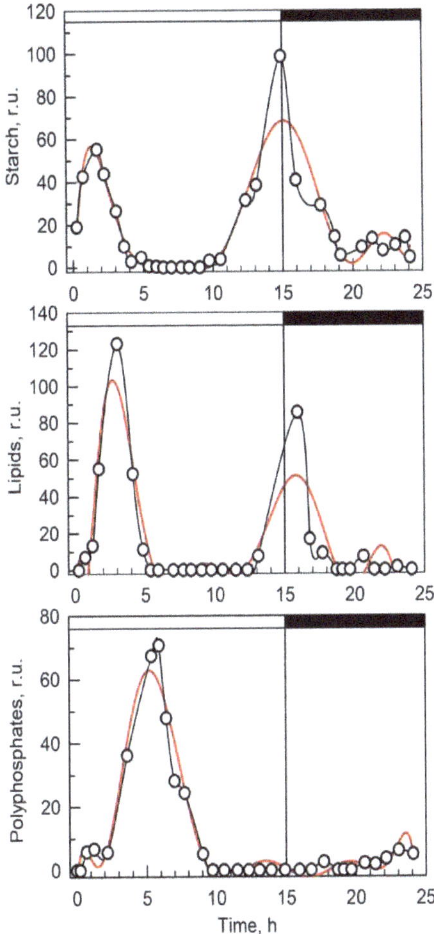

Figure 6. Time evolution of starch (top), lipids (middle), and polyP (bottom) levels in the cells during the cell cycle as determined from the Raman maps. Black lines connect experimental points. Red lines are splines averaging neighboring values.

Both bulk methods and Raman estimates were able to detect starch and polyP. The bulk biochemical analysis identified a steady increase in starch from the beginning of the cell cycle till cell darkening and the start of cell division, when starch started to be consumed. In contrast, Raman analyses detected an increase in the starch level in the first three hours of the cell cycle followed by a complete disappearance. A new increase in starch synthesis was observed by about the 12th hour of the cell cycle and reached a peak at about the 15th hour. The starch was subsequently spent for the ongoing processes of protoplast fission and cell division. Both Raman estimates and conventional fluorescence microscopy revealed similar patterns of polyP localization in both space and time. Both of the methods detected two major groups of polyP granules on each side of a single nucleus, or later, on the outer side of the two newly formed nuclei. PolyP granules gradually disappeared by the time the cells completed division of their nuclei into four. The bulk biochemical analyses detected a rapid increase in both total and polyP content from the beginning of the light period to about the fifth hour of the cell cycle. Thereafter both total and polyP levels remained unchanged.

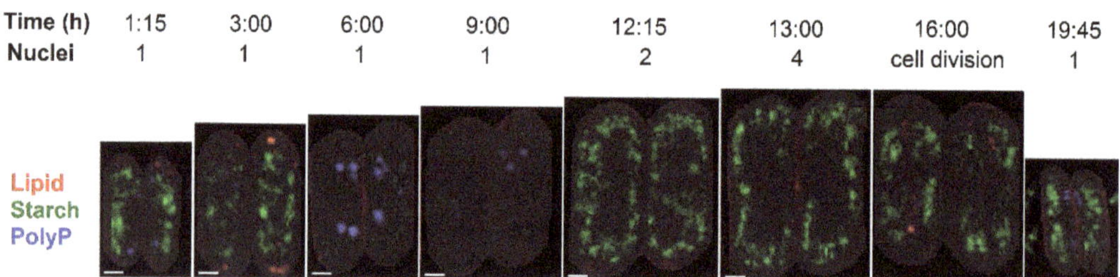

Figure 7. Raman maps showing the distribution of lipid droplets (red), starch bodies (green) and polyP granules (blue) in the same two innermost cells of eight-celled coenobia as presented in Figure 1C. The time counted from the beginning of the cell cycle and the number of nuclei determined by DAPI staining of the culture are indicated in the top row. For a given biomolecule, the color scale is the same for all maps. The white bars correspond to 2 μm.

4. Discussion

Synchronized cultures are a traditional tool for studying the progression of the cell cycle in different organisms. Cell cycle progression, encompassing growth, cell duplication and division events, has been routinely described by biochemical analyzes of macro-molecules (RNA, protein, DNA, starch). Such analyzes are rather time consuming, complicated and mostly require a large quantity of initial sample. Here, we have compared the bulk biochemical analyses of biomolecules with the results of Raman microscopy. Raman microscopy visualizes distributions of the prevalent groups of biomolecules simultaneously, without the need for artificial pre-processing of the sample and with a spatial resolution of a confocal microscope. As Raman microscopy shows the intrinsic chemical content of the sample, even biochemical species not expected in the sample can be seen, provided they are present at sufficient concentrations and that their Raman spectra are correctly interpreted. This can be demonstrated with the example of guanine microcrystals that have been recently observed in two microalgal species [2] without prior knowledge of their existence. In the present study, the guanine microcrystals occurred regularly between approximately the fifth and seventh hours of the cell cycle, i.e., at the time of attainment of the second commitment point and completion of the first nuclear division, and then sporadically during other phases of the cell cycle. The appearance of guanine microcrystals during a natural cell cycle of unstressed cells supports the tentative biological relevance of guanine microcrystals as effective depots of purines [2]. Such biological functions may be related to their localization in the cells. Guanine crystals were localized together with polyP granules, either in the vicinity of the newly formed nuclei, or in the place where nuclei are formed. The two possibilities are impossible to separate based on Raman microscopy, as nuclear localization could not be detected in the algal cells. Both localizations would suggest guanine crystals may serve as purine deposits consumed in the formation of nuclei.

Apart from identifying novel and/or otherwise undetectable biochemical species, Raman microscopy might also be more sensitive to low levels of biochemical species that can be detected and analyzed by fluorescence microscopy and/or bulk analytical techniques. This is demonstrated by the observed presence of lipid bodies at the periphery of some cells during the cell cycle (Figures 4, 7, and S5). These lipid bodies could not be observed by fluorescence microscopy or bulk fluorescent measurements, as they did not increase the cellular lipid content enough to be detected separately. Raman microscopy has been established method for detection of lipids, particularly in stressed cells with increased lipid amount [10–13]. Here, this has been extended to detection of very low lipid levels present in non-stressed cycling dividing cells. Raman microscopy possibly takes advantage of its ability to detect different compounds in dense and solidified, although small, structures.

Raman microscopy and fluorescent microscopy of polyP bodies were in agreement (compare Figures 3 and 5). They visualized polyP granules both in sites where new nuclei will be formed and near them, once they divided. This was particularly true for the first nuclear division. The localization of polyP granules (together with guanine crystals) close to nuclei could suggest they function as energy and phosphate stores for nucleic acid synthesis. This localization is thus in agreement with the classical observation in green alga *Chlorella*, where polyP are preferentially used for the synthesis of DNA (and phosphoproteins), but not RNA [43–45]. Both microscopy methods showed a decrease in polyP granules after the eighth hour of the cell cycle, i.e., at the time when the exponential increase in DNA synthesis starts, further supporting the notion that polyP granules serve as a P source for DNA synthesis. Interestingly, at the same time, the total phosphorus and polyP levels, as determined by bulk biochemical analyses, plateaued. The biochemical method that we used for total phosphorus and polyP was non-specific, and precipitated polyP granules together with DNA and RNA. This could explain the differences between microscopy and biochemical methods. Thus, from the eighth hour, DNA is possibly synthesized at the expense of polyP granules, and it gradually forms a larger proportion of the biochemically detected polyP fraction.

The most striking difference was observed between the bulk biochemical and Raman analyses of starch. The bulk biochemical analysis showed a steady increase in starch, as has been established in the field [1,42,46], whilst Raman microscopy showed a bimodal pattern with an initial increase in starch prior to attainment of the first CP followed by starch disappearance and another increase in starch levels at the time the cells reached the third CP and completed the second nuclear division (Figures 1 and 6). This was really an intriguing observation, probably caused by the difference in light intensities used for cultivation in the biochemical versus Raman experiments. Light intensity is known to affect the accumulation of starch, so that it increases with increasing light intensity [47]. At lower light intensity, less starch would be synthesized; it would be continuously spent for general metabolism and no net starch granules would be visible by Raman microscopy. Although this is a somewhat problematic discrepancy, probably caused by different cultivation conditions, we leave it here because it reveals an interesting phenomenon concerning the complete consumption of accumulated starch during the cell cycle under limited illumination, which would deserve a more detailed study. Due to its high sensitivity, Raman microscopy would be well suited for such studies.

The graphical description of the time evolution of starch, lipid and polyP content during the cell cycle in Figure 4 should be treated as a semi-quantitative proxy. Although recent studies have shown that two-dimensional mapping can be used to quantify the levels of starch [26,48], lipids [26,49], and polyP [50] in microalgal cultures, these studies have averaged the results from Raman maps of several to dozens of cells in order to obtain an accurate quantification when compared to bulk methods. On the other hand, as the cells in the present study were synchronous, variability amongst the cell cultures should be much narrower than in a common non-synchronous culture. Furthermore, the Raman signal is, by the nature of conventional Raman scattering, linearly proportional to the concentration of a compound in a given voxel [18].

The data presented here have proven that Raman microscopy is capable of detecting relatively low concentrations of storage biomolecules present in non-stressed *D. quadri-cauda* cells during a normal cell cycle. Decreasing the limit of detection to levels present in standard dividing cells represents an improvement compared to previous quantification studies [26,48,49], which mostly focused on stressed cells over-accumulating lipids or starch. Although the merits of Raman microscopy have been shown here, it still possesses limitations similar to conventional fluorescence microscopy and bulk biochemical analyses. As concerns the chemical analysis and imaging, the method is well suitable for dense and solidified, although small structures, but rather insensitive for dissolved chemical species, as it requires local concentrations of at least dozens of mM. This precludes the detection of low-concentration metabolites in solution. Furthermore, Raman mapping is of relatively low throughput, and parallelization is problematic, unless one possesses several Raman microscopes. To keep a sampling time of 30–45 min and a high spatial resolution of Raman mapping, in the present study, it was not possible to measure more than two cells of *D. quadricauda* coenobium at a given time-point. Statistically, a more sound set of cells could be acquired by compromising one or both of these requirements. However, lower spatial resolution would impair imaging and quantification of small biomolecular inclusions in Raman maps. On the other hand, increasing intervals between samplings might lead to missing important phases in the cell cycle. Otherwise, there exists the possibility of repeating the same experiment in multiplicates. However, each cell culture, although synchronous, is never exactly the same. Although the specificity of Raman spectrum for a given class of chemical compounds (e.g., lipids, carbohydrates, nucleic acids, proteins) is relatively high (Figures S2–S4), the specificity within particular classes is often much smaller. For example, structurally different proteins of similar mean amino acid composition cannot be discerned. It cannot be used for unknown molecules as their Raman signature is unknown. Furthermore, the excitation energies used for conventional, non-resonant Raman spectroscopy is several orders of magnitude higher than powers required for confocal fluorescence microscopy, which raises concerns about the local heating and photodegradation of some less stable compounds. In photosynthesizing organisms such as microalgae, high fluorescence background, in particular that of chlorophyll, represents the main obstacle for Raman microscopy. Although this issue can be solved by photobleaching prior to the Raman measurement [2,3,28,50–53] this prevents the more desirable time-lapse Raman microscopy of exactly the same cell or coenobium. On the other hand, photobleaching itself does not disqualify the results, as the consequent decay processes take longer than needed for Raman mapping [28]. Despite all the aforementioned limitations and shortcomings, a major advantage of Raman microscopy is the ability to acquire chemical maps of multiple components from a single measurement that provides a unique opportunity to visualize these structures in a mutual context (Figure 7).

5. Conclusions

We have compared the detection of energy-rich compounds (starch, lipid, and polyP) during the cell cycle of *Desmodesmus quadricauda* using two independent methodologies, conventional bulk biochemical analyzes and confocal Raman microscopy. The two methods were comparable in detection of polyP. Fluorescence and Raman microscopies detected the specific localization of polyP granules near the nuclei until the bi-nuclear stage. The two methods differed in detection of starch, possibly suggesting differences in starch localization and/or its solubility during the cell cycle. Raman microscopy was more sensitive in detecting lipids. It also detected guanine crystals; these were localized near polyP granules and nuclei, suggesting their possible role in processes related to nuclear division. Confocal Raman microscopy is capable of detecting even low levels of macromolecules naturally present in the cells during their vegetative development. It is especially suited for the detection of polyP, lipids, and guanine crystals within the cells. The differences in starch detection will require further experiments.

Supplementary Materials: The following are available online at https://www.mdpi.com/2073-4409/10/1/62/s1, Figure S1: Typical Raman spectra of the respective biomolecules detected and quantified in *D. quadricauda* cells. Figure S2: Raman spectra of starch granules and crystalline guanine obtained in situ from *D. quadricauda* compared with the respective chemical references. Figure S3: Raman spectra of small lipid droplets and polyP granules obtained in situ from *D. quadricauda* compared with the respective chemical references. Figure S4: Raman chemical maps of eight-cell coenobium of *D. quadricauda* and the spectral components obtained by a multivariate deconvolution. Figure S5: Raman chemical maps of the cells illustrating in detail the time course of changes in their molecular composition during the cell cycle.

Author Contributions: Conceptualization, K.B., P.M., L.N., and V.Z.; Methodology, Š.M., I.N.I., and M.V.; Validation, K.B., P.M., L.N., and V.Z.; Formal analysis, Š.M. and I.N.I.; Investigation., Š.M., I.N.I., M.V., L.N., V.Z., P.M., and K.B.; Resources, P.M. and K.B.; Data curation, Š.M., I.N.I., M.V., V.Z., P.M., and K.B.; Writing—original draft preparation, Š.M., P.M., V.Z., K.B., and I.N.I.; Writing—review and editing, Š.M., I.N.I., M.V., L.N., V.Z., P.M., and K.B.; Visualization, V.Z., M.V., P.M., and K.B.; Supervision, K.B. and P.M.; Project administration, K.B., P.M., L.N., and V.Z.; Funding acquisition, K.B. and P.M. All authors have read and agreed to the published version of the manuscript.

Funding: This research was funded by Grantová Agentura České Republiky, grant no. 17-06264S.

Institutional Review Board Statement: Not applicable.

Informed Consent Statement: Not applicable.

Data Availability Statement: All data presented in this study are available within this article or Supplementary Materials. There are no special databases associated with this manuscript.

Acknowledgments: We are obliged to technical staff of the Laboratory of Cell Cycles of Algae for excellent technical support. We thank J. D. Brooker for critical reading and language editing of the text.

Conflicts of Interest: The authors declare no conflict of interest. The funders had no role in the design of the study; in the collection, analyses, or interpretation of data; in the writing of the manuscript, or in the decision to publish the results.

References

1. Zachleder, V.; Bišová, K.; Vítová, M. The cell cycle of microalgae. In *The Physiology of Microalgae*; Borowitzka, M.A., Beardall, J., Raven, J.A., Eds.; Springer: Dordrecht, The Netherlands, 2016; Volume 6, pp. 3–46.
2. Moudříková, Š.; Nedbal, L.; Solovchenko, A.; Mojzeš, P. Raman microscopy shows that nitrogen-rich cellular inclusions in microalgae are microcrystalline guanine. *Algal Res.* **2017**, *23*, 216–222. [CrossRef]
3. Mojzeš, P.; Gao, L.; Ismagulova, T.; Pilátová, J.; Moudříková, Š.; Gorelova, O.; Solovchenko, A.; Nedbal, L.; Salih, A. Guanine, a high-capacity and rapid-turnover nitrogen reserve in microalgal cells. *Proc. Natl. Acad. Sci. USA* **2020**, *117*, 32722–32730. [CrossRef]
4. Šetlík, I.; Zachleder, V. The multiple fission cell reproductive patterns in algae. In *The Microbial Cell Cycle*; Nurse, P., Streiblová, E., Eds.; CRC Press Inc.: Boca Raton, FL, USA, 1984; pp. 253–279.
5. Zachleder, V.; van den Ende, H. Cell cycle events in the green alga *Chlamydomonas eugametos* and their control by environmental factors. *J. Cell Sci.* **1992**, *102*, 469–474.
6. Donnan, L.; John, P.C.L. Cell cycle control by timer and sizer in *Chlamydomonas*. *Nature* **1983**, *304*, 630–633. [CrossRef] [PubMed]
7. Šetlík, I.; Berková, E.; Doucha, J.; Kubín, S.; Vendlová, J.; Zachleder, V. The coupling of synthetic and reproduction processes in *Scenedesmus quadricauda*. *Arch. Hydrobiol. Algol. Stud.* **1972**, *7*, 172–217.
8. Zachleder, V.; Šetlík, I. Distinct controls of DNA-replication and of nuclear division in the cell-cycles of the chlorococcal alga *Scenedesmus quadricauda*. *J. Cell Sci.* **1988**, *91*, 531–539.
9. Zachleder, V.; Šetlík, I. Timing of events in overlapping cell reproductive sequences and their mutual interactions in the alga *Scenedesmus quadricauda*. *J. Cell Sci.* **1990**, *97*, 631–638.
10. Zachleder, V. The course of reproductive events in the chloroplast cycle of the chlorococcal alga *Scenedesmus quadricauda* as revealed by using inhibitors of DNA replication. *Plant Cell Physiol.* **1997**, *38*, 56.
11. Tukaj, Z.; Kubínová, A.; Zachleder, V. Effect of irradiance on growth and reproductive processes during the cell cycle in *Scenedesmus armatus* (Chlorophyta). *J. Phycol.* **1996**, *32*, 624–631. [CrossRef]
12. Zachleder, V.; Doucha, J.; Berková, E.; Šetlík, I. The effect of synchronizing dark period on populations of *Scenedesmus quadricauda*. *Biol Plant.* **1975**, *17*, 416–433. [CrossRef]
13. Zachleder, V.; Bišová, K.; Vítová, M.; Kubín, Š.; Hendrychová, J. Variety of cell cycle patterns in the alga *Scenedesmus quadricauda* (Chlorophyta) as revealed by application of illumination regimes and inhibitors. *Eur. J. Phycol.* **2002**, *37*, 361–371. [CrossRef]
14. Vítová, M.; Zachleder, V. Points of commitment to reproductive events as a tool for analysis of the cell cycle in synchronous cultures of algae. *Folia Microbiol.* **2005**, *50*, 141–149. [CrossRef] [PubMed]
15. Xie, B.; Stessman, D.; Hart, J.H.; Dong, H.L.; Wang, Y.J.; Wright, D.A.; Nikolau, B.J.; Spalding, M.H.; Halverson, L.J. High-throughput fluorescence-activated cell sorting for lipid hyperaccumulating *Chlamydomonas reinhardtii* mutants. *Plant Biotechnol. J.* **2014**, *12*, 872–882. [CrossRef] [PubMed]

16. Terashima, M.; Freeman, E.S.; Jinkerson, R.E.; Jonikas, M.C. A fluorescence-activated cell sorting-based strategy for rapid isolation of high-lipid *Chlamydomonas* mutants. *Plant J.* **2014**, *81*, 147–159. [CrossRef] [PubMed]
17. Vítová, M.; Hendrychová, J.; Cepák, V.; Zachleder, V. Visualization of DNA-containing structures in various species of Chlorophyta, Rhodophyta and Cyanophyta using SYBR green I dye. *Folia Microbiol.* **2005**, *50*, 333–340. [CrossRef] [PubMed]
18. Dieing, T.; Hollricher, O.; Toporski, J. *Confocal Raman Microscopy*; Springer: Berlin/Heidelberg, Germany, 2011. [CrossRef]
19. Butler, H.J.; Ashton, L.; Bird, B.; Cinque, G.; Curtis, K.; Dorney, J.; Esmonde-White, K.; Fullwood, N.J.; Gardner, B.; Martin-Hirsch, P.L.; et al. Using Raman spectroscopy to characterize biological materials. *Nat. Protoc.* **2016**, *11*, 664–687. [CrossRef]
20. Koch, M.; Zagermann, S.; Kniggendorf, A.K.; Meinhardt-Wollweber, M.; Roth, B. Violaxanthin cycle kinetics analysed in vivo with resonance Raman spectroscopy. *J. Raman Spectrosc.* **2017**, *48*, 686–691. [CrossRef]
21. Jehlička, J.; Edwards, H.G.M.; Orenc, A. Raman spectroscopy of microbial pigments. *Appl. Environ. Microbiol.* **2014**, *80*, 3286–3295. [CrossRef]
22. Li, K.; Cheng, J.; Ye, Q.; He, Y.; Zhou, J.H.; Cen, K.F. In vivo kinetics of lipids and astaxanthin evolution in Haematococcus pluvialis mutant under 15% CO_2 using Raman microspectroscopy. *Bioresour. Technol.* **2017**, *244*, 1439–1444. [CrossRef]
23. Meksiarun, P.; Spegazzini, N.; Matsui, H.; Nakajima, K.; Matsuda, Y.; Sato, H. In vivo study of lipid accumulation in the microalgae marine diatom *Thalassiosira pseudonana* using Raman spectroscopy. *Appl. Spectrosc.* **2015**, *69*, 45–51. [CrossRef]
24. Samek, O.; Jonáš, A.; Pilát, Z.; Zemánek, P.; Nedbal, L.; Tříska, J.; Kotas, P.; Trtílek, M. Raman microspectroscopy of individual algal cells: Sensing unsaturation of storage lipids in vivo. *Sensors* **2010**, *10*, 8635–8651. [CrossRef] [PubMed]
25. Wu, H.; Volponi, J.V.; Oliver, A.E.; Parikh, A.N.; Simmons, B.A.; Singh, S. In vivo lipidomics using single-cell Raman spectroscopy. *Proc. Natl. Acad. Sci. USA* **2011**, *108*, 3809–3814. [CrossRef] [PubMed]
26. Chiu, L.-D.; Ho, S.-H.; Shimada, R.; Ren, N.-Q.; Ozawa, T. Rapid in vivo lipid/carbohydrate quantification of single microalgal cell by Raman spectral imaging to reveal salinity-induced starch-to-lipid shift. *Biotechnol. Biofuels* **2017**, *10*, 9. [CrossRef] [PubMed]
27. Huang, Y.Y.; Beal, C.M.; Cai, W.W.; Ruoff, R.S.; Terentjev, E.M. Micro-Raman spectroscopy of algae: Composition analysis and fluorescence background behavior. *Biotechnol. Bioenergy* **2010**, *105*, 889–898. [CrossRef] [PubMed]
28. Moudříková, Š.; Mojzeš, P.; Zachleder, V.; Pfaff, C.; Behrendt, D.; Nedbal, L. Raman and fluorescence microscopy sensing energy-transducing and energy-storing structures in microalgae. *Algal Res.* **2016**, *16*, 224–232. [CrossRef]
29. Zachleder, V.; Šetlík, I. Effect of irradiance on the course of RNA synthesis in the cell cycle of *Scenedesmus quadricauda*. *Biol. Plant.* **1982**, *24*, 341–353. [CrossRef]
30. Hlavová, M.; Vítová, M.; Bišová, K. Synchronization of green algae by light and dark regimes for cell cycle and cell division studies. In *Plant Cell Division*; Caillaud, M.-C., Ed.; Springer Science: New York, NY, USA; Berlin/Heidelberg, Germany; Dordrecht, The Netherlands; London, UK, 2016; pp. 3–16.
31. Zachleder, V.; Cepák, V. Visualization of DNA containing structures by fluorochrome DAPI in those algal cells which are not freely permeable to the dye. *Arch. Hydrobiol. Algol. Stud.* **1987**, *47*, 157–168.
32. Ota, S.; Yoshihara, M.; Yamazaki, T.; Takeshita, T.; Hirata, A.; Konomi, M.; Oshima, K.; Hattori, M.; Bišová, K.; Zachleder, V.; et al. Deciphering the relationship among phosphate dynamics, electron-dense body and lipid accumulation in the green alga *Parachlorella kessleri*. *Sci. Rep.* **2016**, *6*, 25731. [CrossRef]
33. Wanka, F. Die Bestimmung der Nucleinsäuren in *Chlorella pyrenoidosa*. *Planta* **1962**, *58*, 594–619. [CrossRef]
34. Lukavský, J.; Tetík, K.; Vendlová, J. Extraction of nucleic acid from the alga *Scenedesmus Quadricauda*. *Arch. Hydrobiol. Algol. Stud.* **1973**, *9*, 416–426.
35. Decallonne, J.R.; Weyns, C.J. A shortened procedure of the diphenylamine reaction for measurement of deoxyribonucleic acid by using light activation. *Anal. Biochem.* **1976**, *74*, 448–456. [CrossRef]
36. Zachleder, V. Optimization of nucleic acids assay in green and blue-green algae: Extraction procedures and the light-activated reaction for DNA. *Arch. Hydrobiol. Algol. Stud.* **1984**, *36*, 313–328. [CrossRef]
37. Miller, G.L. Use of dinitrosalicylic acid reagent for determination of reducing sugar. *Anal. Chem.* **1959**, *31*, 426–428. [CrossRef]
38. Everall, N.J. Confocal Raman microscopy: Performance, pitfalls, and best practice. *Appl. Spectrosc.* **2009**, *63*, 245A–262A. [CrossRef] [PubMed]
39. Eaton, J.W.; Bateman, D.; Hauberg, S.; Wehbring, R. *GNU Octave Version 4.0.0 Manual: A High-Level Interactive Language for Numerical Computations*; CreateSpace Independent Publishing Platform: Boston, MA, USA, 2015.
40. Palacký, J.; Mojzeš, P.; Bok, J. SVD-based method for intensity normalization, background correction and solvent subtraction in Raman spectroscopy exploiting the properties of water stretching vibrations. *J. Raman Spectrosc.* **2011**, *42*, 1528–1539. [CrossRef]
41. Zachleder, V.; Schläfli, O.; Boschetti, A. Growth-controlled oscillation in activity of histone H1 kinase during the cell cycle of *Chlamydomonas reinhardtii* (Chlorophyta). *J. Phycol.* **1997**, *33*, 673–681. [CrossRef]
42. Bišová, K.; Zachleder, V. Cell-cycle regulation in green algae dividing by multiple fission. *J. Exp. Bot.* **2014**, *65*, 2585–2602. [CrossRef]
43. Miyachi, S.; Tamiya, H. Distribution and turnover of phosphate compounds in growing *Chlorella* cells. *Plant Cell Physiol.* **1961**, *2*, 405–414.
44. Miyachi, S.; Tamiya, H. Some observations on the phosphorus metabolism in growing *Chlorella* cells. *Biochim. Biophys. Acta* **1961**, *46*, 200–202. [CrossRef]
45. Miyachi, S.; Kanai, R.; Mihara, S.; Miyachi, S.; Aoki, S. Metabolic roles of inorganic polyphosphates in *Chlorella* cells. *Biochim. Biophys. Acta* **1964**, *93*, 625–634. [CrossRef]

46. Juppner, J.; Mubeen, U.; Leisse, A.; Caldana, C.; Wiszniewski, A.; Steinhauser, D.; Giavalisco, P. The target of rapamycin kinase affects biomass accumulation and cell cycle progression by altering carbon/nitrogen balance in synchronized *Chlamydomonas reinhardtii* cells. *Plant J.* **2018**, *93*, 355–376. [CrossRef] [PubMed]

47. Vítová, M.; Bišová, K.; Umysová, D.; Hlavová, M.; Kawano, S.; Zachleder, V.; Čížková, M. *Chlamydomonas reinhardtii*: Duration of its cell cycle and phases at growth rates affected by light intensity. *Planta* **2011**, *233*, 75–86. [CrossRef] [PubMed]

48. Ji, Y.; He, Y.; Cui, Y.; Wang, T.; Wang, Y.; Li, Y.; Huang, W.E.; Xu, J. Raman spectroscopy provides a rapid, non-invasive method for quantitation of starch in live, unicellular microalgae. *Biotechnol. J.* **2014**, *9*, 1512–1518. [CrossRef] [PubMed]

49. Hosokawa, M.; Ando, M.; Mukai, S.; Osada, K.; Yoshino, T.; Hamaguchi, H.; Tanaka, T. In vivo live cell imaging for the quantitative monitoring of lipids by using raman microspectroscopy. *Anal. Chem.* **2014**, *86*, 8224–8230. [CrossRef] [PubMed]

50. Moudříková, Š.; Sadowsky, A.; Metzger, S.; Nedbal, L.; Mettler-Altmann, T.; Mojzeš, P. Quantification of polyphosphate in microalgae by Raman microscopy and by a reference enzymatic assay. *Anal. Chem.* **2017**, *89*, 12006–12013. [CrossRef]

51. Solovchenko, A.; Khozin-Goldberg, I.; Selyakh, I.; Semenova, L.; Ismagulova, T.; Lukyanov, A.; Mamedov, I.; Vinogradova, E.; Karpova, O.; Konyukhov, I.; et al. Phosphorus starvation and luxury uptake in green microalgae revisited. *Algal Res.* **2019**, *43*, 101651. [CrossRef]

52. Barcytė, D.; Pilátová, J.; Mojzeš, P.; Nedbalová, L. The arctic *Cylindrocystis* (Zygnematophyceae, Streptophyta) green algae are genetically and morphologically diverse and exhibit effective accumulation of polyphosphate. *J. Phycol.* **2020**, *56*, 217–232. [CrossRef]

53. Siebers, N.; Hofmann, D.; Schiedung, H.; Landsrath, A.; Ackermann, B.; Gao, L.; Mojzeš, P.; Jablonowski, N.D.; Nedbal, L.; Amelung, W. Towards phosphorus recycling for agriculture by algae: Soil incubation and rhizotron studies using ^{33}P-labeled microalgal biomass. *Algal Res.* **2019**, *43*, 101634. [CrossRef]

MDPI
St. Alban-Anlage 66
4052 Basel
Switzerland
Tel. +41 61 683 77 34
Fax +41 61 302 89 18
www.mdpi.com

Cells Editorial Office
E-mail: cells@mdpi.com
www.mdpi.com/journal/cells